CISM International Centre for Mechanical Sciences

Courses and Lectures

Volume 583

The series presents lecture notes, monographs, edited works and proceedings in the field of Mechanics, Engineering, Computer Science and Applied Mathematics. Purpose of the series is to make known in the international scientific and technical community results obtained in some of the activities organized by CISM, the International Centre for Mechanical Sciences.

More information about this series at http://www.springer.com/series/76

Guillaume Ovarlez · Sarah Hormozi
Editors

Lectures on Visco-Plastic Fluid Mechanics

Springer

Editors
Guillaume Ovarlez
Laboratory of the Future
University of Bordeaux
Bordeaux
France

Sarah Hormozi
Department of Mechanical Engineering
Ohio University
Athens, OH
USA

ISSN 0254-1971 ISSN 2309-3706 (electronic)
CISM International Centre for Mechanical Sciences
ISBN 978-3-030-07771-6 ISBN 978-3-319-89438-6 (eBook)
https://doi.org/10.1007/978-3-319-89438-6

Softcover re-print of the Hardcover 1st edition 2019

Printed on acid-free paper

This Springer imprint is published by the registered company Springer International Publishing AG part of Springer Nature
The registered company address is: Gewerbestrasse 11, 6330 Cham, Switzerland

Preface

This volume contains the lecture notes from the course *Lectures on Visco-Plastic Fluid Mechanics*, held at the International Centre for Mechanical Sciences (CISM), Udine, Italy, May 30–June 3, 2016. The lectures address fluid flows of yield stress fluids from theoretical, computational and experimental perspectives. The following summarizes the broad subtopic breakdown introduced by the course: (i) analysis of ideal visco-plastic fluid models, (ii) mathematical modelling of flows of visco-plastic fluids, (iii) computing flows of visco-plastic fluids, (iv) rheometry of visco-plastic fluids, (v) rheology of visco-plastic suspensions and (vi) application of visco-plastic fluids in engineering sciences. The lecture notes are designed for advanced graduate students as well as researchers across several disciplines (e.g. mathematics, physics and engineering), with the aim of providing them with tools and techniques that are essential in performing research on the flow problems of visco-plastic fluids. This volume includes six chapters. Each chapter has its own opening that gives a brief introduction.

The organizers of the course, *Lectures on Visco-Plastic Fluid Mechanics*, were Sarah Hormozi (Ohio University, Athens, Ohio, USA) and Guillaume Ovarlez (University of Bordeaux, CNRS, Solvay, LOF, Bordeaux, France), who also acted as scientific editors for these lecture notes. The main lecturers of the course were Ian Frigaard (The University of British Columbia, Vancouver, BC, Canada), Neil Balmforth (The University of British Columbia, Vancouver, BC, Canada), Anthony Wachs (The University of British Columbia, Vancouver, BC, Canada), Ian Wilson (University of Cambridge, Cambridge, UK), Sarah Hormozi (Ohio University, Athens, Ohio, USA) and Guillaume Ovarlez (University of Bordeaux, CNRS, Solvay, LOF, Bordeaux, France). The course was benefited greatly from the assistance and support of Alfredo Soldati (Head of the Department of Fluid Mechanics at CISM), Elisabeth Guazzelli (Rector of CISM), Paolo Serafini (Scientific Editor of CISM) and Marika Minisini (Administrative Assistant at CISM).

Athens, USA
Bordeaux, France
February 2018

Sarah Hormozi
Guillaume Ovarlez

Contents

Background Lectures on Ideal Visco-Plastic Fluid Flows

I. A. Frigaard

Abstract These notes are intended to supplement a short lecture course covering the theoretical background of the dynamics of ideal visco-plastic fluids, e.g. Bingham fluids, Herschel-Bulkley fluids. They are targeted at an applied mathematics or engineering audience. The intention is to give a non-rigorous introduction to those parts of the theory that: (a) appear to have use in applications; (b) are needed for computational methods; (c) mark out visco-plastic fluids from purely viscous generalised Newtonian fluids.

1 General Aspects of Visco-Plastic Fluid Flows

On continuum scales visco-plastic fluids are governed by the Navier-Stokes equations, with constitutive law specific to the type of fluid considered. Here we consider only *ideal* visco-plastic fluids. By this we mean models such as the Bingham, Herschel-Bulkley and Casson fluid models. These models have a yield stress $\hat{\tau}_Y$ and satisfy the von Mises yield criterion.

The objective of these notes is to develop the theoretical background of these fluids, targeted at an applied mathematics/engineering practitioner. The intention is to give a non-rigorous introduction (only covering proofs formally, if at all). We also have a specific target which is to cover those parts of the theory that: (a) appear to have use in applications; (b) are needed for computational methods; (c) distinguish visco-plastic fluids from purely viscous generalised Newtonian fluids.

This research is funded by the NSERC Discovery grant programme which is gratefully acknowledged. Parts of these notes contain results from ongoing and recent work with my group. I would like to thank Emad Chaparian, Ida Karimfazli and Ali Roustaei for their help with computed examples.

I. A. Frigaard (✉)
University of British Columbia, Vancouver, B.C., Canada
e-mail: frigaard@math.ubc.ca

G. Ovarlez and S. Hormozi (eds.), *Lectures on Visco-Plastic Fluid Mechanics*,
CISM International Centre for Mechanical Sciences 583,
https://doi.org/10.1007/978-3-319-89438-6_1

1.1 General Setup

In general we will refer to the flow domain as Ω, with boundary $\partial\Omega$, and consider flows in two or three dimensions. Most of what follows is described in terms of dimensionless variables. Where needed, dimensional variables will be denoted with a “hat” symbol, e.g. $\hat{\tau}_Y$ for the yield stress.

It is supposed that there is a representative length-scale $\hat{L}$ attached to the flow domain. There is also a velocity scale $\hat{U}_0$, relevant to the problem at hand. This velocity scale typically comes either from imposed boundary conditions on the velocity (e.g. flow along a pipe at a fixed flow rate), or is deduced from the forces or boundary traction. These two scales are used to make the Navier-Stokes equations dimensionless as follows:

$$\mathbf{x} = \frac{\hat{\mathbf{x}}}{\hat{L}}, t = \frac{\hat{t}\hat{U}_0}{\hat{L}}, \mathbf{u} = \frac{\hat{\mathbf{u}}}{\hat{U}_0}, p = \frac{\hat{p}\hat{L}}{\hat{\mu}_0\hat{U}_0}, \tau_{ij} = \frac{\hat{\tau}_{ij}\hat{L}}{\hat{\mu}_0\hat{U}_0}, f_i = \frac{\hat{f}_i\hat{L}^2}{\hat{\mu}_0\hat{U}_0}, \tag{1}$$

leading to:

$$Re\frac{\partial u_i}{\partial t} + u_j\frac{\partial u_i}{\partial x_j} = -\frac{\partial p}{\partial x_i} + \frac{\partial \tau_{ij}}{\partial x_j} + f_i, \tag{2}$$

$$0 = \frac{\partial u_i}{\partial x_i}. \tag{3}$$

The fluid density is $\hat{\rho}$, $\hat{\mu}_0$ represents a viscosity scale and $\hat{\mathbf{f}}$ is a body force (per unit volume). We see that the stresses and body forces have been scaled to balance one another, using a viscous stress scale: $\hat{\mu}_0\hat{U}_0/\hat{L}$.

Scaled constitutive laws are:

$$\tau_{ij}(\mathbf{u}) = \left[\eta_v(\dot{\gamma}(\mathbf{u})) + \frac{B}{\dot{\gamma}(\mathbf{u})}\right]\dot{\gamma}_{ij}(\mathbf{u}) \iff \tau(\mathbf{u}) > B, \tag{4}$$

$$\dot{\gamma}(\mathbf{u}) = 0 \iff \tau(\mathbf{u}) \le B, \tag{5}$$

where the components of the strain rate tensor are:

$$\dot{\gamma}_{ij} = \frac{\partial u_i}{\partial x_j} + \frac{\partial u_j}{\partial x_i},$$

and where

$$\dot{\gamma}(\mathbf{u}) = \left[\frac{1}{2}\sum_{i,j=1}^{3}[\dot{\gamma}_{ij}(\mathbf{u})]^2\right]^{1/2} \qquad \tau(\mathbf{u}) = \left[\frac{1}{2}\sum_{i,j=1}^{3}[\tau_{ij}(\mathbf{u})]^2\right]^{1/2}. \tag{6}$$

The 2 dimensionless parameters appearing above are firstly the Reynolds number:

$$Re = \frac{\hat{\rho}\hat{L}\hat{U}_0}{\hat{\mu}_0}, \tag{7}$$

which represents the balance of inertial and viscous stresses. Secondly the Bingham number

$$B = \frac{\hat{\tau}_Y \hat{L}}{\hat{U}_0 \hat{\mu}_0}, \tag{8}$$

is representative of the balance between the yield stress and the viscous stress scale. In the case of a Bingham or Casson fluid, one would use the plastic viscosity, i.e. $\hat{\mu}_0 = \hat{\mu}_p$. For fluids such as the Herschel-Bulkley one would commonly include the representative strain rate in the definition, e.g. $\hat{\mu}_0 = \hat{\kappa}\hat{U}_0^{n-1}/\hat{L}^{n-1}$. The function $\eta_v(\dot{\gamma})$ represents only the viscous part of the effective viscosity: e.g. $\eta_v = 1$ for Bingham or Newtonian fluids, $\eta_v = \dot{\gamma}^{n-1}$ for a Herschel-Bulkley fluids and $\eta_v = 1 + 2\sqrt{B/\dot{\gamma}}$ for a Casson fluid. The effective viscosity is simply $\eta(\dot{\gamma}) = \eta_v(\dot{\gamma}) + B/\dot{\gamma}$.

1.2 Classical Formulation

There are a number of analytical and semi-analytical solutions to flow problems involving the Bingham fluid, (slightly less so for the Herschel-Bulkley and Casson fluid). However, most are one-dimensional (1D) and derived under assumptions of symmetry, full development, etc. to simplify the equations. For more complex flows, one would also commonly try to solve the Navier-Stokes equations in their *classical* formulation. However, for ideal visco-plastic fluids we soon encounter some difficulties in doing this. Certainly (2)–(4) would be fine for any yielded region of flow, supplemented of course with appropriate boundary conditions. However, in regions where (5) is satisfied the fluid motion is less clear.

Let us consider briefly the question of whether it is is viable to model such a region directly. Suppose that $\Omega_p(t) \subseteq \Omega$ contains unyielded fluid. The momentum equations are satisfied in $\Omega_p(t)$, but because (5) holds in such a *plug* region the stresses are indeterminate. Specifically, this means that there are more equations of motion than there are momentum equations, i.e. the stresses cannot be determined fully (at least in 2D and 3D).

Kinematically, (5) implies that a plug moves as a rigid solid. However, a plug is not exactly a rigid solid, since mass may either enter or leave the plug through a yield surface. Determining the yield surfaces and their movement represent the key difficulties in dealing with visco-plastic fluid flows in a classical formulation. For illustration suppose that the domain Ω_p is bounded by a yield surface Γ, with outward normal $\mathbf{n}$ and we suppose that the yield surfaces of the plug move instantaneously in the direction of $\mathbf{n}$ with speed u_p.

As with any rigid body motion, the velocity within the plug can be decomposed instantaneously into a linear motion and a rotation about a point. Let $\mathbf{x}_c(t) = (x_c(t), y_c(t), z_c(t))$ denote a point that remains within the plug over the time period of interest. The plug velocity is then described instantaneously by the linear motion of the point $\mathbf{x}_c(t)$, moving with velocity $\mathbf{u}_c(t)$, and by a rotation $\omega_c(t)$ about $\mathbf{x}_c(t)$, i.e. $\mathbf{u} = \mathbf{u}_c(t) + (\mathbf{x} - \mathbf{x}_c) \wedge \omega_c(t)$, or componentwise:

$$u_i = u_{c,i} + \epsilon_{ijk}(x_j - x_{c,j})\omega_{c,k},$$

where ϵ_{ijk} is the permutation symbol. The 6 unknowns: $u_{c,i}$, $\omega_{c,i}$, $i = 1, 2, 3$ must be determined from the 6 equations corresponding to conservation of linear and angular momentum. These 6 equations are, (for $i = 1, 2, 3$):

$$\begin{aligned} 0 = &\int_{\Omega_p(t)} (\dot{u}_{c,i} + \epsilon_{ijk}(x_j - x_{c,j})\dot{\omega}_{c,k}) \, \mathrm{d}\mathbf{x} - \int_{\Gamma(t)} \sigma_{ij} n_j \, \mathrm{d}s \\ &- \int_{\Omega_p(t)} f_i \, \mathrm{d}\mathbf{x} + \int_{\Gamma(t)} (u_{c,i} + \epsilon_{ijk}(x_j - x_{c,j})\omega_{c,k}) \times \\ &\qquad [(u_{c,l} + \epsilon_{lmn}(x_m - x_{c,m})\omega_{c,n})n_l - u_p] \, \mathrm{d}s \end{aligned} \tag{9}$$

$$\begin{aligned} 0 = &\int_{\Omega_p(t)} \epsilon_{ijk}[x_j - x_{c,j}][\dot{u}_{c,k} + \epsilon_{klm}(x_l - x_{c,l})\dot{\omega}_{c,m}] \, \mathrm{d}\mathbf{x} - \\ &\int_{\Omega_p(t)} \epsilon_{ijk}(x_j - x_{c,j}) f_k \, \mathrm{d}\mathbf{x} + \\ &\int_{\Gamma(t)} \epsilon_{ijk}[x_j - x_{c,j}][u_{c,k} + \epsilon_{klm}(x_l - x_{c,l})\omega_{c,m}] \times \\ &[(u_{c,l} + \epsilon_{lmn}(x_m - x_{c,m})\omega_{c,n})n_l - u_p] \, \mathrm{d}s. \\ &- \int_{\Gamma(t)} \epsilon_{ijk}(x_j - x_{c,j})\sigma_{kl} n_l \, \mathrm{d}s \end{aligned} \tag{10}$$

In (9) and (10) we have denoted the time derivatives of $u_{c,i}$, $\omega_{c,i}$ by $\dot{u}_{c,i}$, $\dot{\omega}_{c,i}$, $i = 1, 2, 3$, and the stress tensor is $\sigma_{ij} = -p\delta_{ij} + \tau_{ij}$. The surface integrals involving σ_{ij} are understood to mean the limits of these integrals taken from the yielded fluid region, i.e. from the side on which the stress is determinate. The classical problem consists of (2)–(4) with (9) and (10), associated boundary conditions and continuity conditions at the yield surface, i.e. the velocity and traction are continuous across a yield surface (6 conditions).

Equations (9) and (10) constitute 6 nonlinear ordinary (integro-)differential equations for $u_{c,i}$, $\omega_{c,i}$, $i = 1, 2, 3$, which are coupled to the flow outside the plug, through continuity of the traction and velocity vectors. Note that (9) and (10) are integro-differential equations since $x_{c,i}$, $i = 1, 2, 3$ are obtained by integrating $u_{c,i}$, $i = 1, 2, 3$.

This classical formulation is clearly complex. Although it may appear that we have sufficient equations for the unknowns, this is actually not clear. The speed of propagation of the yield surface in the normal direction, u_p, appears above but is determined wholly by changes in the deviatoric stress field, τ_{ij}, which itself is fully determined only outside of the plug. This is a key difference between this and the simpler problem of an interface between the fluid an another phase: in such cases a kinematic condition governs movement of the interface. Here the yield surface is not a *material* surface and we have no such condition to model evolution of the yield surface.

To our knowledge, no solutions of such general problems have ever been computed in multiple dimensions, without some kind of simplifying assumptions (symmetry, steady flow, etc), and little is known about the regularity of classical solutions to this problem. Where advances have been made are in simpler 1D scenarios where plug motion and stress fields can be determined.

Fortunately, we are able to derive a different formulation of the flow problem which does allow for the solution of quite general flows. This is a weak or variational formulation, which has the advantage of not needing to know a priori where the plug regions are. The key differences in this formulation can be examined through the simpler problem of Stokes flow, which we now look at.

2 Stokes Flow

Fluids with any measurable yield stress (e.g. $\geq$0.1 Pa) tend to be relatively viscous. Therefore, flows in which inertial effects are small are more typical than when dealing with common Newtonian fluids. Equally, the most interesting aspects of visco-plastic fluids (as opposed to purely viscous generalised Newtonian fluids) are related to yielding, the occurrence of plugs, yield surfaces etc., and all these flow features are more prevalent in flows that are slow. Thus, the study of Stokes flows of yield stress fluids is both of practical interest and allows us to understand the some of key dynamical differences that this particular rheological feature introduces.

We will consider a general problem in which the boundary of Ω is assumed smooth enough for whatever follows (e.g. Lipschitz continuous), and is divided in two parts: $\partial\Omega_v$ on which the velocity is specified and $\partial\Omega_t$ on which the traction is specified:

$$\mathbf{u} = \mathbf{u}_s \text{ on } \partial\Omega_v, \tag{11}$$

$$\boldsymbol{\sigma} \cdot \mathbf{n} = \mathbf{g}_s \text{ on } \partial\Omega_t, \tag{12}$$

for $\boldsymbol{\sigma} = -p\boldsymbol{\delta} + \boldsymbol{\tau}$ where $\boldsymbol{\delta}$ is the Kronecker delta tensor. The Stokes equations are:

$$0 = \nabla p + \nabla \cdot \boldsymbol{\tau} + \mathbf{f}, \text{ in } \Omega, \tag{13}$$

$$0 = \nabla \cdot \mathbf{u}, \text{ in } \Omega. \tag{14}$$

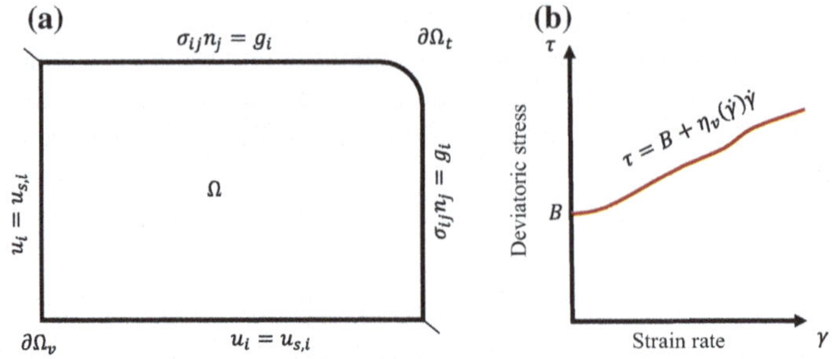

Fig. 1 **a** flow domain and boundary conditions; **b** flowcurves considered

Scaled constitutive laws are:

$$\boldsymbol{\tau}(\mathbf{u}) = \left[\eta_v(\dot{\gamma}(\mathbf{u})) + \frac{B}{\dot{\gamma}(\mathbf{u})}\right]\dot{\boldsymbol{\gamma}}(\mathbf{u}) \iff \tau(\mathbf{u}) > B, \tag{15}$$

$$\dot{\gamma}(\mathbf{u}) = 0 \iff \tau(\mathbf{u}) \leq B, \tag{16}$$

and we'll consider that $\eta_v(\dot{\gamma})$ is any strictly positive function of $\dot{\gamma}$, such that $\eta_v(\dot{\gamma})\dot{\gamma}$ is increasing. In other words the flow curve of the fluid is increasing; see Fig. 1.

For what follows, we assume implicit summation of repeated indices and define a tensor inner product:

$$\boldsymbol{a} : \boldsymbol{b} = \frac{1}{2}a_{ij}b_{ij},$$

so that our preceding definitions of $\dot{\gamma}$ and τ are simply:

$$\dot{\gamma} = \sqrt{\dot{\boldsymbol{\gamma}} : \dot{\boldsymbol{\gamma}}}, \qquad \tau = \sqrt{\boldsymbol{\tau} : \boldsymbol{\tau}}, \tag{17}$$

i.e. the norms associated to the tensor inner product.

We now introduce two spaces of admissible functions. The space $\mathcal{V}$ of admissible velocities is:

$$\mathcal{V} = \{\mathbf{v} : \ \mathbf{v} = \mathbf{u}_s \text{ on } \partial\Omega_v; \ \ \nabla \cdot \mathbf{v} = 0, \ \text{in } \Omega\}. \tag{18}$$

Additionally, $\mathcal{V}$ is regarded as a subspace of the relevant Sobolev space for the solution. Which space is appropriate depends on the dimension of the problem and the behaviour of $\eta_v(\dot{\gamma})$ at large strain rates, e.g. for Bingham fluids we are typically in $[W^{1,2}(\Omega)]^d$, for dimension $d = 2, \ 3$. We shall proceed assuming that the functions are sufficiently smooth for what follows. The space $\mathcal{S}$ of admissible stress tensors is:

$$\mathcal{S} = \{\tilde{\boldsymbol{\sigma}} = -\tilde{p}\boldsymbol{\delta} + \tilde{\boldsymbol{\tau}} : \ \tilde{\boldsymbol{\sigma}} \cdot \mathbf{n} = \mathbf{s}_s \text{ on } \partial\Omega_t; \ \ -\nabla\tilde{p} + \nabla \cdot \tilde{\boldsymbol{\tau}} + \mathbf{f} = \mathbf{0} \text{ in } \Omega\}. \tag{19}$$

Again $\mathcal{S}$ will also lie within an appropriate closed function space, that is dual to that of the strain rate. We bypass these aspects here. Physically, the space $\mathcal{V}$ contains velocities that are divergence free and satisfy the velocity boundary conditions, while the space $\mathcal{S}$ contains stress tensors that satisfy the Stokes momentum equations and the traction boundary conditions.

Most of what follows below was known (at least formally) by the 1970s. The variational principles were initially developed by Prager for Bingham fluids in the late 1950s; see e.g. Prager (1954). Broader frameworks were developed for wider classes of generalised Newtonian fluids in the 1960 and 1970s; see e.g. Johnson (1960, 1961), Yoshioka and Adachi (1971a, b). These variational principles developed initially in a formal way and only later were subjected to more rigorous analysis, e.g. Mosolov and Miasnikov (1965, 1966, 1967), Duvaut and Lions (1976). A more detailed review can be found in Huilgol (2015), which we partly follow below in terms of presentation.

2.1 *Principle of Virtual Power*

Suppose we take $\mathbf{v} \in \mathcal{V}$ and $\tilde{\boldsymbol{\sigma}} \in \mathcal{S}$, which consequently satisfies (13). Taking the dot product of (13) with $\mathbf{v}$, integrating over Ω and using the divergence theorem leads to:

$$0 = \int_{\Omega} [\nabla \cdot \tilde{\boldsymbol{\sigma}} + \mathbf{f}] \cdot \mathbf{v} \, \mathrm{d}\mathbf{x} = \int_{\Omega} \frac{\partial}{\partial x_j}[\tilde{\sigma}_{ij} v_i] - \frac{\partial v_i}{\partial x_j}\tilde{\sigma}_{ij} + f_i v_i \, \mathrm{d}\mathbf{x}$$
$$= \int_{\partial\Omega} v_i \tilde{\sigma}_{ij} n_j \, \mathrm{d}s + \int_{\Omega} f_i v_i \, \mathrm{d}\mathbf{x} + \int_{\Omega} \frac{\partial v_i}{\partial x_j} \tilde{p}\delta_{ij} \, \mathrm{d}\mathbf{x} - - \int_{\Omega} \frac{\partial v_i}{\partial x_j} \tilde{\tau}_{ij} \, \mathrm{d}\mathbf{x}.$$

We now substitute from (11) and (12) to deal with the first term. The third term disappears by virtue of (14) and for the last term we use the symmetry of the stress:

$$0 = \int_{\partial\Omega_v} u_{s,i} \tilde{\sigma}_{ij} n_j \, \mathrm{d}s + \int_{\partial\Omega_t} v_i g_i \, \mathrm{d}s + \int_{\Omega} f_i v_i \, \mathrm{d}\mathbf{x} - \int_{\Omega} \tilde{\boldsymbol{\tau}} : \dot{\boldsymbol{\gamma}}(\mathbf{v}) \, \mathrm{d}\mathbf{x} \qquad (20)$$

This expression is mainly useful in deriving a range of other results (as we will do).

It is known as the *principle of virtual power* (in dimensional form each term has units of power). It represents an energy balance in which the first 3 terms are the work done by the surface forces (due to either imposed velocity or traction) and the work done by the body forces. These are balanced by the last term which represents the visco-plastic dissipation rate of energy.

Two other forms of (20) that we will use are as follows.

(i) If $\boldsymbol{\sigma}(\mathbf{u})$ is the stress associated with the solution $\mathbf{u}$ and if $\mathbf{v} \in \mathcal{V}$, then

$$\int_{\Omega} \boldsymbol{\tau}(\mathbf{u}) : [\dot{\boldsymbol{\gamma}}(\mathbf{v}) - \dot{\boldsymbol{\gamma}}(\mathbf{u})] \, \mathrm{d}\mathbf{x} = \int_{\partial\Omega_t} (\mathbf{v} - \mathbf{u}) \cdot \mathbf{g} \, \mathrm{d}s + \int_{\Omega} \mathbf{f} \cdot (\mathbf{v} - \mathbf{u}) \, \mathrm{d}\mathbf{x}. \qquad (21)$$

(ii) If $\tilde{\boldsymbol{\sigma}} \in \mathcal{S}$:

$$\int_{\Omega} [\boldsymbol{\tau}(\mathbf{u}) - \tilde{\boldsymbol{\tau}}] : \dot{\boldsymbol{\gamma}}(\mathbf{v}) \, \mathrm{d}\mathbf{x} = \int_{\partial\Omega_v} u_{s,i}(\sigma_{ij} - \tilde{\sigma}_{ij}) n_j \, \mathrm{d}s. \tag{22}$$

2.2 Mechanical Energy Balance

The mechanical energy balance is simply (20) evaluated at the solution:

$$\int_{\Omega} \boldsymbol{\tau}(\mathbf{u}) : \dot{\boldsymbol{\gamma}}(\mathbf{u}) \, \mathrm{d}\mathbf{x} = \int_{\partial\Omega_v} u_{s,i} \sigma_{ij} n_j \, \mathrm{d}s + \int_{\partial\Omega_t} \mathbf{g} \cdot \mathbf{u} \, \mathrm{d}s + \int_{\Omega} \mathbf{f} \cdot \mathbf{u} \, \mathrm{d}\mathbf{x}. \tag{23}$$

As commented before, this balances the work done on the fluid at the boundary and by the body forces. Later we shall look at the transient version of this, in which the difference between the above power terms governs growth/decay of the kinetic energy of the flow.

2.3 Velocity Minimization

Recall that we consider visco-plastic fluids with flowcurve of form $\tau(\dot{\gamma}) = B + \dot{\gamma}\eta_v(\dot{\gamma})$; see Fig. 2a. We define $\phi(\dot{\gamma})$ as follows:

$$\phi(\dot{\gamma}) = \int_{0^+}^{\dot{\gamma}} \tau(s) \, \mathrm{d}s = \int_{0^+}^{\dot{\gamma}} \eta_v(s) \, \mathrm{d}s + B\dot{\gamma} = \phi_v(\dot{\gamma}) + B\dot{\gamma}. \tag{24}$$

Note that $\phi''(\dot{\gamma})$ is the slope of the flowcurve, which we assume is increasing. Therefore, $\phi(\dot{\gamma})$ is a convex function of $\dot{\gamma}$; the viscous part $\phi_v(\dot{\gamma})$ is strictly convex. Examples of $\phi(\dot{\gamma})$ for some common models are:

$$\begin{aligned}
\text{Newtonian}: \quad & \phi(\dot{\gamma}) = \frac{\dot{\gamma}^2}{2}, \\
\text{Bingham}: \quad & \phi(\dot{\gamma}) = \frac{\dot{\gamma}^2}{2} + B\dot{\gamma}, \\
\text{Herschel-Bulkley}: \quad & \phi(\dot{\gamma}) = \frac{\dot{\gamma}^{n+1}}{n+1} + B\dot{\gamma}, \\
\text{Casson}: \quad & \phi(\dot{\gamma}) = \frac{\dot{\gamma}^2}{2} + \frac{4B^{1/2}\dot{\gamma}^{3/2}}{3} + B\dot{\gamma}.
\end{aligned}$$

For any $\mathbf{v} \in \mathcal{V}$ the velocity potential $\Phi(\mathbf{v})$ is defined as:

$$\Phi(\mathbf{v}) = \int_{\Omega} \phi(\dot{\gamma}(\mathbf{v})) \, \mathrm{d}\mathbf{x} - \int_{\Omega} \mathbf{f} \cdot \mathbf{v} \, \mathrm{d}\mathbf{x} - \int_{\partial\Omega_t} \mathbf{v} \cdot \mathbf{g} \, \mathrm{d}s. \tag{25}$$

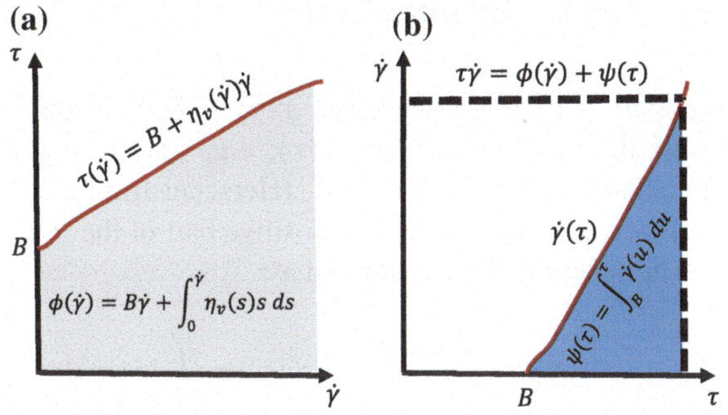

Fig. 2 **a** $\phi(\dot\gamma)$; **b** $\psi(\tau)$ and the relation $\phi(\dot\gamma) + \psi(\tau) = \dot\gamma\tau$

Theorem 2.1 *The solution* **u** *minimizes* $\Phi(\mathbf{v})$ *over* $\mathcal{V}$.

Proof We need to show that $\Phi(\mathbf{v}) \geq \Phi(\mathbf{u})$, $\forall \mathbf{v} \in \mathcal{V}$.

$$\begin{aligned}
\Phi(\mathbf{v}) - \Phi(\mathbf{u}) &= \int_\Omega \phi(\dot\gamma(\mathbf{v})) - \phi(\dot\gamma(\mathbf{u}))\,\mathrm{d}\mathbf{x} - \int_\Omega \mathbf{f}\cdot(\mathbf{v}-\mathbf{u})\,\mathrm{d}\mathbf{x} \\
&\quad - \int_{\partial\Omega_t} (\mathbf{v}-\mathbf{u})\cdot\mathbf{g}\,\mathrm{d}s \\
&= \int_\Omega \phi(\dot\gamma(\mathbf{v})) - \phi(\dot\gamma(\mathbf{u}))\,\mathrm{d}\mathbf{x} - \int_\Omega \boldsymbol{\tau}(\mathbf{u}) : [\dot{\boldsymbol{\gamma}}(\mathbf{v}) - \dot{\boldsymbol{\gamma}}(\mathbf{u})]\,\mathrm{d}\mathbf{x} \\
&\geq \int_\Omega \phi(\dot\gamma(\mathbf{v})) - \phi(\dot\gamma(\mathbf{u})) - \tau(\mathbf{u})\dot\gamma(\mathbf{v}) + \tau(\mathbf{u})\dot\gamma(\mathbf{u})\,\mathrm{d}\mathbf{x} \\
&= \int_\Omega \left(\int_{\dot\gamma(\mathbf{u})}^{\dot\gamma(\mathbf{v})} [\tau(s) - \tau(\mathbf{u})]\,\mathrm{d}s \right) \mathrm{d}\mathbf{x} \geq 0.
\end{aligned}$$

The second line follows from (21) and the third line uses the Cauchy-Schwarz inequality. In the last line the integrand is positive whenever $\dot\gamma(\mathbf{v}) \geq \dot\gamma(\mathbf{u})$ and negative otherwise (see Fig. 2a). □

Note that by using (20) the velocity potential can be rewritten as:

$$\Phi(\mathbf{v}) = \int_\Omega \phi(\dot\gamma(\mathbf{v}))\,\mathrm{d}\mathbf{x} - \int_\Omega \boldsymbol{\tau}(\mathbf{u}) : \dot{\boldsymbol{\gamma}}(\mathbf{v})\,\mathrm{d}\mathbf{x} + \int_{\partial\Omega_v} u_{s,i}\sigma_{ij}(\mathbf{u})n_j\,\mathrm{d}s. \tag{26}$$

This has an interesting geometric interpretation. Of all admissible **v** with the same $\dot\gamma(\mathbf{v})$, $\Phi(\mathbf{v})$ is minimised locally by having $\dot{\boldsymbol{\gamma}}(\mathbf{v})$ parallel to $\boldsymbol{\tau}(\mathbf{u})$. From the constitutive law we know that this is a property satisfied by the solution.

2.4 Stress Maximization

We start with the flow curve $\tau(\dot{\gamma})$, which is uniquely defined for $\tau > B$ and invert this relation to give $\dot{\gamma}(\tau)$, e.g. $\dot{\gamma}(\tau) = \tau$ (Newtonian), $\dot{\gamma}(\tau) = (\tau - B)_+$ (Bingham), $\dot{\gamma}(\tau) = (\tau - B)_+^{1/n}$ (Herschel-Bulkley), $\dot{\gamma}(\tau) = (\tau^{1/2} - B^{1/2})_+^2$ (Casson), where $(f)_+$ denotes the positive part of the function f.

The function $\psi(\tau)$ is defined as:

$$\psi(\tau) = \int_B^\tau \dot{\gamma}(s)\,\mathrm{d}s, \tag{27}$$

and is illustrated graphically as the shaded area in Fig. 2b. For the most common models we have:

$$\begin{aligned}
\text{Newtonian}: \quad \psi(\tau) &= \frac{\tau^2}{2}, \\
\text{Bingham}: \quad \psi(\tau) &= \frac{(\tau - B)_+^2}{2}, \\
\text{Herschel-Bulkley}: \quad \psi(\tau) &= \frac{(\tau - B)_+^{1/n+1}}{1/n+1}, \\
\text{Casson}: \quad \psi(\tau) &= \left(\frac{\tau^2}{2} - \frac{4B^{1/2}\tau^{3/2}}{3} + B\tau - \frac{B^2}{6}\right)_+.
\end{aligned}$$

Note for fixed τ and $\dot{\gamma}$ related via any of these constitutive laws, we have:

$$\tau\dot{\gamma} = \psi(\tau) + \phi(\dot{\gamma}), \tag{28}$$

as is evident graphically from Fig. 2b, by considering the area of the rectangle marked with the broken line.

The second variational principle involves the set of admissible stress tensors, $\tilde{\boldsymbol{\sigma}} \in \mathcal{S}$. For any admissible stress tensor, $\tilde{\boldsymbol{\sigma}} \in \mathcal{S}$, the stress potential $\Psi(\tilde{\boldsymbol{\sigma}})$ is defined as:

$$\Psi(\tilde{\boldsymbol{\sigma}}) = -\int_\Omega \psi(\tilde{\tau})\,\mathrm{d}\mathbf{x} + \int_{\partial\Omega_v} u_{s,i}\tilde{\sigma}_{ij}n_j\,\mathrm{d}s. \tag{29}$$

Theorem 2.2 *The solution stress tensor, $\boldsymbol{\sigma}$ maximizes $\Psi(\tilde{\boldsymbol{\sigma}})$ over $\mathcal{S}$.*

Proof Let the velocity solution be $\mathbf{u}$ with associated stress $\boldsymbol{\sigma} = -p\boldsymbol{\delta} + \boldsymbol{\tau}$. We need to show that $\Psi(\boldsymbol{\sigma}) \geq \Psi(\tilde{\boldsymbol{\sigma}})$, $\forall\tilde{\boldsymbol{\sigma}} \in \mathcal{S}$.

$$\begin{aligned}
\Psi(\boldsymbol{\sigma}) - \Psi(\tilde{\boldsymbol{\sigma}}) &= \int_{\Omega} \psi(\tilde{\tau}) - \psi(\tau)\,\mathrm{d}\mathbf{x} + \int_{\partial\Omega_v} u_{s,i}(\sigma_{ij} - \tilde{\sigma}_{ij})n_j\,\mathrm{d}s, \\
&= \int_{\Omega} \psi(\tilde{\tau}) - \psi(\tau) + (\boldsymbol{\tau}(\mathbf{u}) - \tilde{\boldsymbol{\tau}}) : \dot{\boldsymbol{\gamma}}(\mathbf{u})\,\mathrm{d}\mathbf{x}, \\
&\geq \int_{\Omega} \psi(\tilde{\tau}) - \psi(\tau) + (\tau(\mathbf{u}) - \tilde{\tau})\dot{\gamma}(\mathbf{u})\,\mathrm{d}\mathbf{x} \\
&= \int_{\Omega} \psi(\tilde{\tau}) - \psi(\tau) + (\tau - \tilde{\tau})\dot{\gamma}(\tau)\,\mathrm{d}\mathbf{x} \\
&= \int_{\Omega} \left(\int_{\tau}^{\tilde{\tau}} [\dot{\gamma}(s) - \dot{\gamma}(\tau)]\,\mathrm{d}s \right) \mathrm{d}\mathbf{x} \geq 0.
\end{aligned}$$

The second line follows from (22) and the third line uses the Cauchy-Schwarz inequality. In the last line the integrand is positive when $\tilde{\tau} \geq \tau$ and negative otherwise. □

Relationship of the two principles The velocity minimization is the primal problem and the stress maximization is the associated dual problem. The two variational problems given are however not equivalent. The velocity minimization is the basis of existence and uniqueness results for these flows, as we'll briefly touch on later. The stress maximization does not admit a unique solution as the stress may be non-unique where $\tau \leq B$. The two potential functions are however related as follows.

Theorem 2.3 *Let the velocity solution be* $\mathbf{u}$ *with associated stress* $\boldsymbol{\sigma} = -p\boldsymbol{\delta} + \boldsymbol{\tau}$. *Then for all* $\tilde{\boldsymbol{\sigma}} \in \mathcal{S}$ *and all* $\mathbf{v} \in \mathcal{V}$ *we have:*

$$\Psi(\tilde{\boldsymbol{\sigma}}) \leq \Psi(\boldsymbol{\sigma}) = \Phi(\mathbf{u}) \leq \Phi(\mathbf{v}). \tag{30}$$

Proof This result combines the velocity minimization and stress maximization, so all that remains to be shown is that $\Psi(\boldsymbol{\sigma}) = \Phi(\mathbf{u})$.

$$\begin{aligned}
\Phi(\mathbf{u}) - \Psi(\boldsymbol{\sigma}) &= \int_{\Omega} \phi(\dot{\gamma}(\mathbf{u}))\,\mathrm{d}\mathbf{x} - \int_{\Omega} \mathbf{f}\cdot\mathbf{u}\,\mathrm{d}\mathbf{x} - \int_{\partial\Omega_t} \mathbf{u}\cdot\mathbf{g}\,\mathrm{d}s \\
&\quad + \int_{\Omega} \psi(\tau)\,\mathrm{d}\mathbf{x} - \int_{\partial\Omega_v} u_{s,i}\sigma_{ij}n_j\,\mathrm{d}s \\
&= \int_{\Omega} \phi(\dot{\gamma}(\mathbf{u})) + \psi(\tau)\,\mathrm{d}\mathbf{x} - \int_{\Omega} \mathbf{f}\cdot\mathbf{u}\,\mathrm{d}\mathbf{x} - \int_{\partial\Omega} u_i\sigma_{ij}n_j\,\mathrm{d}s \\
&= \int_{\Omega} [\dot{\gamma}\tau](\mathbf{u})\,\mathrm{d}\mathbf{x} - \int_{\Omega} \mathbf{f}\cdot\mathbf{u} + \frac{\partial}{\partial x_j}[u_i\sigma_{ij}]\,\mathrm{d}\mathbf{x} \\
&= \int_{\Omega} [\dot{\gamma}\tau](\mathbf{u})\,\mathrm{d}\mathbf{x} - \int_{\Omega} \boldsymbol{\tau}(\mathbf{u}) : \dot{\boldsymbol{\gamma}}(\mathbf{u})\,\mathrm{d}\mathbf{x} = 0.
\end{aligned}$$

□

3 Using Variational Methods for Stokes Flows

Here we outline some of the main uses of the variational principles of the previous section.

3.1 Existence, Uniqueness and Properties of Solutions

Theorem 2.1 is the basis for proving existence and uniqueness of the velocity $\mathbf{u} \in \mathcal{V}$, using general results from convex optimization. As well as the minimization formulation, an equivalent formulation that is often used is as a variational inequality, e.g. Duvaut and Lions (1976), Glowinski et al. (1981).

It can be shown that the solution $\mathbf{u}$ satisfies:

$$\langle \Phi_v'(\mathbf{u}), \mathbf{v} - \mathbf{u} \rangle + B \int_\Omega \dot{\gamma}(\mathbf{v}) - \dot{\gamma}(\mathbf{u}) \, \mathrm{d}\mathbf{x} \geq L(\mathbf{v} - \mathbf{u}), \tag{31}$$

$\forall \mathbf{v} \in \mathcal{V}$. Here $\Phi_v(\mathbf{u})$ denotes the viscous part of $\Phi(\mathbf{u})$ and $\Phi_v'(\mathbf{u})$ denotes the Gateaux derivative, given by the following inner product:

$$\langle \Phi_v'(\mathbf{u}), \mathbf{v} - \mathbf{u} \rangle = \int_\Omega \eta_v(\dot{\gamma}(\mathbf{u}))\dot{\boldsymbol{\gamma}}(\mathbf{u}) : [\dot{\boldsymbol{\gamma}}(\mathbf{v}) - \dot{\boldsymbol{\gamma}}(\mathbf{u})] \, \mathrm{d}\mathbf{x}. \tag{32}$$

The linear operator $L(\mathbf{v})$ is defined for $\mathbf{v} \in \mathcal{V}$ as:

$$L(\mathbf{v}) = \int_{\partial\Omega_t} \mathbf{v} \cdot \mathbf{g} \, \mathrm{d}s + \int_\Omega \mathbf{f} \cdot \mathbf{v} \, \mathrm{d}\mathbf{x}. \tag{33}$$

The inequality (31) is an elliptic inequality of the second type. Equivalence of (31) with the velocity minimization is addressed in e.g. Proposition 2.2 in Chap. 5 of Glowinski (1984).

With the above framework, we can show that $\Phi_v(\mathbf{v}) + L(\mathbf{v})$ is strictly convex and Gateaux-differentiable. The plastic term $B \int_\Omega \dot{\gamma}(\mathbf{u}) \, \mathrm{d}\mathbf{x}$ contained in $\Phi(\mathbf{u})$ is non-differentiable, but can be shown to be convex (not strictly). Standard results are then applied show that there exists a unique solution; e.g. Theorem 2.1 in Chap. 5 of Glowinski (1984).

For steady inertial problems (31) is extended by the addition of a term $Reb(\mathbf{u}, \mathbf{u}, \mathbf{v} - \mathbf{u})$, where $b(\cdot, \cdot, \cdot)$ is a trilinear form. One approach here is to formally replace $b(\mathbf{u}, \mathbf{u}, \mathbf{v} - \mathbf{u})$ with $b(\mathbf{u}, \zeta, \mathbf{v} - \mathbf{u})$. For sufficiently small $Re||\zeta||$, this problem remains elliptic and has a unique solution $\mathbf{u}_\zeta$. We then consider $Re||\zeta||$ small enough that $\zeta \mapsto \mathbf{u}_\zeta$ is a contraction mapping. This leads to the (usual) result that for inertial flows, provided Re and the data are small enough there is a unique steady solution.

More general transient Navier-Stokes problems were considered by Duvaut and Lions (1976), also within a variational setting. The mathematical theory for weak solutions and consideration of regularity has evolved, but is not considered further here. The reader is referred to Cioranescu (1976), Fuchs and Seregin (2000), Malek et al. (2005) for further insight.

3.2 *Monotonicity and Continuity*

Various results concerning monotonicity and continuity can be derived straightforwardly from the variational inequality (31) and the velocity minimization. For this section, to keep things simple we assume that

$$\partial\Omega_t = \emptyset, \quad \partial\Omega_v = \partial\Omega,$$

and that $\mathbf{u}_s = 0$. Additionally we assume the Bingham fluid model, for which:

$$\Phi(\mathbf{u}) = \int_\Omega \frac{1}{2}\dot{\gamma}(\mathbf{u}) : \dot{\gamma}(\mathbf{u}) + B\dot{\gamma}(\mathbf{u})\, \mathrm{d}\mathbf{x} = \frac{1}{2}\langle \mathbf{u}, \mathbf{u}\rangle + Bj(\mathbf{u}),$$

with obvious notation.

For the Bingham model the first term $\langle \mathbf{u}, \mathbf{u}\rangle$ is an elliptic bilinear form, which makes the manipulations below easier. Much of what follows can also be established for the other models (using monotonicity and convexity of Φ_v) and for different boundary conditions. Essentially what is needed is Poincaré's and Korn's inequality, general results on embedding of function spaces (Adams 1975; Dacorogna 2008), and for the boundary conditions to be such that if $\mathbf{v} \in \mathcal{V}$ is a rigid motion on the whole of Ω then $\mathbf{v} = 0$. This latter condition allows for the study of zero flow conditions via the results of Temam and Strang (1980), which characterize the space of bounded deformations.

Flow rates and bulk motions

In general the operator $L(\mathbf{v})$ denotes an integral over the flow domain, taken as a scalar product with a body force $\mathbf{f}$. Supposing that $\mathbf{f}_1$ and $\mathbf{f}_2$ are 2 body forces, with corresponding solutions $\mathbf{u}_1$ and $\mathbf{u}_2$, the variational inequality (31) holds for each solution, with the other taken as a test function. Summing the variational inequalities leads to

$$0 \le \langle \mathbf{u}_1 - \mathbf{u}_1, \mathbf{u}_2 - \mathbf{u}_1\rangle \le \int_\Omega (\mathbf{u}_2 - \mathbf{u}_1) \cdot (\mathbf{f}_2 - \mathbf{f}_1)\, \mathrm{d}\mathbf{x}. \tag{34}$$

The above inequality is strict unless $\mathbf{u}_2 = \mathbf{u}_1$, which is only possible if $\mathbf{u}_2 = \mathbf{u}_1 = 0$. What this inequality says is that the change in $\mathbf{u}$ in the direction of $\mathbf{f}$ is positive (integrated over Ω). There are a number of situations in which this is used.

- In simple unidirectional flows, e.g. flow along a duct or pipe of arbitrary cross-section, f would for example denote the (constant negative) pressure gradient

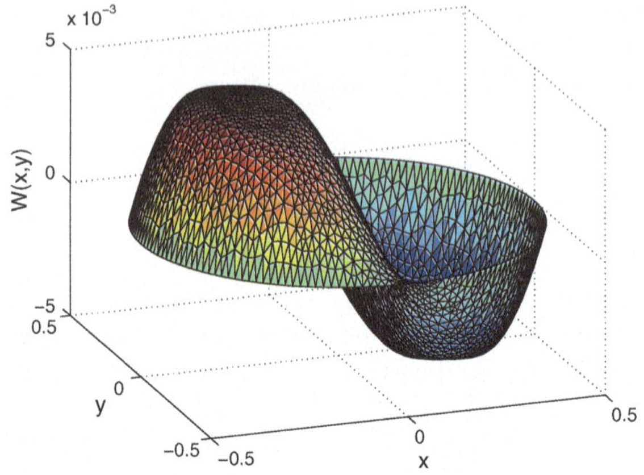

Fig. 3 Example of the velocity computed for an exchange flow in a circular pipe. Here the straight interface is fixed in the centre of the pipe and the rheological properties are the same in the 2 Bingham fluids

imposed in the flow direction and the integral of **u** is the flow rate. Thus, increasing the size of the pressure gradient results in an increase in the flow rate.

- This also extends to multiple fluids in ducts, as is relevant for example in thin-film type of models of displacement flows, e.g. Frigaard (1998).
- In multi-dimensions, by taking increases in any single component of f we get corresponding monotone increases in the flow in that same direction.
- In problems involving particle settling, see e.g. Putz and Frigaard (2010), this type of inequality shows that particles move in the direction of the applied force (i.e. the scalar product of the particle velocity with the force on the particle is positive), and for suitable symmetric particles there settling speed is monotone with the body force (heavy particles fall faster).
- This type of monotonicity also allows us to iteratively solve some problems, either with imposed force or with imposed motion. For example, in Frigaard and Scherzer (1998, 2000), Moyers-Gonzalez and Frigaard (2004) exchange flow problems are solved both numerically and semi-analytically: varying the pressure gradient to ensure a zero net flow between 2 yield stress fluids of different densities flowing axially in a duct. An example of the velocity field for one such flow is shown in Fig. 3. Here the straight interface is fixed in the centre of the pipe and the rheological properties are the same in the 2 Bingham fluids.
  Without iterating to find a pressure gradient that balances buoyancy, the net flow is in one direction, as illustrated in Fig. 4 (here one of the fluids is Newtonian).
- Variational methods were used extensively by Yoshioka, Adachi and co-workers in the 1970s, e.g. Adachi and Yoshioka (1973), Yoshioka and Adachi (1973). These include estimates of particle drag and results that to some extent generalise classical results of Faxen. Some of these results are less interesting in the modern era of computations, where Stokes flows are readily computed, but are worth reviewing for those working in this area.

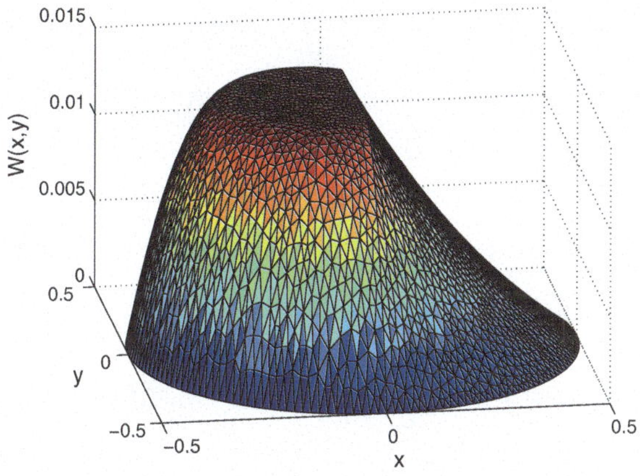

Fig. 4 Example of the velocity computed for a flow of a Newtonian fluid and a Bingham fluid in a circular pipe, separated by a straight interface

Continuity with respect to parameters

Solutions are generally continuous with respect to the data.
Exercise 1: Show that the solutions are continuous with respect to $\mathbf{f}$.
Exercise 2: For Bingham numbers B_1 and B_2, with corresponding solutions $\mathbf{u}_1$ and $\mathbf{u}_2$, show that:

$$\langle \mathbf{u}_1 - \mathbf{u}_1, \mathbf{u}_2 - \mathbf{u}_1 \rangle \leq C_1 |B_1 - B_2|^2, \tag{35}$$

$$\langle \dot{\gamma}(\mathbf{u}) \rangle \leq C_2 |B_1 - B_2|, \tag{36}$$

$$||\mathbf{u}_1 - \mathbf{u}_1||_{[H^1(\Omega)]^d} \leq C_3 |B_1 - B_2|, \tag{37}$$

for positive constants $C_1,\ C_2,\ C_3$ depending on Ω.

Note: In the dimensionless formulation considered, we have scaled away the plastic viscosity in the constitutive law. Otherwise, one can demonstrate similar continuity with respect to the plastic viscosity. This is essentially contained in the above continuity results.

Dissipation

Exercise 3: From the variational inequality (31) show that the plastic dissipation $\langle \dot{\gamma}(\mathbf{u}) \rangle$ decreases monotonically with B, and strictly unless $\mathbf{u} = 0$.

Further monotonicity results come from the velocity minimization. Suppose that we vary B, with say $B_1 < B_2$, and corresponding solutions: $\mathbf{u}_1$ and $\mathbf{u}_2$. As both solutions are in $\mathcal{V}$, we have:

$$\Phi(\mathbf{u}_1; B_1) = \min_{\mathbf{v} \in \mathcal{V}} \Phi(\mathbf{v}; B_1) \leq \min_{\mathbf{v} \in \mathcal{V}} \Phi(\mathbf{v}; B_2) = \Phi(\mathbf{u}_2; B_2). \tag{38}$$

Now, considering that here $\mathbf{u}_s = 0$, on using (26) we find

$$\Phi(\mathbf{u}) = \int_\Omega \phi(\dot{\gamma}(\mathbf{u})) - \boldsymbol{\tau}(\mathbf{u}) : \dot{\boldsymbol{\gamma}}(\mathbf{u}) \, d\mathbf{x}$$
$$= \int_\Omega \phi_v(\dot{\gamma}(\mathbf{u})) - \eta_v(\dot{\gamma}(\mathbf{u}))\dot{\gamma}(\mathbf{u}) \, d\mathbf{x} \leq 0. \qquad (39)$$

The last statement can be seen graphically and is easily calculated for common models:

$$\text{Bingham}: \quad \int_\Omega \phi_v(\dot{\gamma}(\mathbf{u})) - \eta_v(\dot{\gamma}(\mathbf{u}))\dot{\gamma}(\mathbf{u}) \, d\mathbf{x} = -\frac{1}{2}\int_\Omega \dot{\gamma}^2(\mathbf{u}) \, d\mathbf{x},$$
$$\text{Herschel-Bulkley}: \quad \int_\Omega \phi_v(\dot{\gamma}(\mathbf{u})) - \eta_v(\dot{\gamma}(\mathbf{u}))\dot{\gamma}(\mathbf{u}) \, d\mathbf{x} = -\frac{n}{n+1}\int_\Omega \dot{\gamma}^{n+1}(\mathbf{u}) \, d\mathbf{x},$$
$$\text{Casson}: \quad \int_\Omega \phi_v(\dot{\gamma}(\mathbf{u})) - \eta_v(\dot{\gamma}(\mathbf{u}))\dot{\gamma}(\mathbf{u}) \, d\mathbf{x} = -\int_\Omega \frac{\dot{\gamma}^2(\mathbf{u})}{2} + \frac{2B^{1/2}\dot{\gamma}^{3/2}(\mathbf{u})}{3} \, d\mathbf{x}.$$

Thus, for example we see that $\int_\Omega \dot{\gamma}^{n+1}(\mathbf{u}) \, d\mathbf{x}$ decreases with B for the Herschel-Bulkley fluid flow.

Note that although for the Bingham model the above expression happens to be (minus) the viscous dissipation functional, in general this is not the same. Instead it is equal to (minus) the stress functional Ψ evaluated at $\tau(\dot{\gamma}(\mathbf{u}))$.

Exercise 4: In the dimensional setting the minimization approach above can be used to show monotonic decrease in $-\Psi$ with an increase in consistency (or plastic viscosity).

4 Static Solutions to the Stokes Flow Problem

A unique feature of an ideal visco-plastic fluid flow is that it may withstand a shear stress while at rest, i.e. contrary to the definition of a fluid. In this section we look at situations in which the velocity may be zero everywhere in Ω. Therefore, we again assume that $\mathbf{u}_s = 0$. For simplicity we also assume that

$$\partial\Omega_t = \emptyset, \quad \partial\Omega_v = \partial\Omega,$$

although potentially some of what follows could be extended to include suitable surface tractions. Additionally, although not strictly necessary we will assume the Bingham fluid model for simplicity.

4.1 Physical Intuition of the Zero Flow

In the situation described, with essentially no boundary forcing, the flow can only be driven by the body force $\hat{\mathbf{f}}$, which induces stresses of size $\hat{\mathbf{f}}_0 \hat{L}$, where $\hat{\mathbf{f}}_0 = ||\hat{\mathbf{f}}||$, for some appropriately chosen norm of $\hat{\mathbf{f}}$. It is intuitive that the velocity scale be chosen to balance this driving stress with the viscous stress, i.e.

$$\hat{U}_0 = \frac{\hat{\mathbf{f}}_0 \hat{L}^2}{\hat{\mu}_0}.$$

It is also intuitive that if the driving stress is insufficient to exceed the yield stress, then the strain rate will be zero and, by virtue of the boundary conditions the velocity will also be zero. This balance is captured in:

$$\frac{\hat{\tau}_Y}{\hat{\mathbf{f}}_0 \hat{L}} = \frac{\hat{\tau}_Y \hat{L}}{\hat{\mu}_0 \hat{U}_0} = B.$$

In other words, with the velocity scale as defined, we expect zero velocity for large enough B. For a given geometry and body force, we may expect that there is a minimal value of B for which the velocity is zero and we call this the critical value B_{cr}.

4.2 Analytical Derivation and Definition

The above intuition is easily confirmed analytically, starting with the mechanical energy balance (23), which we may write as:

$$\langle \dot{\gamma}^2(\mathbf{u}) \rangle + B \langle \dot{\gamma}(\mathbf{u}) \rangle = \langle \mathbf{u} \cdot \mathbf{f} \rangle. \tag{40}$$

The two terms on the left-hand side represent the rates of viscous and plastic dissipation, respectively. The right-hand side represents the work done by the body force. By the Hölder inequality

$$\langle \mathbf{u} \cdot \mathbf{f} \rangle \leq ||\mathbf{u}||_p ||\mathbf{f}||_q$$

where $|| \cdot ||_p$ represents the L^p norm over Ω and $1/p + 1/q = 1$. Secondly, due the no-slip condition, the only rigid body motion possible is $\mathbf{u} = 0$. For $\mathbf{u} \neq 0$, following Temam and Strang (1980), there exists a constant $C_{TS} > 0$, depending on Ω, such that

$$C_{TS} ||\mathbf{u}||_p \leq \langle \dot{\gamma}(\mathbf{u}) \rangle,$$

provided that $p \leq d/(d-1)$ for dimension $d = 2, 3$. Thirdly, for Dirichlet boundary conditions we have $\langle \dot{\gamma}^2(\mathbf{u}) \rangle = \langle |\nabla \mathbf{u}|^2 \rangle$ and then the Poincaré inequality gives:

$$C_P \langle \mathbf{u}^2 \rangle = C_P ||\mathbf{u}||^2 \leq \langle |\nabla \mathbf{u}|^2 \rangle = \langle \dot{\gamma}^2(\mathbf{u}) \rangle,$$

again for $C_P > 0$, depending on Ω.

Using the above inequalities and assuming $\mathbf{u} \neq 0$, together with the mechanical energy balance gives:

$$C_P\|\mathbf{u}\|^2 + C_{TS}B||\mathbf{u}||_p \leq ||\mathbf{u}||_p||\mathbf{f}||_q,$$

which yields

$$\|\mathbf{u}\|^2 \leq \frac{||\mathbf{f}||_q - C_{TS}B}{C_p}||\mathbf{u}||_p. \tag{41}$$

We see that when $B \geq ||\mathbf{f}||_q/C_{TS}$ the assumption that $\mathbf{u} \neq 0$ is contradicted. Consequently, $\mathbf{u} = 0$ for $B \geq ||\mathbf{f}||_q/C_{TS}$. For $d = 2$ we have $p = q = 2$ and for $d = 3$ we have $p = 3/2$, $q = 3$.

Note: the inequality (41) gives a sufficient condition and not a necessary one: it is quite possible that the steady regime is motionless at lower values of B. The above analysis establishes that there is a critical B, by focusing on the importance of the size of $\mathbf{f}$, measured here via $\|\mathbf{f}\|_q$. However, the distribution of $\mathbf{f}$ plays an equally important role in determining fluid motion, as the following two simple examples (for $d = 2$) show.

(i) if $\mathbf{f} = -y\mathbf{e}_y$ then the following motionless state satisfies the governing equations and therefore is a steady solution for all values of B:

$$\tau_{ij} = 0, \quad u_i = 0, \quad p = \frac{1}{2}y^2 + P_0.$$

Exercise: Show that a small y-dependent $\tau_{yy} = -\tau_{xx}$ can also admit static solutions—how large can $|\tau_{yy}|$ be?

(ii) Alternatively consider: $\mathbf{f} = -x\mathbf{e}_y$, which only gives a motionless solution for sufficiently large B. A simple 1D example of such a flow is considered in Karimfazli and Frigaard (2013), in the context of natural convection.

Suppose that we consider only $\mathbf{f}$ for which there is a Helmholtz decomposition $\mathbf{f} = -\nabla\phi + \nabla \times \mathbf{A}$ where $\nabla\phi$ and $\nabla \times \mathbf{A}$ are the irrotational and divergence free components of $\mathbf{f}$. Using the divergence theorem we can easily show that $\langle \mathbf{u} \cdot \nabla\phi \rangle = 0$ and thus

$$\langle \mathbf{u} \cdot \mathbf{f} \rangle = \langle \mathbf{u} \cdot (\nabla \times \mathbf{A}) \rangle \tag{42}$$

Equation (42) illustrates that the irrotational component of $\mathbf{f}$ is conservative and does no work. This also suggests how one should select a representative scale for the body force, i.e. $\hat{\mathbf{f}}_0$. Since only $(\nabla \times \mathbf{A})$ drives the flow, a suitable choice would be:

$$\hat{\mathbf{f}}_0 = \|\nabla \times \mathbf{A}\|_q. \tag{43}$$

When $\mathbf{A} = 0$, then there exists a motionless steady regime for all values of B. If instead $\mathbf{A} \neq 0$, the steady state cannot be motionless unless B is less than a critical value, B_{cr}. The critical Bingham number is analyzed as follows: assuming that $\mathbf{u} \neq 0$,

$$\langle \dot{\gamma}^2(\mathbf{u}) \rangle = -\langle \dot{\gamma}(\mathbf{u}) \rangle \left[B - \frac{\langle \mathbf{u} \cdot \mathbf{f} \rangle}{\langle \dot{\gamma}(\mathbf{u}) \rangle} \right] \leq -\langle \dot{\gamma}(\mathbf{u}) \rangle \left[B - \sup_{\mathbf{v} \in \mathcal{V}, \mathbf{v} \neq 0} \left\{ \frac{\langle \mathbf{v} \cdot \mathbf{f} \rangle}{\langle \dot{\gamma}(\mathbf{v}) \rangle} \right\} \right]. \quad (44)$$

Therefore, defining B_{cr} by:

$$B_{cr} = \sup_{\mathbf{v} \in \mathcal{V}, \mathbf{v} \neq 0} \left\{ \frac{\langle \mathbf{v} \cdot \mathbf{f} \rangle}{\langle \dot{\gamma}(\mathbf{v}) \rangle} \right\} = \sup_{\mathbf{v} \in \mathcal{V}, \mathbf{v} \neq 0} \left\{ \frac{\langle \mathbf{v} \cdot (\nabla \times \mathbf{A}) \rangle}{\langle \dot{\gamma}(\mathbf{v}) \rangle} \right\}, \quad (45)$$

we see that $B \geq B_{cr}$ enforces that $\langle \dot{\gamma}^2(\mathbf{u}) \rangle = 0$, and hence $\mathbf{u} = 0$.

Two-dimensional flows: If we consider only 2D flows, then we consider $\mathbf{f}$ to comprise only the (x, y)-components of a 3D force. Note that for a 2D flow that is independent of z, the vector potential is $\mathbf{A} = (\cdot, \cdot, A(x, y))$, and any admissible $\mathbf{v} \in \mathcal{V}$ may be expressed in terms of its stream function $\psi_{\mathbf{v}}$, (set arbitrarily to 0 on $\partial\Omega$). We see that:

$$\langle \mathbf{v} \cdot (\nabla \times \mathbf{A}) \rangle = \langle \nabla \psi_{\mathbf{v}} \cdot \nabla A \rangle = -\langle A \nabla^2 \psi_{\mathbf{v}} \rangle = -\langle \psi_{\mathbf{v}} \nabla^2 A \rangle,$$

so that optimality in (45) promotes alignment of the streamlines with contours of the vector potential A, in so far as is possible. Alternately, we can note that $-\nabla^2 \psi_{\mathbf{v}}$ is the vorticity $\omega_{\mathbf{v}}$ associated to $\mathbf{v}$. Hence optimality maximizes $\langle A \omega_{\mathbf{v}} \rangle$.

Other models: If we consider models other than the Bingham, (e.g. Casson, Herschel-Bulkley), it is worth noting that they all have the same critical limit B_{cr}. The above analysis can be repeated and in place of showing $\langle \dot{\gamma}^2(\mathbf{u}) \rangle = 0$, we would have that $\langle \phi_v(\mathbf{u}) \rangle = 0$, which is also a positive functional, equal to zero only when $\mathbf{u} = 0$. Alternatively, one could use the stress maximization formulation. If for one fluid type we find $\mathbf{u} = 0$, then the stress field does not exceed B and must be admissible for other fluids (i.e. the Stokes equations are satisfied). Thus, $\Psi = 0$ for the other fluid, which also does not yield.

Sharpness of the bound (45) Using the previous results on continuity and monotonicity, it can be shown that the viscous dissipation functional converges to zero faster than either the plastic dissipation or work term, as $B \to B_{cr}^-$. Using this it can be established that the condition (45) is also a sharp bound determining static flow, i.e. in the limit $B \to B_{cr}$ the solution $\mathbf{u}$ is a maximizer.

Other settings: Other than the flows considered above, which are internal flows, this type of flow problem has also been considered by Ionescu and co-workers in the contexts of landslides, e.g. Hild et al. (2002), Ionescu and Lachand-Robert (2005), Hassani et al. (2005). These problems generally have a free surface and an unknown surface internally which the material may yield, under the weight of gravity and the resulting shear stresses. Given the geophysical length-scales involved, spatial variation of the yield stress is also common and/or some variation in the force (which depends on the loading and topography). In this context, the terminologies "load limit analysis" and "blocking solutions" have been used.

4.3 Examples of Zero Flow Limits

A number of zero flow limits are known exactly. Generally, these are for flows with a single velocity component.

1. Flow through a plane channel of unit half-width, driven by a unit pressure gradient has $B_{cr} = 1$, (exercise).
2. Flow through a pipe of unit radius, driven by a unit pressure gradient has $B_{cr} = 1/2$, (exercise).
3. Flow in one direction along a duct with unit square cross-section, driven by a unit pressure gradient has $B_{cr} = 1/(2 + \sqrt{\pi}) \approx 0.2650...$; see Mosolov and Miasnikov (1965) and discussion later of similar examples.

Some other flows that are two-dimensional or axisymmetric have critical limits that have been computed and are regarded as accurate.

1. Settling of a disc under gravity in an infinite fluid. Here the density difference of the solid disc and surrounding fluid drives the flow. The balance of settling force and yield stresses is captured in the plastic drag coefficient, defined as:

$$C_D^p = \frac{\hat{F}_s}{\hat{A}_\perp \hat{\tau}_Y} = \frac{\Delta\hat{\rho}\hat{g}\pi\hat{D}^2/4}{\hat{D}\hat{\tau}_Y}.$$

   This represents the ratio of driving force ($\hat{F}_s$) to cross-sectional area perpendicular to the flow ($\hat{A}_\perp$), all divided by the yield stress ($\hat{D}$ is the disc diameter and unit cylinder length can be assumed in the 3rd dimension). As the yield stress is increased the particle becomes static at a critical value $C_{d,c}^p$. The value $C_{d,c}^p$ has been computed by Tokpavi et al. (2008) and appears to coincide with the analytical prediction for perfectly plastic yielding of a cylindrical pile pushed laterally through soil; see Randolph and Houlsby (1984): $C_{d,c}^p = 4\sqrt{2} + 2\pi \approx 11.94....$ Figure 5 shows the limiting flow around a settling circular disc, from Tokpavi et al. (2008). Here Od denotes a Bingham number based on the particle velocity and the limit of no-flow is approached as $Od \to \infty$.
2. Settling of a sphere under gravity in an infinite fluid. Again the density difference of the solid disc and surrounding fluid drives the flow and the plastic drag coefficient, is:

$$C_D^p = \frac{\hat{F}_s}{\hat{A}_\perp \hat{\tau}_Y} = \frac{\Delta\hat{\rho}\hat{g}\pi\hat{D}^3/6}{\pi\hat{D}^2/4\hat{\tau}_Y}.$$

   The critical limit was computed by Beris et al. (1985) and found to be $C_{d,c}^p \approx 6.993$.

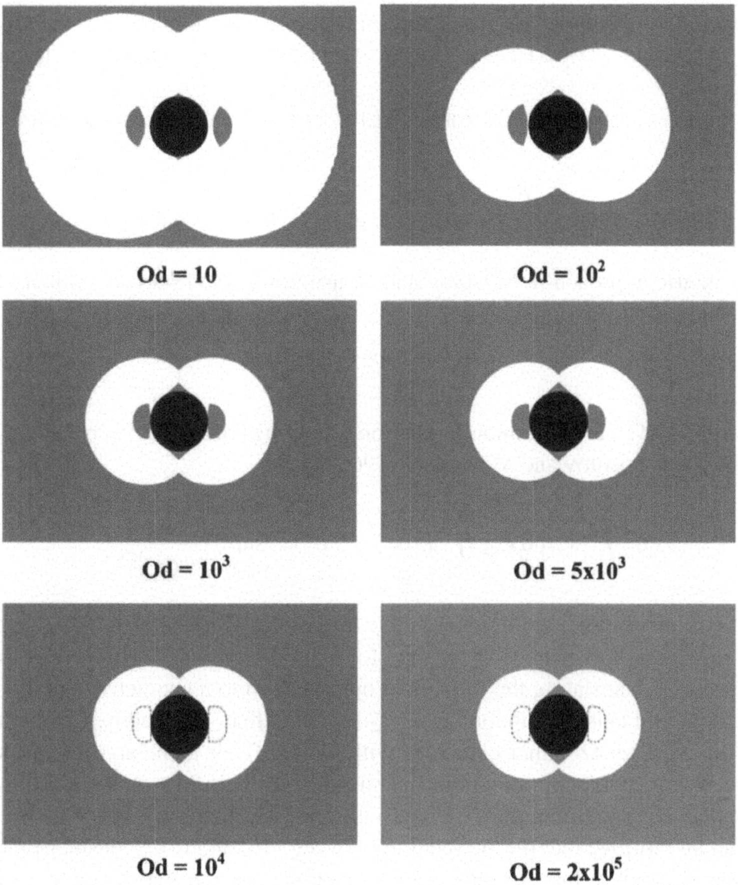

Fig. 5 Limiting flow around a settling circular disc. Reproduced with permission from Tokpavi et al. (2008)

4.4 *Exact Methods*

Aside from one-dimensional flows for which the solution may be computed analytically and the limit of zero flow analysed directly, there are two classes of flows about which more can be said in any generality: (i) anti-plane shear flows in closed regions; (ii) two-dimensional planar flows.

Anti-plane shear flows in closed regions

Here Ω is a 2-dimensional closed region in the (x, y)-plane and the velocity $\mathbf{u} = (0, 0, w(x, y))$, i.e. this is flow along a cylinder of arbitrary cross-section. General consideration of this type of flow/no-flow threshold in a mathematical framework has been undertaken in Mosolov and Miasnikov (1965, 1966, 1967), which contain a wealth of interesting results and insights, but are quite technical. Within these papers is a prescriptive methodology for determining the critical limit and this is explained clearly in Huilgol (2006). Here we just give an introduction to the geometric aspect of this work.

Assuming a constant body force (in the z-direction), the analysis presented leads to:

$$B_{cr} = \sup_{u \in H_0^1(\Omega); u \neq 0} \frac{\int_\Omega u \, \mathbf{dx}}{\int_\Omega |\nabla u| \, \mathbf{dx}}, \tag{46}$$

The geometric approach in Mosolov and Miasnikov (1965) starts from there being no flow in Ω if

$$\int_\Omega B|\nabla u| - u \, \mathbf{dx} \geq 0 \tag{47}$$

for all $u \in H_0^1(\Omega)$. If u is a smooth function satisfying the boundary conditions, then it is shown by Mosolov and Miasnikov (1965) that

$$B' \int_\Omega |\nabla u| \, \mathbf{dx} \geq \int_\Omega u \, \mathbf{dx} : \qquad B' = \sup_{\Omega' \subseteq \Omega} \frac{\text{meas}(\Omega')}{\text{meas}(\partial\Omega')} \tag{48}$$

where Ω' is an arbitrary sub-domain of the domain Ω, with boundary $\partial\Omega'$, i.e. effectively the largest ratio of area to perimeter for a sub-domain of Ω. This is the essential geometric step, translating the functional optimisation to a geometric one. It is also shown that there exists a subdomain $\Omega' \subseteq \Omega$ that defines B' exactly. The boundary $\partial\Omega'$ of the *optimal* Ω' either coincides with that of Ω, or is the arc of a circle that is tangent with $\partial\Omega$ at common points. Thus, according to Mosolov and Miasnikov (1965) the yield condition, for a non-zero flow in Ω, is that: $B \geq B'$.

To get an intuitive feel for the relation between (46) and (48), consider functions that approximately minimise (46). These functions are constant on as large a sub-domain as is possible, but must drop to zero in a narrow ring close to the boundary. For such functions, the quotient in (46) can be approximately evaluated and is seen to relate to the geometric ratio in (48). Assuming the equivalence of B' and B_{cr} it is possible to calculate various limiting flows, as in Mosolov and Miasnikov (1965), Huilgol (2006). Some of the known geometries are: (a) square duct; (b) rectangular duct; (c) L-shaped duct; (d) circular pipe; (e) concentric annulus; (f) equilateral triangular duct; (g) elliptical pipe.
Exercise: Derive the critical limit for the flow through a square pipe and through an equilateral triangular duct. Assume sides of unit length and assume the problem is scaled so that a unit pressure gradient is applied in the flow direction.

Two-dimensional planar flows

The continuity and monotonicity results suggest that the viscous dissipation plays no role in the limiting flows. On neglecting the viscous terms in the constitutive law, in yielded regions we have approximately:

$$\tau \approx B \frac{\dot{\gamma}}{\|\dot{\gamma}\|}. \tag{49}$$

which appears to approximate the constitutive law for perfectly plastic materials (Hill 1950). Consequently, it is natural to ask if the theory of perfect plasticity can determine the yield limit?

For many problems we may switch between formulations in which either a body force or a boundary motion is is prescribed (see Sect. 4.6 below). For the latter we have no body force and the system of equations (assuming (49) to be exact) is:

$$\nabla \cdot \tau - \nabla p = 0, \tag{50}$$

$$\|\tau\| = B. \tag{51}$$

In such 2D flows the steady momentum equations plus the von Mises yield criterion, form a closed set of hyperbolic equations for the plastic regions of flow; see e.g. Chakrabarty (2012). The stress at each point in the plastic region can be represented by a shear stress, an angle ϕ and the pressure p:

$$\sigma_{xx} = -p - B\sin(2\phi), \quad \sigma_{yy} = -p + B\sin(2\phi), \quad \tau_{xy} = B\cos(2\phi).$$

We may find a pair of orthogonal curves, at each point along which the magnitude of the shear stress is equal to B. The principle stress directions bisect the right angle between these curves, which are called the sliplines (α and β families). Thus, conventionally ϕ represents the counterclockwise orientation which an α-line makes with the x-axis. The sliplines are the characteristic lines of the hyperbolic system (50)–(51). Consequently, there may be discontinuity in normal derivatives of the stress (and velocity) components across these curves.

Computation of the sliplines is carried out by integrating the Hencky equations, which state that along each line of a family of sliplines, a quantity is conserved (Riemann invariant):

$$p + 2B\phi = const. \qquad \text{along an } \alpha\text{-line}, \tag{52}$$

$$p - 2B\phi = const. \qquad \text{along a } \beta\text{-line}. \tag{53}$$

The stresses and slipline field can thus be resolved independently of the velocity. The velocity can itself be evaluated using a similar set of equations—the Geiringer equations. Other features of the sliplines are described in Hill (1950), Chakrabarty (2012).

The precise relation between (50)–(51) and the analogous limiting 2D planar viscoplastic flows is not fully understood. The slipline method can be used to give admissible velocity and stress fields, which may then approximate Φ and Ψ. Certainly there are cases where the slipline method appears to give B_{cr} exactly (e.g. flows around a circular cylinder), but this is probably not generally true.

Note that the slipline method creates an envelope of characteristics that delineate an area within which $\tau = B$, but does not give information outside this envelope. This envelope may or may not correspond to the envelope of yielded fluid for the analogous visco-plastic problem. Secondly, the visco-plastic problem often contains

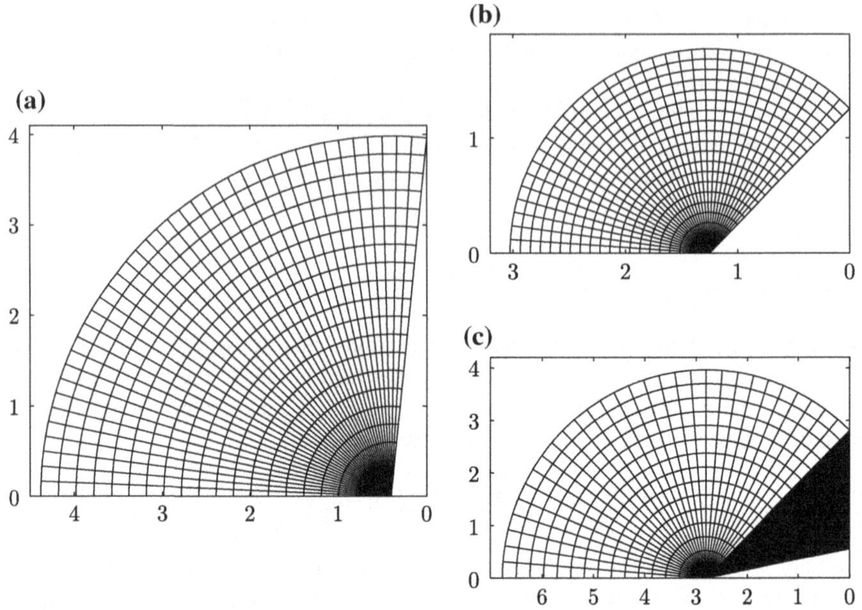

Fig. 6 Slipline network around the diamond: **a** $\chi = 10$; **b** $\chi = 1$; **c** $\chi = 0.2$. Due to symmetry only 1/4 of the flow is shown. Reproduced with permission from Chaparian et al. (2017)

regions that are below the yield stress as well as those at just above the yield stress. Nevertheless, the slipline method is a useful tool in some cases.

An example of where the slipline method approximates the limiting flows very effectively is shown below in Figs. 6 and 7, from Chaparian et al. (2017). Here a dense diamond shaped particle is settling along its axis of symmetry. Figure 6 shows the characteristic slipline network computed for 3 diamonds with different aspect ratios χ, i.e. assuming the surrounding medium is a perfectly plastic fluid. Figure 7 presents the analogous computations for the diamond-shaped particle settling through a Bingham fluid, close to the limit of zero flow. The outer yield surfaces in both the slipline analysis and the computed visco-plastic solution are very similar and the calculated critical plastic drag coefficients coincide with those calculated from the slipline analysis.

4.5 Approximation Methods

In specific problems we may be able to approximate B_{cr} directly using a variety of methods, some examples of which follow.

1. *Crude approximations*: For example, in a 2D flow it is easy to deduce that $\dot{\gamma} \geq 2|u_x| = 2|v_y|$. Lets consider a continuous $\mathbf{f} = (f_1, f_2)$ in a domain with maximal

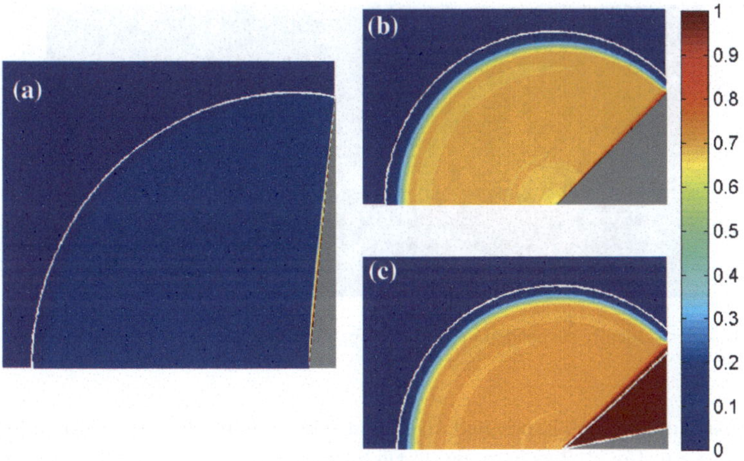

Fig. 7 Calculated speed colourmap around diamond, close to the critical limit: **a** $\chi = 10$; **b** $\chi = 1$; **c** $\chi = 0.2$. The speed is normalized with the particle settling speed. Due to symmetry only 1/4 of the flow is shown. Reproduced with permission from Chaparian et al. (2017)

linear dimensions in x and y-directions ℓ_x and ℓ_y, respectively. We can bound as follows:

$$|f_1(x, y)u(x, y)| = |f_1(x, y) \int_{x_{\partial\Omega}}^{x} u_x(s, y)\, \mathrm{d}s|$$
$$\leq \frac{\max_\Omega\{|f_1(x, y)|\}}{2} \int_{x_{\partial\Omega}}^{x} \dot{\gamma}(s, y)\, \mathrm{d}s,$$

where $x_{\partial\Omega}$ denotes the nearest position to x on $\partial\Omega$. An analogous bound follows for $|f_2(x, y)v(x, y)|$. We then integrate over Ω, noting that no point is more than $\ell_x/2$ or $\ell_y/2$ from the nearest boundary, to give:

$$\int_\Omega \mathbf{f} \cdot \mathbf{u}\, \mathrm{d}\mathbf{x} \leq \frac{||\mathbf{f}||_\infty(\ell_x + \ell_y)}{4} \int_\Omega \dot{\gamma}(\mathbf{u})\, \mathrm{d}\mathbf{x} \quad \Rightarrow \quad B_{cr} \leq \frac{||\mathbf{f}||_\infty(\ell_x + \ell_y)}{4}.$$

Estimates such as this are fairly easy to derive. Except in cases of particular symmetry the (upper) bounds for B_{cr} derived tend to be fairly conservative, and this is for the simple reason that the geometric nature of the limiting flows is completely ignored.

What then is the point of such estimates? Firstly, these are instructive and may give a practical engineering answer where Ω is complex. Secondly, the techniques can be applied in more complex problems. An example is where we may have a media with high geometric diversity and it is impossible to say much about a specific feature, except in statistical terms. As an example, consider bubbles held suspended in a visco-plstic media. For a given yield stress there is a wide

Fig. 8 Bubbles statically suspended in a carbopol gel held within a tube

diversity of bubble shapes and sizes that may remain statically suspended in a gel, see e.g. Fig. 8. To characterize such situations it is perhaps less valuable to target a particular idealized bubble shape, than to offer conservative but general predictions.
Examples of the use of bounds such as the above, applied to bubbles, can be found in Dubash and Frigaard (2004). When compared with accurate computations of a single bubble rising in an infinite medium these bounds are shown to be very conservative. See for example the very nice computational study of Tsamopoulos et al. (2008).

2. *Newtonian solution*: In any flow situation the solution for the Newtonian fluid gives a stress field which is admissible (and is known for many simple problems). When this stress field does not exceed the yield stress, the stress maximization principle can be used to provide an upper bound for B_{cr}.
3. *Linear elastic stress field*: As with the above, linear elasticity may be used to generate an admissible stress field, which can be inserted into the stress maximization principle to provide an upper bound for B_{cr}. More clearly, in such a test-problem if B exceeds the maximum deviatoric stress over Ω then $\Psi = 0$. Therefore, the maximum deviatoric stress of any admissible stress field gives an upper bound for B_{cr}.
4. *Slipline method*: As discussed, the slipline method leads to admissible stress and velocity fields that may be good approximations.

4.6 Computational Methods

There are a number of different ways to compute the limit of zero flow.

1. ***Direct method***: this refers to the obvious method of computing the solution, say $\mathbf{u}_B$, at given B, measuring some functional of $\mathbf{u}_B$ (typically a norm, a dissipation functional or flow rate), then increasing B iteratively until that measure becomes zero. As we generally have monotonicity results on $\mathbf{u}_B$, we know that $\mathbf{u}_B = 0$

for $B \geq B_{cr}$. This gives us a "one-sided" zero to determine. We should be a bit cautious with using methods that rely on the gradient, but other simple methods such as the bisection method work effectively. If high precision is needed in determining B_{cr} then one should know the rate at which $\mathbf{u}_B \to 0$ as $B \to B_{cr}$, so that a tolerance can be assigned to the iterative procedure.

Depending on the solution method for $\mathbf{u}_B$, it may be significantly quicker to find B_{cr} by iterating from above. This is the case if e.g. augmented Lagrangian methods are used, which can solve for $\mathbf{u}_B = 0$ correctly, whereas regularization methods cannot. The volocity and strain rate iterations can quickly converge in such situations (being zero). It is recommended that a method such as the augmented Lagrangian method be used for any direct iteration.

Note however that this is not always sufficient for an accurate evaluation of B_{cr}. For example, Moyers-Gonzalez and Frigaard (2004) found order 1 errors in computing the critical limit for flow through a square duct. The errors arose due to the strongly geometrical nature of the limiting solutions as $B \to B_{cr}$, which are not easily represented on fixed meshes. In general some form of mesh adaptivity is advisable in computing these limits.

2. ***Re-scaling the problem***: Note that since we take $\mathbf{u}_B \to 0$ as $B \to B_{cr}$ and we compute $\mathbf{u}_B$ numerically at each iteration only to some specified tolerance, it is sometimes easier to study $B \to B_{cr}$ through a transformed problem for which the velocity does not vanish. Typically this involves re-scaling the problem, e.g. using an imposed velocity rather than a specified force.

   Example, channel flow: As a simple example of this, consider plane Poiseuille flow of a Bingham fluid along a plane channel of width $2\hat{L}$, driven by a constant pressure gradient $-\hat{G}$ in the flow direction. From integrating the 1-dimensional momentum balance, we would find that the velocity profile has a central plug region of width: $2\hat{L}y^*$, where

$$\hat{L}y^*\hat{G} = \hat{\tau}_Y,$$

   and as $y^* \to 1$ the flow stops. This is captured by the critical ratio: $\hat{\tau}_Y/(\hat{L}\hat{G}) \geq 1$, which can be thought of as a critical B in the sense we have been discussing. In this case, we might define a velocity scale from the pressure gradient as $\hat{U} = \hat{G}\hat{L}^2/\hat{\mu}_p$, and hence a Bingham number $B = \hat{\tau}_Y\hat{L}/(\hat{U}\hat{\mu}_p) = \hat{\tau}_Y/(\hat{L}\hat{G})$. We see that as $B \to 1 = B_{cr}$ the flow stops.

   Alternatively, if we do not want to compute small velocities in the limit $B \to B_{cr}^-$, we may instead suppose that $\hat{G}$ is unknown, but that the mean speed $\hat{U}_0$ is known. We now define a new Bingham number $\tilde{B}$ as:

$$\tilde{B} = \frac{\hat{\tau}_Y\hat{L}}{\hat{U}_0\hat{\mu}_p},$$

   and can easily find the dimensionless velocity (scaled with $\hat{U}_0$) is:

$$\tilde{W}(y) = \begin{cases} \frac{\tilde{B}}{2y^*}(1 - y^*)^2 & 0 \le |y| \le y^*, \\ \frac{\tilde{B}}{2y^*}[(1 - y^*)^2 - (|y| - y^*)^2] & y^* < |y| \le 1, \end{cases} \tag{54}$$

where $y = \pm y^*$ are the dimensionless positions of the yield surfaces. Since the mean velocity has been used to scale the flow, it follows that:

$$1 = \int_0^1 \tilde{W}(y)\, dy, \tag{55}$$

which leads to:

$$0 = y^{*3} - 3y^*\left[1 + \frac{2}{\tilde{B}}\right] + 2, \tag{56}$$

which must be solved for y^*. Having found y^*, we may compute $\hat{G}$ using $\hat{L}y^*\hat{G} = \hat{\tau}_Y$.

The re-scaled problem is equivalent to the first one for any non-zero velocity. The limit $B \to B_{cr}^-$ is equivalent to $\tilde{B} \to \infty$, but the $\tilde{B}$-problem always has order 1 velocities (which may be preferable numerically). The polynomial (56) is easily solved to find the unique root which satisfies $0 < y^* < 1$. As $\tilde{B} \to \infty$, the unyielded plug regions widen towards the walls of the channel. An asymptotic expansion for large $\tilde{B}$ gives:

$$y^*(\tilde{B}) \sim 1 - \frac{\sqrt{2}}{\tilde{B}^{1/2}} + \frac{2}{3\tilde{B}} + O(\tilde{B}^{-3/2}), \tag{57}$$

which gives the relation between the 2 limits as: $B = y^* \to B_{cr} = 1$ as $\tilde{B} \to \infty$. More explicitly, we observe that $\tilde{B}^{-1} \sim (B_{cr} - B)^2/2$ as $B \to B_{cr}^-$. From its definition, we also see that $\tilde{B}^{-1} = \hat{U}_0/\hat{U}$, which is the mean velocity, scaled with $\hat{U}$. Thus, in the unscaled problem, if we take

$$Q(B) = \int_0^1 W(y)\, dy,$$

as our measure of the velocity, we see that $Q(B) \sim (B_{cr} - B)^2/2$ as $B \to B_{cr}^-$ and evidently $Q(B) = 0$ for $B \ge B_{cr}$.

Re-scaling as above is common in many problems. Note that the functional in (45) is invariant to rescaling the velocity *and* the supremum is attained by the solution in this limit. Computationally, one might solve e.g. the $\tilde{B}$-problem for successively large $\tilde{B}$, then substitute the solution $\tilde{\mathbf{u}}$ directly into (45) to approximate B_{cr}. Numerically, as $||\tilde{\mathbf{u}}|| \sim O(1)$, we may often find B_{cr} in a more precise and stable way than with the direct method for $\mathbf{u} \sim 0$. Having found B_{cr} in this way, the asymptotic behaviour could be studied using either method.

An example of a more complex version of the channel flow problem is illustrated below, from Roustaei et al. (2016). Here the flow is through a 2D affine fracture, of minimal width $2\hat{L}$. Computations are made for the $\tilde{B}$-problem for 3 different complex geometries at successively large $\tilde{B} = 100,\ 1000,\ 10000$ (only one is shown below in Fig. 9). Convergence of B to B_{cr} as $\tilde{B}$ increases is plotted in Fig. 10

3. ***Informed guessing***: The main alternative to the direct method (with or without re-scaling of the problem), is to find the limit exactly by inferring the form of the limiting solution, from computation or intuition. Often a few computations at large $\tilde{B}$ suffice to expose the main structure of the solution in the limit of no flow. Having understood this, we may construct a test function with the same limiting behaviour and insert this directly into (45), which is invariant to re-scaling. For sufficiently simple problems we can then evaluate B_{cr}.

   As an example of this, Fig. 11 shows the speed colourmap from a differentially heated square cavity (hot wall to the left, cold wall to the right, insulated top and bottom); see Karimfazli et al. (2015). The driving force here is buoyancy, which drives a single convection cell. For sufficiently large B the steady velocity solution appears to rotate in a central plug region, separated by thin shear layers from the walls and regions of unyielded fluid in each corner, where $\mathbf{u} = 0$. Having understood that as $B \to B_{cr}$, the shear layers become progressively thin, we can postulate a test function that approximates this behaviour. By doing this we are able to evaluate $B_{cr} = 1/32$ for the square cavity.

   It is worth commenting that for this same problem, with other aspect ratios cavities we are unable to evaluate B_{cr} exactly, although parts of their final structure are easily predicted. For the limiting case of an infinitely tall cavity we can solve the corresponding 1D problem and evaluate B_{cr} directly, but intermediate aspect ratios still require direct computation and iteration.

4. ***Using regularization methods***: It has been common to use regularization methods in solving visco-plastic flow problems numerically. In computing general flows this is a reasonable strategy and is typically faster than the augmented Lagrangian method. However, for the problem of zero flow all regularization methods lead to motion at sub-yield stress values of the deviatoric stress, and this is simply because these methods become Newtonian-like at low strain rates. For this reason we don't feel that these methods are ideally suitable for problems of zero flow.

   Notwithstanding the above, there are many instances these regularized fluid models are used in situations where the stress falls below B. The difficulty here is that although the velocity field of the regularized model should converge to that of the exact model (as the regularization parameter $\varepsilon \to 0$, there is no guarantee that the stress will converge. This is a general issue with understanding the meaning of yield surface positions computed using these methods.

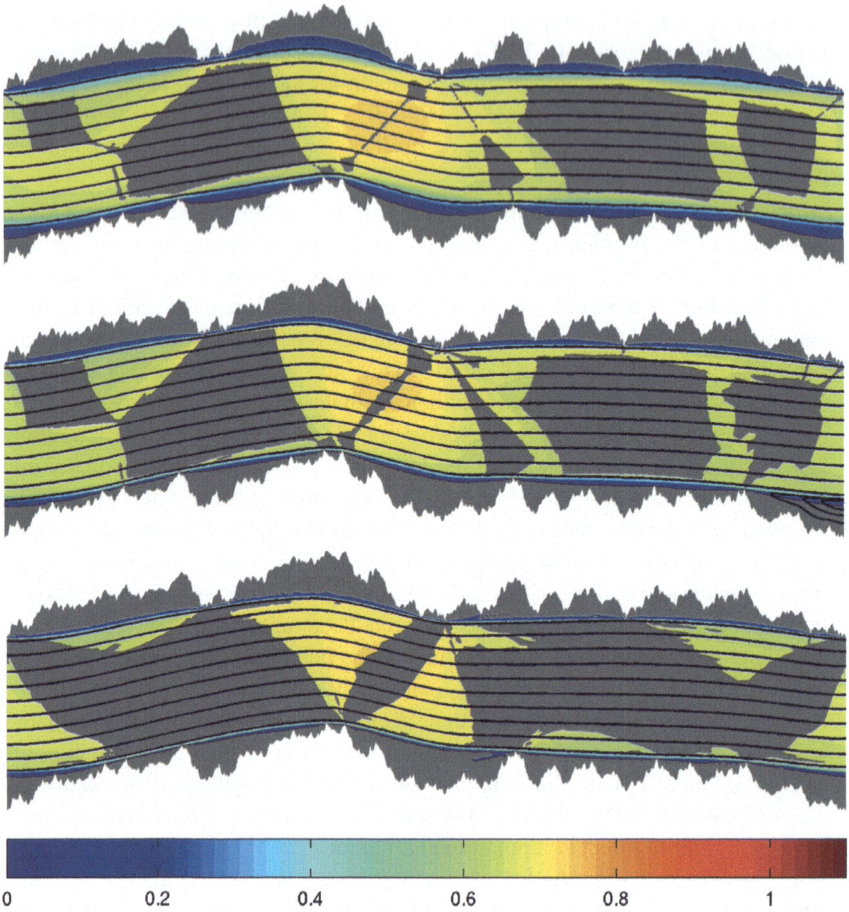

Fig. 9 Computed examples of speed $|\mathbf{u}|$ and streamlines for a fracture formed from two affine surfaces. Parameters are: $\tilde{B} = 100,\ 1000,\ 10000$, from top to bottom, with gray denoting unyielded plug regions. Reproduced with permission from Roustaei et al. (2016)

However, the treatment of zero flow problems with regularization could be improved significantly by using the stress maximization principle. The stress field computed from the regularized constitutive law is still admissible. This means that if the stress of the regularized model lies below B everywhere then the actual stress must also lie below B, i.e. the regularized model computations when used with care can predict sufficient conditions for the flow to stop.

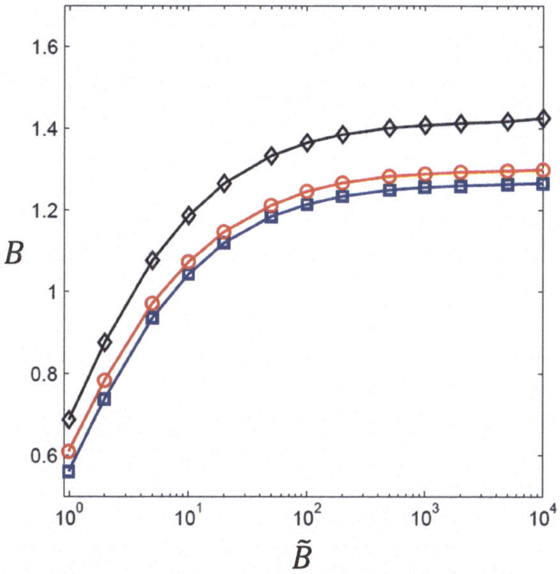

Fig. 10 Computed Oldroyd number as function of the Bingham number for the affine fracture of Fig. 9 and two other similar fractures. Blue squares: the two surface are symmetrical. Red circles: the two surfaces are uncorrelated. Black lozenges: the two surfaces are identical, shifted laterally (Fig. 9). Reproduced with permission from Roustaei et al. (2016)

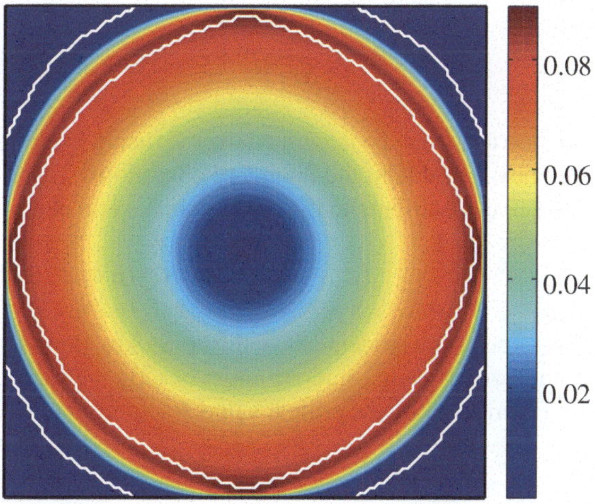

Fig. 11 Contours of the computed speed as $B \to B_{cr}^- = 1/32$. Here computed at $B = 0.0311$. The white curves represent yield surfaces. Reproduced with permission from Karimfazli et al. (2015)

5 Flow Stability

We now look at flows which are transient and inertial, keeping to the same setup as before. We assume that $\mathbf{u}_s = 0$ and that

$$\partial\Omega_t = \emptyset, \quad \partial\Omega_v = \partial\Omega.$$

Fluid motion in the domain is governed by the Navier-Stokes equations

$$\frac{\partial u_i}{\partial t} + Re\, u_j \frac{\partial u_i}{\partial x_j} = -\frac{\partial P}{\partial x_i} + \frac{\partial \tau_{ij}}{\partial x_j} + f_i, \tag{58}$$

and the continuity equation

$$\frac{\partial u_i}{\partial x_i} = 0. \tag{59}$$

The two dimensionless groups governing the fluid dynamics are B and the Reynolds number

$$Re = \frac{\hat{\rho}\hat{U}_0\hat{L}}{\hat{\mu}_0} \left(= \frac{\hat{\rho}\hat{f}_0\hat{L}^3}{\hat{\mu}_0^2} \right). \tag{60}$$

The last expression follows if the velocity scale is chosen to balance the driving force with a viscous stress. Thus as usual, Re represents the ratio of inertial stresses to viscous stresses.

Steady inertial flows Before considering transient flows, lets consider inertial steady state solutions to (58). Setting the time derivative to zero, multiplying by u_i and integrating over Ω results in

$$Re\langle u_j \frac{\partial u_i}{\partial x_j} u_i \rangle + \langle \dot{\gamma}^2(\mathbf{u}) \rangle + B\langle \dot{\gamma}(\mathbf{u}) \rangle = \langle \mathbf{u} \cdot \mathbf{f} \rangle. \tag{61}$$

The first term may be integrated over Ω, using the divergence theorem and (59), and becomes zero due to the boundary conditions. Therefore, the inertial terms do not themselves contribute to the steady mechanical energy balance, which becomes:

$$\langle \dot{\gamma}^2(\mathbf{u}) \rangle + B\langle \dot{\gamma}(\mathbf{u}) \rangle = \langle \mathbf{u} \cdot \mathbf{f} \rangle.$$

This expression is identical with the Stokes flow. Consequently, the inertial flow has non-zero steady solutions under exactly the same conditions as the Stokes flow, i.e. determined by the critical value B_{cr}.

5.1 Stability of Static Flows

In the case that $B \geq B_{cr}$, the only steady solution is the static solution $\mathbf{U}(\mathbf{x}) = 0$. The velocity $\mathbf{u}(\mathbf{x}, t)$, (which is also the perturbation from the steady solution $\mathbf{U}(\mathbf{x})$), satisfies the unsteady mechanical energy equation:

$$\frac{d}{dt}\frac{\|\mathbf{u}\|^2}{2} + \langle \dot{\gamma}^2(\mathbf{u}) \rangle + B\langle \dot{\gamma}(\mathbf{u}) \rangle = \langle \mathbf{u} \cdot \mathbf{f} \rangle, \tag{62}$$

where $\|\mathbf{u}\|$ denotes the $L^2(\Omega)$. Assuming that $\mathbf{u} \neq 0$ initially, we proceed as in (44):

$$\begin{aligned}\frac{d}{dt}\frac{\|\mathbf{u}\|^2}{2} &= -\langle\dot{\gamma}^2(\mathbf{u})\rangle - \langle\dot{\gamma}(\mathbf{u})\rangle\left[B - \frac{\langle\mathbf{u}\cdot\mathbf{f}\rangle}{\langle\dot{\gamma}(\mathbf{u})\rangle}\right] \\ &\le -\langle|\nabla\mathbf{u}|^2\rangle - \langle\dot{\gamma}(\mathbf{u})\rangle\,[B - B_{cr}] \\ &\le -C_P\|\mathbf{u}\|^2 - C_{TS}\|\mathbf{u}\|_p\,[B - B_{cr}]\,. \end{aligned} \tag{63}$$

In the case that $d \le 2$, we may write $\|\mathbf{u}\|_p = \|\mathbf{u}\|$. Then, from the differential inequality (63), we can establish that $\|\mathbf{u}\|(t) \to 0$ in a finite time. Explicitly, this means that $\|\mathbf{u}\|(t) = 0$ for $t \ge t_{stop}$:

$$t_{stop} = \frac{1}{C_P}\ln\left(1 + \frac{C_P\|\mathbf{u}\|(0)}{C_{TS}(B - B_{cr})}\right).$$

Exercise: Use the differential inequality (63) to show that $\|\mathbf{u}\|(t) = 0$ for $t \ge t_{stop}$.

Stability analysis of the above style, leading to the finite stopping time bound, was first developed by Bristeau (1975), Glowinski et al. (1981), applied to 1D duct flows as an illustrative test example, The analysis applies equally to general flows in 2D enclosures, as considered here. Note that t_{stop} is an upper bound for the time at which the flow will stop.

For $d = 3$ we have $p = 3/2$, so cannot establish finite time decay to zero although the decay is at least exponential (due to the viscous dissipation). In practice however, we are still likely to observe finite time decay in 3D, as for computational solutions we deal with a discretized solution and for all finite dimensional subspaces $\|\mathbf{u}\|_{3/2}$ and $\|\mathbf{u}\|$ are equivalent norms.

Somewhat remarkably, we see that the Reynolds number Re does not enter the analysis above. This is explained by noticing that there is no background flow to transfer energy to $\mathbf{u}$ and there is no flux of energy across the boundary. Thus, condition (45) is sufficient to ensure the global energy stability of the static steady solution, i.e. for all initial $\mathbf{u}_0$ and all Re. This is completely different to the situation for Newtonian fluids.

This type of analysis can be applied to flows with slightly different boundary conditions. For example, in a Poiseuille flow along a duct we may include a constant pressure drop as part of the body force and then impose periodicity of the solution in the axial direction.

Example 1 One-dimensional flows for which analytical steady state solutions can be found include plane Couette, Poiseuille flows and pipe flow. For these Poiseuille flow solutions it is fairly straightforward to find the critical B_{cr}. In Chatzimina et al. (2005) the authors consider these steady flows as initial conditions, reducing the driving force to zero for $t > 0$, and track evolution of the velocity (in the plane Couette flow problem this means that the wall velocity is set to zero for $t > 0$). The solutions decay to zero, as shown below in one example from Chatzimina et al. (2005); see Fig. 12. As soon as the forcing is removed these flows all have steady state $\mathbf{U} = 0$, for all B, so that $B_{cr} = 0$. As B is increased above B_{cr} the stopping times in Chatzimina et al. (2005) decrease like B^{-1} and the estimate of t_{stop} becomes progressively sharp.

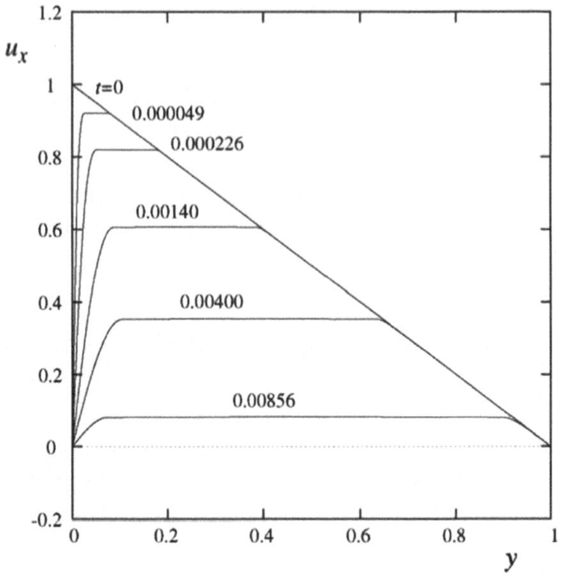

Fig. 12 Cessation of 1D Couette flow for $B = 20$. At $t = 0$ the initial condition is the linear profile between parallel walls. For $t > 0$ the moving wall become stationary and the velocity decays. Reproduced with permission from Chatzimina et al. (2005)

Example 2 This second example, from Karimfazli et al. (2015) concerns a differentially heated square cavity (hot wall to the left, cold wall to the right, insulated top and bottom), for which $B_{cr} = 1/32$; see Fig. 11 earlier for the static limit. Here we initially assume $B = 0$ and that the flow has attained the associated Newtonian steady flow. For $t > 0$ a large yield stress is introduced $B > B_{cr}$ and the flow decays to zero. Figure 13 tracks this decay over successive times for flows at different Rayleigh numbers Ra (all for Prandtl number $Pr = 1$). The final column of figures on the right in each of these simulations represents the solution just before it becomes static. Note that the decay times here vary with Ra. Note also that the final non-zero velocity fields shown are very similar to that in Fig. 11.

Figure 14 illustrates the development of velocity and temperature norms due to introduction of the yield stress $B > B_{cr}$ for $t > 0$. The analysis is slightly more involved than that presented above, but still results in finite time decay of the kinetic energy to zero within a finite time. This decline is more rapid for larger B. The decay of the temperature perturbation (here $||\theta||(t)$) experiences a brief period where convection is still active, followed by a longer period of pure exponential (conductive) decay after the velocity becomes zero.

Example 3 A third example, from Wachs and Frigaard (2016) concerns a heavy circular disc settling in a Bingham fluid confined between two parallel vertical walls. For sufficiently large $B \geq B_{cr}$ the disc does not settle in steady state. A similar stability analysis can be performed to show that the transient flow should be energy stable for $B > B_{cr}$. Figure 15 presents a computed example of this situation, showing the evolution of the strain rate at successive times. Here the disc is initially settling

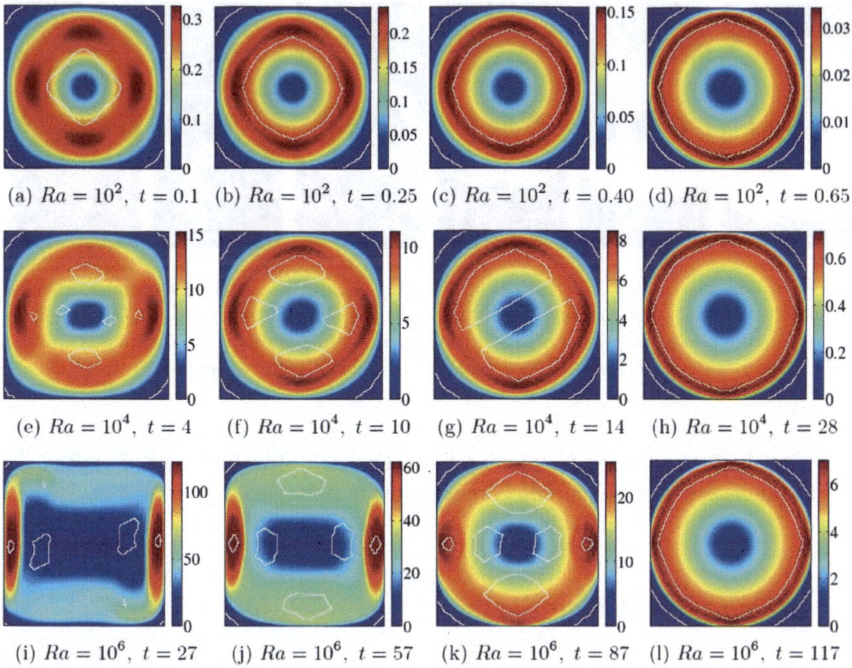

Fig. 13 Colour maps of the speed at intermediate stages in the transition from the steady Newtonian flow to a completely static state, due to the introduction of a large yield stress $B = 0.075$ for $t > 0$. White curves represent the yield surfaces; $Pr = 1$. Reproduced with permission from Karimfazli et al. (2015)

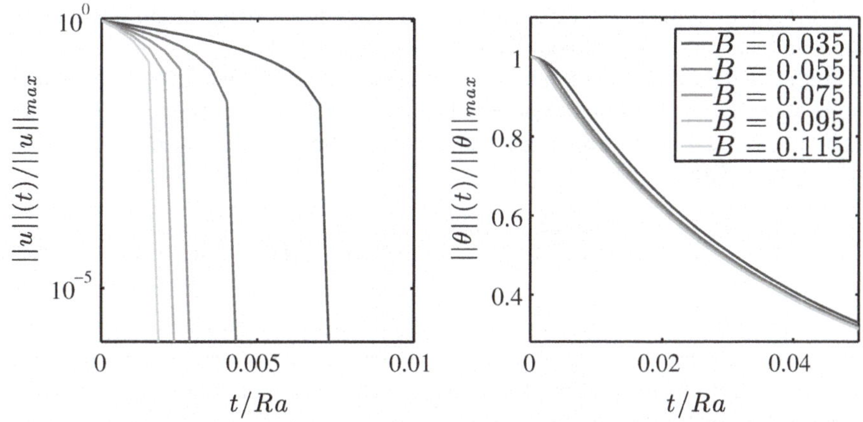

Fig. 14 Development of $\|\mathbf{u}\|$ and $\|\theta\|$ after the introduction of $B > B_{cr}$ for $t > 0$, all for $Ra = 10^4$ (decay trends are quite similar at lower and higher Ra). Reproduced with permission from Karimfazli et al. (2015)

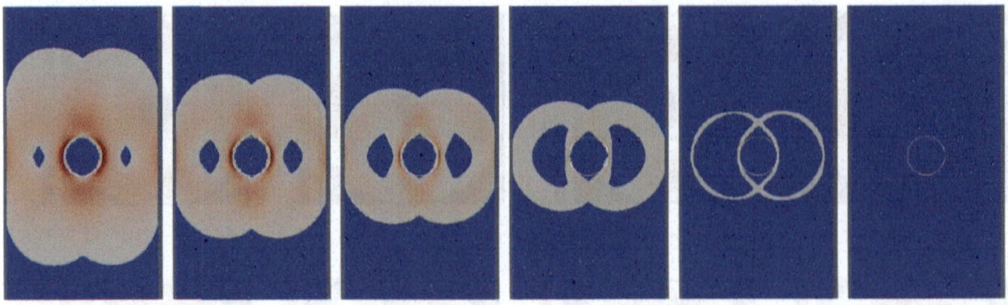

Fig. 15 Colourmap of the strain rate around a settling circular disc at successive times. The disc has initially the same steady settling velocity as in a Newtonian fluid. Here $B > B_{cr}$ for $t > 0$ and the motion comes to a complete stop in a finite time. The frame of reference is translated upwards as the particle falls. Reproduced with permission from Wachs and Frigaard (2016)

at the steady state rate for a Newtonian fluid. The yield stress $B > B_{cr}$ is applied for $t > 0$ and we see that the particle comes to rest in finite time.

5.2 *Stability of Moving Flows*

In the case that $B < B_{cr}$ we may have a non-zero steady solution $\mathbf{U}(\mathbf{x})$. We assume that is the case and consider the evolution of a time varying perturbation: $\mathbf{U}(\mathbf{x}) + \mathbf{u}(\mathbf{x}, t)$. Here we are back in the more usual realms of hydrodynamic stability. Although it is possible to conduct linear and weakly nonlinear analyses of these flows, the presence of the unyielded regions complicates and those studies that exist studies tend to be quite specific to each flow configuration. Here we outline the fully nonlinear analysis, which although conservative is general.

Both $\mathbf{U}$ and the perturbed solution $\mathbf{U} + \mathbf{u}$ satisfy (58)–(59). We subtract the steady momentum equations from the perturbed equations, take the dot product with $\mathbf{u}$ and integrate over Ω. Using the divergence theorem, no-slip condition and (59), we derive the Reynolds-Orr equation:

$$\frac{d}{dt}\frac{\|\mathbf{u}\|^2}{2} + Re\langle \mathbf{u} \cdot [(\mathbf{u} \cdot \nabla)\mathbf{U}]\rangle + \langle (\tau(\mathbf{U} + \mathbf{u}) - \tau(\mathbf{U})) : \dot{\gamma}(\mathbf{u})\rangle = 0. \tag{64}$$

Note that the first term represents evolution of the perturbation kinetic energy only, (unlike (63) above). The second term reflects the transfer of energy from the base flow to the perturbation and the last term gives the visco-plastic dissipation of the perturbation kinetic energy.

This last term is positive definite, as follows from convexity. For the Bingham fluid it is possible to bound the visco-plastic dissipation below by the purely viscous part of the dissipation, e.g. as in Nouar and Frigaard (2001). This leads to the following energy inequality:

$$\frac{d}{dt}\frac{\|\mathbf{u}\|^2}{2} \leq -Re\langle \mathbf{u} \cdot [(\mathbf{u} \cdot \nabla)\mathbf{U}]\rangle - \langle \dot{\gamma}^2(\mathbf{u})\rangle. \tag{65}$$

Interestingly, except for having inequality rather than equality, the right-hand side of the above is analogous to the mechanical energy equation for a purely viscous Newtonian fluid. We do not know of any way in which the influence of B can be retained explicitly in the dissipative terms, for general flows, although in specific examples this may be possible.

Note that B does not appear explicitly in (65), although evidently B influences $\mathbf{U}$. Similarly, although $\mathbf{f}$ influences the base flow, it does not appear in (65) explicitly. Lastly, note that although Re does not appear in the steady mechanical energy equation, which $\mathbf{U}$ satisfies, it does influence the base flow in the momentum balance and also appears here in the Reynolds-Orr equation. Conventionally, the further analysis of (65) leads to the definition of Re_E, such that for $Re < Re_E$ the viscous dissipation dominates the energy transfer term and $\|\mathbf{u}\| \to 0$ exponentially. For brevity, we simply sketch the simplest general analysis.

We start from (44) assuming $B < B_{cr}$:

$$\langle \dot{\gamma}^2(\mathbf{U})\rangle \leq [B_{cr} - B]\,\langle \dot{\gamma}(\mathbf{U})\rangle \leq [B_{cr} - B]\,|\Omega|^{1/2}\langle \dot{\gamma}^2(\mathbf{U})\rangle^{1/2}, \tag{66}$$

from the Cauchy-Schwarz inequality. Therefore,

$$\langle |\nabla \mathbf{U}|^2\rangle = \langle \dot{\gamma}^2(\mathbf{U})\rangle \leq [B_{cr} - B]^2\,|\Omega|.$$

The inertial term in (65) can be treated using the Cauchy-Schwarz inequality and Sobolev embedding theorem:

$$-\langle \mathbf{u} \cdot [(\mathbf{u} \cdot \nabla)\mathbf{U}]\rangle \leq C_b \|\mathbf{u}\|^2_{H^1}\,\|\mathbf{U}\|_{H^1};$$

where $C_b > 0$ depends only on Ω; see e.g. Frigaard et al. (2001). Now we may use the general bound on $\langle |\nabla \mathbf{U}|^2\rangle$:

$$-\langle \mathbf{u} \cdot [(\mathbf{u} \cdot \nabla)\mathbf{U}]\rangle \leq C_b \left(1 + \frac{1}{C_P}\right)(B_{cr} - B)\,|\Omega|^{1/2}\|\mathbf{u}\|^2_{H^1}.$$

Thus, the Reynolds-Orr equation becomes:

$$\frac{d}{dt}\frac{\|\mathbf{u}\|^2}{2} \leq \left[ReC_b \left(1 + \frac{1}{C_P}\right)(B_{cr} - B)\,|\Omega|^{1/2} - 1\right]\langle |\nabla \mathbf{u}|^2\rangle.$$

We see that the right-hand side is negative provided that:

$$Re < \frac{1}{C_b \left(1 + \frac{1}{C_P}\right)(B_{cr} - B)\,|\Omega|^{1/2}} = Re_E. \tag{67}$$

Assuming $Re < Re_E$, then the perturbation decays exponentially to zero:

$$\|\mathbf{u}\|(t) \leq \|\mathbf{u}\|(0)\mathrm{e}^{-C_P(Re_E - Re)t}.$$

Note that although this appears to be a viscous decay, the yield stress has influence. Firstly, the energy Reynolds number Re_E scales like $(B_{cr} - B)^{-1}$, so that $Re_E \to \infty$ as $B \to B_{cr}^-$, recovering the unconditional decay of the previous section. Secondly, note that as $Re_E \to \infty$ the exponential decay timescale decreases to zero.

Comments

1. The above analysis can of course be refined significantly. In Newtonian energy stability theory, in classical shaped domains and for specific $\mathbf{U}$, one commonly computes an optimal bound from a variational problem, instead of C_b and the general bound on $\langle|\nabla\mathbf{U}|^2\rangle$. Many examples of this form are found in e.g. Joseph (1976a, b).
2. The bound (67) is new as far as we know. It extends general results, such as those of Serrin (1959), to yield stress fluids. There seem to be relatively few examples of nonlinear stability that have been tackled. Partly this is because they do not fall into the above framework readily, i.e. there are few 2D steady flows with velocity imposed on a closed region. One example would be the driven cavity flow, which has been well studied as a benchmark problem.
3. Other classical flows studied include the Poiseuille flow. Here a different analysis route is taken because: (i) the flow domain is infinite, with fixed flow rate instead of a body force; (ii) the base flow velocity is known analytically. The latter allows one to analyze the inertial energy transfer term in the Orr-Sommerfeld problem more directly; see Nouar and Frigaard (2001).

References

Adachi, K., & Yoshioka, N. (1973). On creeping flow of a visco-plastic fluid past a circular cylinder. *Chemical Engineering Science, 28*, 215–226.

Adams, R. A. (1975). *Sobolev spaces*. New York: Academic.

Beris, A. N., Tsamopoulos, J. A., Armstrong, R. C., & Brown, R. A. (1985). Creeping motion of a sphere through a Bingham plastic. *Journal of Fluid Mechanics, 158*, 219–244.

Bristeau, M. O. (1975). *Application de la mthode des lments finis la rsolution numrique d'inquations variationnelles de type Bingham*. These de 3me cycle, Universite de Paris VI, Juin.

Chakrabarty, J. (2012). *Theory of plasticity*. Butterworth-Heinemann.

Chaparian, E., Balmforth, N., & Frigaard, I. A. (2017). Yield limit analysis of symmetric particle sedimentation in a bingham fluid. preprint.

Chatzimina, M., Georgiou, G. C., Argyropaidas, I., Mitsoulis, E., & Huilgol, R. R. (2005). Cessation of Couette and Poiseuille flows of a Bingham plastic and finite stopping times. *Journal of Non-Newtonian Fluid Mechanics, 129*, 117–127.

Cioranescu, D. (1976). Sur une classe de fluides non-newtoniens. *Applied Mathematics and Optimization, 3*, 263–282.

Dacorogna, B. (2008). Applied mathematical sciences series. In *Direct methods in the calculus of variations* (Vol. 78). Springer.

Dubash, N., & Frigaard, I. A. (2004). Conditions for static bubbles in viscoplastic fluids. *Physics of Fluids*, *16*, 4319–4330.

Duvaut, G., & Lions, J. L. (1976). *Inequalities in mechanics and physics*. Springer.

Frigaard, I. A. (1998). Stratified exchange flows of two Bingham fluids in an inclined slot. *Journal of Non-Newtonian Fluid Mechanics*, *78*, 61–87.

Frigaard, I. A., & Scherzer, O. (1998). Uniaxial exchange flows of two Bingham fluids in a cylindrical duct. *IMA Journal of Applied Mathematics*, *61*, 237–266.

Frigaard, I. A., & Scherzer, O. (2000). The effects of yield stress variation in uniaxial exchange flows of two Bingham fluids in a pipe. *SIAM Journal on Applied Mathematics*, *60*, 1950–1976.

Frigaard, I. A., Scherzer, O., & Sona, G. (2001). Uniqueness and non-uniqueness in the steady displacement of two visco-plastic fluids. *ZAMM*, *81*, 99–118.

Fuchs, M., & Seregin, G. (2000). Lecture notes in mathematics. In *Variational methods for problems from plasticity theory and for generalized newtonian fluids* (Vol. 1749). Springer

Glowinski, R. (1984). *Numerical methods for nonlinear variational problems*. Springer.

Glowinski, R., Lions, J. L., & Trémolières, R. (1981). *Numerical analysis of variational inequalities*. Studies in mathematics and its applications. (trans: from French version of 1976). North-Holland.

Hassani, R., Ionescu, I. R., & Lachand-Robert, T. (2005). Shape optimization and supremal minimization approaches in landslides modelling. *Applied Mathematics and Optimization*, *52*, 349–364.

Hild, P., Ionescu, I. R., Lachand-Robert, T., & Rosca, I. (2002). The blocking property of an inhomogeneous Bingham fluid. applications to landslides. *Mathematical Modelling and Numerical Analysis (M2AN)*, *36*, 1013–1026.

Hill, R. (1950). *The mathematical theory of plasticity*. Oxford University Press.

Huilgol, R. R. (2006). A systematic procedure to determine the minimum pressure gradient required for the flow of viscoplastic fluids in pipes of symmetric cross-section. *Journal of Non-Newtonian Fluid Mechanics*, *136*, 140–146.

Huilgol, R. R. (2015). *Fluid mechanics of viscoplasticity*. Springer.

Ionescu, I. R., & Lachand-Robert, T. (2005). Generalized cheeger's sets related to landslides. *Calculus of Variations and PDEs*, *23*, 227–249.

Johnson, M. W. (1960). Some variational theorems for non-newtonian flow. *Physics of Fluids*, *3*, 871–878.

Johnson, M. W. (1961). On variational principle for non-newtonian fluids. *Transactions. Society of Rheology*, *5*, 9–21.

Joseph, D. D. (1976a). Springer tracts in natural philosophy. *Stability of fluid motions II*. Springer, Heidelberg.

Joseph, D. D. (1976b). Springer tracts in natural philosophy. *Stability of fluid motions I*. Springer, Heidelberg.

Karimfazli, I., & Frigaard, I. A. (2013). Natural convection flows of a bingham fluid in a long vertical channel. *Journal of Non-Newtonian Fluid Mechanics*, *201*, 39–55.

Karimfazli, I., Frigaard, I. A., & Wachs, A. (2015). A novel heat transfer switch using the yield stress. *Journal of Fluid Mechanics*, *783*, 526–566.

Malek, J., Ruzicka, M., & Shelukhin, V. V. (2005). Herschel-Bulkley fluids: existence and regularity of steady flows. *Mathematical Models and Methods in Applied Sciences*, *15*, 1845–1861.

Mosolov, P. P., & Miasnikov, V. P. (1965). Variational methods in the theory of the fluidity of a viscous-plastic medium. PPM. *Journal of Mechanics and Applied Mathematics*, *29*, 468–492.

Mosolov, P. P., & Miasnikov, V. P. (1966). On stagnant flow regions of a viscous-plastic medium in pipes. PPM. *Journal of Mechanics and Applied Mathematics*, *30*, 705–717.

Mosolov, P. P., & Miasnikov, V. P. (1967). On qualitative singularities of the flow of a viscoplastic medium in pipes. PPM. *Journal of Mechanics and Applied Mathematics*, *31*, 581–585.

Moyers-Gonzalez, M. A., & Frigaard, I. A. (2004). Numerical solution of duct flows of multiple visco-plastic fluids. *Journal of Non-Newtonian Fluid Mechanics*, *122*, 227–241.

Nouar, C., & Frigaard, I. A. (2001). Nonlinear stability of Poiseuille flow of a Bingham fluid: Theoretical results and comparison with phenomenological criteria. *Journal of Non-Newtonian Fluid Mechanics*, *100*, 127–149.

Prager, W. (1954). Studies in mathematics and mechanics. In *On slow visco-plastic flow* (pp. 208–216). New York: Academic Press Inc. Presented to Richard von Mises by Friends, Colleagues, and Pupils.

Putz, A., & Frigaard, I. A. (2010). Creeping flow around particles in a Bingham fluid. *Journal of Non-Newtonian Fluid Mechanics*, *165*, 263–280.

Randolph, M. F., & Houlsby, G. T. (1984). The limiting pressure on a circular pile loaded laterally in cohesive soil. *Géotechnique*, *34*, 613–623.

Roustaei, A., Chevalier, T., Talon, L., & Frigaard, I. A. (2016). Non-darcy effects in fracture flows of a yield stress fluid. submitted to *Journal of Fluid Mechanics*.

Serrin, J. (1959). On the stability of viscous fluid motions. *Archive for Rational Mechanics and Analysis*, *3*, 1–13.

Temam, R., & Strang, G. (1980). Functions of bounded deformation. *Archive for Rational Mechanics and Analysis*, *75*, 7–21.

Tokpavi, D., Magnin, A., & Jay, P. (2008). Very slow flow of Bingham viscoplastic fluid around a circular cylinder. *Journal of Non-Newtonian Fluid Mechanics*, *154*, 65–76.

Tsamopoulos, J., Dimakopoulos, Y., Chatzidai, N., Karapetsas, G., & Pavlidis, M. (2008). Steady bubble rise and deformation in Newtonian and viscoplastic fluids and conditions for bubble entrapment. *Journal of Fluid Mechanics*, *601*, 123–164.

Wachs, A., & Frigaard, I. A. (2016). Particle settling in yield stress fluids: limiting time, distance and applications. submitted to *Journal of Non-Newtonian Fluid Mechanics*.

Yoshioka, N., & Adachi, K. (1971a) On variational principles for a non-newtonian fluid. *Journal of Chemical Engineering of Japan*, *4*, 217–220.

Yoshioka, N., & Adachi, K. (1971b). Applications of the extremum principles for non-newtonian fluids. *Journal of Chemical Engineering of Japan*, *4*, 221–226.

Yoshioka, N., & Adachi, K. (1973). Some deductions from the extremum principles for non-newtonian fluids. *Journal of Chemical Engineering of Japan*, *6*, 134–140.

Viscoplastic Asymptotics and Other Analytical Methods

Neil J. Balmforth

1 Introduction

The goal of these lectures is to outline some of the ideas behind the use of asymptotic analysis and other analytical methods in viscoplastic fluid mechanics (other than variational and bounding techniques, which are covered in lectures by I. Frigaard). General texts on the relevant mathematical methodologies include Bender and Orszag (1978), Hinch (1991), Whitham (1974) and Sneddon (1957). Asymptotic analysis surrounds the identification and use of a small parameter to simplify and then solve mathematical problems to build useful approximate solutions. Often, the requirement that a particular parameter is small limits the settings for asymptotic analysis, but the asymptotic developments usually provide key insight by identifying and distilling down the critical physical details and processes. Important phenomenological properties and the scaling with the dimensionless parameters of the problem can then emerge. At the same time, the asymptotic solutions furnish useful limiting test cases for numerical schemes and compact physical predictions.

2 Viscoplastic Lubrication Theory

2.1 Mathematical Formulation

Consider a two-dimensional incompressible viscoplastic fluid slowly flowing down a relatively narrow conduit; flow is driven either by a pressure gradient, the motion of the bounding surfaces, or both. The conduit is described by an arc-length-based

N. J. Balmforth (✉)
University of British Columbia, Vancouver, B.C., Canada
e-mail: njb@math.ubc.ca

G. Ovarlez and S. Hormozi (eds.), *Lectures on Visco-Plastic Fluid Mechanics*,
CISM International Centre for Mechanical Sciences 583,
https://doi.org/10.1007/978-3-319-89438-6_2

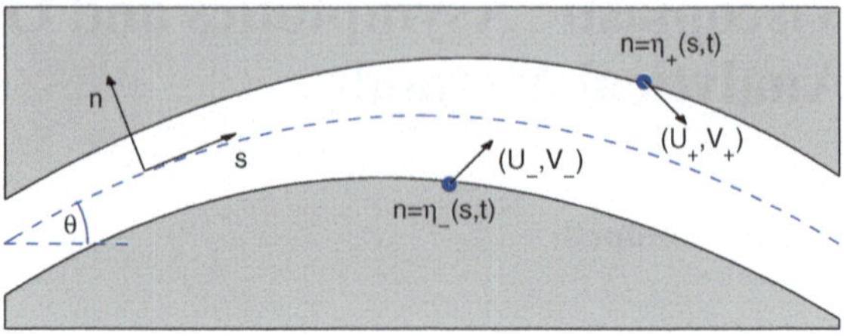

Fig. 1 Sketch showing the geometry of a narrow conduit sandwiched between two moving surfaces and filled by a lubricating viscoplastic fluid, together with the arc-length-based coordinates

curvilinear coordinated system (s, n) following a fixed curve threaded along the length; s is arc-length and n is the normal coordinate (see Fig. 1). We define $\mathbf{u} = (u, v)$ as the velocity in these coordinates (i.e. with respect to the (s, n) axes). Similarly, the deviatoric stress tensor τ_{ij} is also referred to this system. Inertia is neglected.

Conservation of mass and force balance can be expressed in the form (Hewitt and Balmforth 2012)

$$\frac{\partial u}{\partial s} + (1 - \kappa n)\frac{\partial v}{\partial n} - \kappa v = 0, \tag{1}$$

$$\frac{\partial \tau_{ss}}{\partial s} + (1 - \kappa n)\frac{\partial \tau_{sn}}{\partial n} - 2\kappa\tau_{sn} = \frac{\partial p}{\partial s}, \tag{2}$$

$$\frac{\partial \tau_{sn}}{\partial s} + (1 - \kappa n)\frac{\partial \tau_{nn}}{\partial n} + \kappa(\tau_{ss} - \tau_{nn}) = \frac{\partial p}{\partial n}, \tag{3}$$

where κ denotes the curvature.

The strain rate tensor has components,

$$\dot{\gamma}_{ss} = \frac{2}{1 - \kappa n}\left(\frac{\partial u}{\partial s} - \kappa v\right), \quad \dot{\gamma}_{nn} = 2\frac{\partial v}{\partial n}, \quad \dot{\gamma}_{sn} = \frac{1}{1 - \kappa n}\left(\frac{\partial v}{\partial s} + \kappa u\right) + \frac{\partial u}{\partial n}, \tag{4}$$

which can be fed into the constitutive law:

$$\tau_{ij} = \mu(\dot{\gamma})\dot{\gamma}_{ij} + \tau_Y \frac{\dot{\gamma}_{ij}}{\dot{\gamma}}, \quad \text{if} \sqrt{\tau_{ss}^2 + \tau_{sn}^2} > \tau_Y, \tag{5}$$

and $\dot{\gamma}_{ij} = 0$ otherwise, where $\dot{\gamma} \equiv \sqrt{\dot{\gamma}_{ss}^2 + \dot{\gamma}_{sn}^2}$.

The upper and lower surfaces are located at $n = \eta_\pm$. Here, the kinematic conditions demand that

$$v = V_\pm = \frac{\partial \eta_\pm}{\partial t} + \frac{U_\pm}{1 - \kappa\eta_\pm}\frac{\partial \eta_\pm}{\partial s} \qquad \text{with } u = U_\pm. \tag{6}$$

To identify the principal balance of terms in the limit of a slender gap, one can introduce a characteristic thickness, $\mathcal{H}$, length, $\mathcal{L}$, speed, $\mathcal{U}$, and pressure, $\mathcal{P}$, and then non-dimensionalize the variables:

$$\hat{s} = \frac{s}{\mathcal{L}}, \quad (\hat{n}, \hat{\eta}_\pm) = \frac{1}{\mathcal{H}}(n, \eta_\pm), \quad \hat{u} = \frac{u}{\mathcal{U}}, \quad \hat{v} = \frac{v\mathcal{L}}{\mathcal{U}\mathcal{H}} \tag{7}$$

and

$$\hat{t} = \frac{\mathcal{U}t}{\mathcal{L}}, \quad \hat{p} = \frac{p}{\mathcal{P}}, \quad (\hat{\tau}, \hat{\sigma}) = \frac{\mathcal{L}}{\mathcal{H}\mathcal{P}}(\tau_{\rm sn}, \tau_{\rm ss}), \tag{8}$$

The aspect ratio, $\epsilon = \mathcal{H}/\mathcal{L}$, is small when the gap is slender, and this parameter can be used to establish the leading-order relations from the governing equations, and design expansions to proceed to higher order if needed.

The disparate scalings of the two velocity components is guided by the requirement that the gap remains thin and the main balance (to $O(\epsilon)$) in the conservation of mass equation,

$$\frac{\partial \hat{u}}{\partial \hat{s}} + \frac{\partial \hat{v}}{\partial \hat{n}} = 0. \tag{9}$$

The scaling of the pressure and stress components leads to the main force balances,

$$\frac{\partial \hat{p}}{\partial \hat{s}} = \frac{\partial \hat{\tau}}{\partial \hat{n}}, \quad \frac{\partial \hat{p}}{\partial \hat{n}} = 0. \tag{10}$$

That is, the pressure is largely constant across the slot, but its gradient down that conduit is balanced by the resistance due to the shear stress.

The usual strategy at this stage is to drop the hat decoration that clutters our dimensionless notation and streamline the formulae by exploiting (s, n, t) subscripts to denote partial derivatives (except in the case of the components of the stress and strain-rate tensors, for which we use different font in any case). Following this suit, we write

$$p = p(s, t) \qquad \& \qquad \tau = \tau_-(s, t) + (n - \eta_-)p_s, \tag{11}$$

using the shear stress on the lower surface $\tau_- = \tau(s, \eta_-, t)$. Moreover,

$$\tau_+ - \tau_- = hp_s \tag{12}$$

where $h = \eta_+ - \eta_-$ is the local gap thickness.

The strain rates, scaled by $\mathcal{U}/\mathcal{H}$, become

$$[\dot{\gamma}_{\rm ss}, \dot{\gamma}_{\rm sn}] = [2\epsilon u_s, u_n + \epsilon\kappa u] + O(\epsilon^2). \tag{13}$$

At first sight, the disparity in scaling here suggests that the shear rate $\dot{\gamma}_{\rm sn}$ and therefore the shear stress τ dominate the state of the material to furnish

$$\dot{\gamma} \sim |\dot{\gamma}_{\rm sn}| \sim |u_n| \qquad \& \qquad \tau \sim \mu(\dot{\gamma})u_n + B\ \mathrm{sgn}(u_n), \tag{14}$$

where $\mu(\dot{\gamma})$ is the plastic viscosity scaled by $\mathcal{U}/\mathcal{H}\mathcal{P}$ and we (somewhat loosely) refer to the yield-stress parameter $B = \tau_Y/(\epsilon\mathcal{P})$ as a Bingham number.

However, there is another type of solution for the stress state in which $u_n = O(\epsilon)$. In this case, we must have that

$$u \sim u_p(s,t) + \epsilon u_1(s,n,t) + \ldots \qquad \& \qquad (\dot{\gamma}_{ss}, \dot{\gamma}_{sn}) = \epsilon(2u_{ps}, u_{1n} + \kappa u_p) + O(\epsilon^2). \tag{15}$$

The viscous part of the shear stress is then small, but the yield stress contribution demands the stress state is dictated by

$$(\tau, \sigma) = \frac{(\dot{\gamma}_{ns}, \dot{\gamma}_{ss})}{\dot{\gamma}} \qquad \& \qquad \tau^2 + \sigma^2 = B^2. \tag{16}$$

The existence of these two types of stress solutions has in the past caused undue confusion: the so-called "lubrication paradox" (Lipscomb and Denn 1984) surrounds how the use of the first type of solution leads to an apparent inconsistency in the lubrication theory. More specifically, the blind use of this approximation leads to a solution of the problem in which τ appears to fall below B suggesting that the fluid is not yielded and one should then take $u_n = 0$ over a moving "plug" region. But the full solution of the problem (and, in particular, the imposition of the boundary conditions) subsequently indicates that these plugs often cannot be rigid and u still depends on s. i.e. the plugs are in extension or compression in the direction of the slot. The resolution of this "paradox" is not that the theory is internally inconsistent, but that one has simply chosen the wrong solution to the stress state, and the second one in (15)–(16) is needed with $u_n = O(\epsilon)$ rather than zero. The plugs are not therefore truly rigid but held close to the yield stress in the fashion of a perfectly plastic material. Walton and Bittleston (1991) introduced the terminology "pseudo-plugs" to emphasize this feature of the lubrication problem. The only inconsistency is to implement the first solution in (14) over regions where u_n is not order one.

Despite Walton & Bittleston's paper, and several others since (e.g. Balmforth and Craster (1999), Smyrnaios and Tsamopoulos (2001), Putz et al. (2009)), an unfortunate failing of the literature on viscoplastic flows is to refer to the "lubrication paradox" as though there really were one. This has led to some recent attempts to revise the pseudo-plug solution in an ad hoc and superfluous fashion to try to make that region into a genuinely rigid zone (e.g. Fusi et al. (2012)). Worse, there is even a vein of literature in which it is argued that there are problems with the constitutive law itself and one must regularize to surmount the paradox (e.g. Wilson (1999)). No such deviants are necessary.

The main consequence of all this in the lubrication theory is that one may supplement the fully yielded part of the flow where $u_n = O(1)$ with a region over which u_n is plug-like to complete the asymptotic solution. In most situations u_n is $O(\epsilon)$ and there is a pseudo-plug. In others, one can consistently take $u_n = 0$ and connect the fully yielded region to a true plug. In both cases, this procedure renders no change to the leading-order results and the asymptotic solution is completed in what is effectively the naive manner without worry. We next illustrate all this using a specific model problem, squeeze flow of Bingham fluid between two plates, to

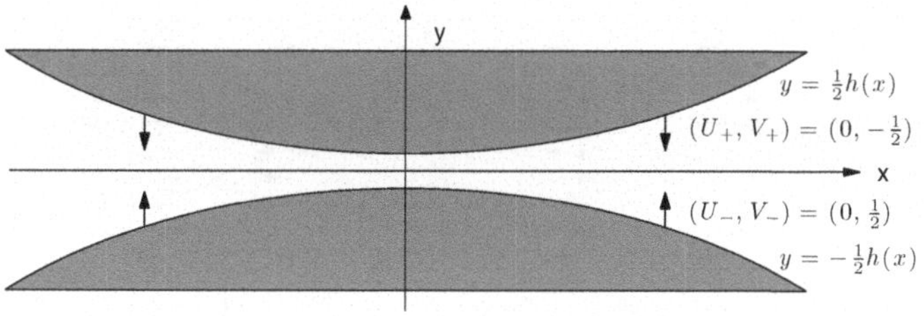

Fig. 2 Sketch of a Bingham squeeze flow

make the analysis more transparent and avoid the clutter and opacity of the general formulation. At the same time, this leads us to iron out some finer details of how the fully yielded region fits together (matches) with the pseudo-plug.

2.2 *Squeeze Flow of Bingham Fluid*

For the planar problem of squeeze flow of Bingham fluid between two symmetrically approaching surfaces (See Fig. 2), our curvilinear coordinate system reduces to a Cartesian one: $(s, n) \rightarrow (x, y)$. We then have

$$p = p(x, t) \qquad \& \qquad \tau = y p_x, \tag{17}$$

and

$$u(x, \eta_\pm = \pm\frac{1}{2}h, t) = 0 \qquad \& \qquad v(x, \eta_\pm = \pm\frac{1}{2}h, t) = \mp\frac{1}{2}, \tag{18}$$

in view of the symmetry of the problem about $y = 0$, and on fixing the velocity scale $\mathcal{U} = V/\epsilon$ in terms of the closure speed V of the separation of the plates.

If we set $\mathcal{P} = \mu V/\epsilon^2\mathcal{H}$, with $\mathcal{H}$ the initial plate separation, then the constitutive law implies

$$[\tau, \sigma] = (1 + B\dot{\gamma}^{-1})[u_y, 2\epsilon u_x] + O(\epsilon^2), \tag{19}$$

provided $\tau^2 + \sigma^2 < B^2$. Over the fully yield part of the flow, where $\tau = u_y + B\ \mathrm{sgn}(u_y) \gg \sigma$, we now find

$$u_y = (|y| - Y)p_x\ \mathrm{sgn}(y), \qquad Y = \frac{B}{|p_x|}, \tag{20}$$

or

$$u = -\frac{1}{2}p_x\left(\frac{1}{2}h - |y|\right)\left(\frac{1}{2}h - 2Y + |y|\right) \tag{21}$$

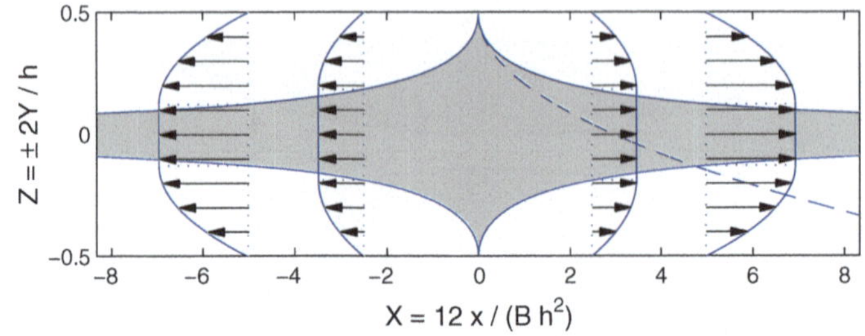

Fig. 3 The velocity profile and pseudoplug (shaded) in Bingham squeeze flow. The dashed line shows the approximation, $Y \sim \frac{1}{2}h - \sqrt{x/B}$, for $B \gg 1$

(cf. Fig. 3). Moreover, this region occupies $Y < |y| < \frac{1}{2}h$. It is inconsistent to continue this solution into $|y| < Y$ where, instead, $u_y = O(\epsilon)$. In fact, here we must have

$$u = u_p(x,t) + \epsilon u_1(x,y,t), \qquad \text{with} \qquad u_p = -\frac{1}{2}p_x\left(\frac{1}{2}h - Y\right)^2, \tag{22}$$

in order that the fully yielded solution in (21) be continuous with the pseudo-plug solution $u \sim u_p$.

The $y-$integral of the continuity equation, $u_x + v_y = 0$, in conjunction with the boundary conditions (18), now implies

$$\frac{\partial}{\partial x}\int_{-\frac{1}{2}h}^{\frac{1}{2}h} u\mathrm{d}y = 1. \tag{23}$$

Given the fact that $u(x = 0, y, t) = 0$ and

$$\int_{-\frac{1}{2}h}^{\frac{1}{2}h} u\mathrm{d}y \equiv -\int_{-\frac{1}{2}h}^{\frac{1}{2}h} yu_y\mathrm{d}y \equiv -\frac{1}{3}p_x(h+Y)\left(\frac{1}{2}h - Y\right)^2, \tag{24}$$

we therefore arrive at an algebraic problem for p_x (recalling that $Y = B/|p_x|$):

$$-\frac{1}{3}p_x[h(x) + Y]\left[\frac{1}{2}h(x) - Y\right]^2 = x. \tag{25}$$

From (23), it is clear that the right-hand side corresponds to the net flux down the gap, Q. Hence, Eq. (25) is the flux-pressure-gradient relation of the lubrication flow: $Q = \frac{1}{3}S(h+Y)(\frac{1}{2}h - Y)^2$, if $S = |p_x|$ denotes the magnitude of the pressure gradient (cf. Sect. 2.7 below).

Evidently, p_x and therefore Y and u_p depend on position x, highlighting how the flow in $|y| < Y$ is only a pseudo-plug and the curves $|y| = Y(x)$ are fake yield surfaces (in the terminology of Walton & Bittleston). Introducing the variables $Z = Y/h$ and $X = 12x/(Bh^2)$, we may write (25) as a simple cubic for $Z = Z(X)$; the relevant solution for flat surfaces ($h = constant$) is shown in Fig. 3, and reproduces one derived by Covey and Stanmore (1981).

One application of squeeze flow theory is to compute the lubrication pressure arising between two colliding spheres in a suspension. For the two-dimensional version of this problem, the colliding objects are disks and we may write $h \sim 1 + \frac{1}{2}x^2$ as an approximation of the geometry of the gap when the approach is close. For Newtonian fluid, $Y \to 0$ and we find the classical solution,

$$p_x = -\frac{12x}{h^3}, \qquad \text{or} \qquad p = p_a + \frac{6}{h^2}, \tag{26}$$

where p_a is the ambient fluid pressure. One can then compute the force on the disks, which is to leading order, $\int_{-\infty}^{\infty}(p - p_a)\,dx$. For a viscoplastic fluid, the situation is rather different: $h \to \frac{1}{2}x^2$ as one moves out of the narrowest part of the gap and the relevant solution of the cubic is

$$Y \to \frac{1}{2}h \;\;\&\;\; |p_x| \to \frac{2B}{h}, \;\; \text{or} \;\; p \sim p_a + 2\sqrt{2}B\left[\frac{1}{2}\pi - \tan^{-1}\frac{|x|}{\sqrt{2}}\right]. \tag{27}$$

The physical significance of this solution is that the yield stress dominates as one moves out of the gap where the shear rates are in decline. The fake yield surfaces then approach the disk surfaces. Critically, the lubrication pressure, $p - p_a$, is not integrable over the gap, which indicates that the force between the colliding objects is no longer dominated by the lubrication forces over the narrowest parts of the gap. Instead, the force is dictated by the yield stress elsewhere, which complicates the macroscopic description of particle interaction forces in viscoplastic suspensions.

2.3 The Pseudo-Plug Solution and Finer Matching Details

To look at the pseudo-plug more carefully and study how it joins on to the fully yielded flow, we focus on the case of a squeeze flow between flat plates and consider only the first quadrant $(x, y) > 0$ to avoid having to keep track of various sign changes. Given $\sigma^2 + \tau^2 = B^2$ and $\sigma > 0$ because $u_x > 0$, we find that

$$\tau = yp_x \equiv \frac{Bu_{1y}}{\sqrt{4u_{px}^2 + u_{1y}^2}} \quad \& \quad \sigma = -p_x\sqrt{Y^2 - y^2} \equiv \frac{2Bu_{px}}{\sqrt{4u_{px}^2 + u_{1y}^2}} \tag{28}$$

over the pseudo-plug. Hence

$$u_{1y} = -\frac{2yu_{px}}{\sqrt{Y^2 - y^2}}. \tag{29}$$

Evidently, $u \to u_p$ from below $y = Y$ and u_{1y} diverges as one approaches that level from the pseudoplug, thus allowing for a larger velocity gradient. These observations do not, however, yet constitute an exercise in matched asymptotics as we appear to be patching two solutions together at a particular point in y.

It turns out that the join is effected over a narrower region surrounding the fake yield surface. To resolve this layer we put

$$y = Y + \delta\zeta \qquad \& \qquad u = u_p + \delta^a U(s, \zeta, t), \tag{30}$$

where $\delta \ll 1$, but its relation with ϵ, and the exponent a both remain to be found. Now, for the fully yielded region we have $\dot{\gamma} \sim |u_y|$, whereas we know that $\dot{\gamma} \sim \epsilon\sqrt{4u_{px}^2 + u_{1y}^2}$ over the pseudo-plug. To connect these two limits, we therefore introduce the approximation $\dot{\gamma} \approx \sqrt{\delta^{2a-2}U_\zeta^2 + 4\epsilon u_{px}^2}$ into the constitutive law:

$$\tau \equiv -B + \delta\zeta p_x \approx \delta^{a-1}U_\zeta + \frac{B\delta^{a-1}U_\zeta}{\sqrt{\delta^{2a-2}U_\zeta^2 + \epsilon^2 u_{px}^2}}. \tag{31}$$

Given $U_\zeta < 0$, the main balance of terms in this relation is just $-B \approx -B$; demanding that the next-order corrections all have the same size implies $a = 2$ and $\delta = \epsilon^{2/3}$. Hence, we arrive at

$$U_\zeta^3 - \zeta p_x U_\zeta^2 + 2Bu_{px}^2 \approx 0. \tag{32}$$

This cubic has a real solution that converges to the limit $U_\zeta \sim \zeta p_x$ for $\zeta \to \infty$, and to the limit $U_\zeta \sim -u_{px}\sqrt{2B/\zeta p_x}$ for $\zeta \to -\infty$; i.e. the limits of the fully yielded and pseudo-plug solutions for $y \to Y$. In other words, the pseudo-plug can be matched to the fully yielded flow over a thin layer of thickness $\epsilon^{2/3}$.[1]

2.4 Strategy for More General Solutions

Returning to more general situations, we express the shear rate u_n in terms of the shear stress using the inverse of the constitutive law,

$$u_n = \Gamma(\tau) \tag{33}$$

(see Fig. 4). Unlike the original law, which is multi-valued for $\tau < \tau_Y$, this inverse is unambiguous with $u_n = 0$ for that range of stress (although the interpretation is not that the fluid is rigid, but that it may be a pseudo-plug). We may integrate this equation over the gap and change the integration variable from

[1]The corresponding analysis presented in Balmforth and Craster (1999) contains erroneous powers; Putz et al. (2009) previously provide corrected formulae.

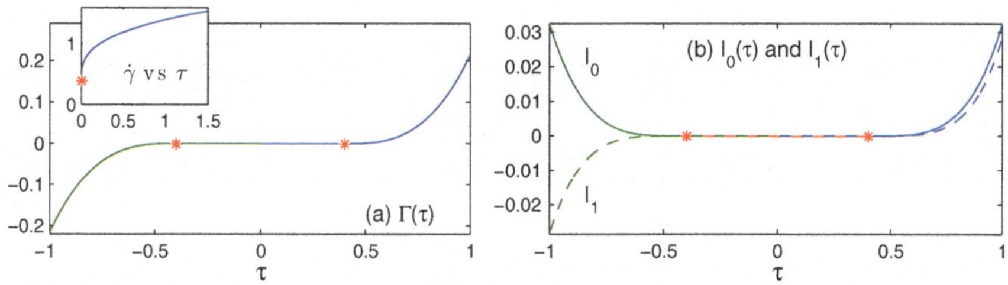

Fig. 4 Illustration of the inverse of the Herschel-Bulkley law ($\tau = K\dot{\gamma}^n + \tau_Y$, where K is the consistency and n the power-law index) and the $I_j(\tau)$ functions it generates, for power-law index $n = \frac{1}{3}$

$$n \equiv \frac{\tau - \tau_-}{p_s} + \eta_- \equiv h\frac{\tau - \tau_-}{\tau_+ - \tau_-} + \eta_- \tag{34}$$

to τ:

$$U_+ - U_- = \int_{\eta_-}^{\eta_+} u_n \mathrm{d}n = h\frac{I_0(\tau_+) - I_0(\tau_-)}{\tau_+ - \tau_-}, \tag{35}$$

where, as illustrated in Fig. 4,

$$I_j(\tau) = \int^{\tau} z^j \Gamma(z)\mathrm{d}z. \tag{36}$$

We may also integrate our continuity relation, $u_s + v_n = 0$, in n and use the kinematic conditions on the bounding surfaces to arrive at

$$h_t + \frac{\partial}{\partial s}\int_{\eta_-}^{\eta_+} u \ \mathrm{d}n = 0. \tag{37}$$

Hence,

$$Q(t) = \int_0^s h_t(s', t)\mathrm{d}s' + U_+\eta_+ - U_-\eta_- - h^2\frac{I_1(\tau_+) - I_1(\tau_-)}{(\tau_+ - \tau_-)^2} + h^2\tau_-\frac{I_0(\tau_+) - I_0(\tau_-)}{\tau_+ - \tau_-}, \tag{38}$$

where Q is an integration constant. Equations (35) and (38) constitute an algebraic problem to solve for the surface shear stresses $\tau_\pm(s, t)$ given the constitutive functions $I_j(\tau)$, the instantaneous geometry of the gap and the motion of the bounding surfaces. A further relation is needed to determine $Q(t)$. In some problems (like in the squeeze flow problem) a spatial symmetry can be used to fix this quantity; in others, the pressure drop Δp across the slot is known and so we have an additional relation,

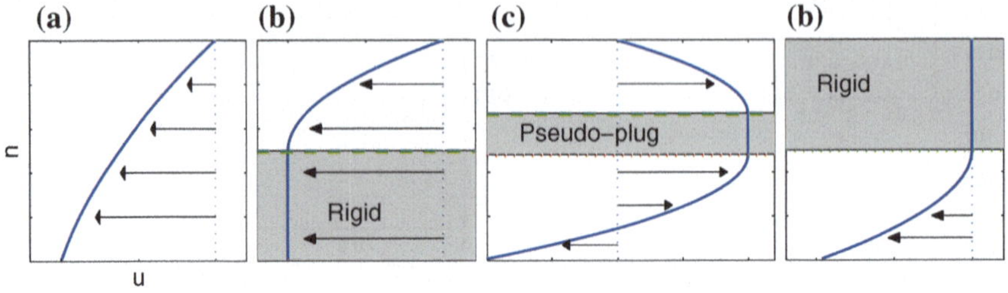

Fig. 5 Examples of the four possible flow configurations

$$\Delta p = \int p_s \mathrm{d}s = \int (\tau_+ - \tau_-) \frac{\mathrm{d}s}{h}. \tag{39}$$

Note that the problem here is parameterized in terms of the surface shear stresses $\tau_\pm$, which hides and avoids an otherwise complicating feature of the problem: at no stage is there a need to recognise the detailed flow configuration. Instead, one constructs after $\tau_\pm$ are determined. The issue that one would otherwise face is that with general motions allowed for the boundaries and an initially unkown pressure gradient along the gap, there are many possibilities for the form of the flow within the conduit: there may be moving pseudo-plugs, genuinely rigid zones attached to the walls or neither. If one were to work with the velocity field directly, the solution strategy would necessarily require a first effort to find the relevant flow configuration. However, by working with $\tau_\pm$, and because the shear stress itself varies only linearly across the gap, this task is avoided. After determining the surface stresses, one can identify the flow configuration based on how they compare to B (see Fig. 5):

- A: Fully yielded zone; $|\tau_-|, |\tau_+| > B$ and $\mathrm{sgn}(\tau_-) = \mathrm{sgn}(\tau_+)$.
- B: Lower plug; $|\tau_-| < B$.
- C: Central pseudo-plug; $|\tau_-|, |\tau_+| > B$ and $\mathrm{sgn}(\tau_-) = -\mathrm{sgn}(\tau_+)$.
- D: Upper plug; $|\tau_+| < B$.

The preceding formulation is a slight generalization of that presented by Hewitt and Balmforth (2012). They provide a number of examples for illustration, including viscoplastic journal and slider bearings (Pinkus and Sternlicht 1961), and indicate further how the construction can be made computationally efficient and be used to dynamically evolve the gap or film (if one the surfaces is free). Figure 6 shows one of their journal bearing solutions in which the two cylinders of the bearing have prescribed rotation rates and translation speeds with respect to one another. Here, the arc-length coordinate here simply corresponds to angle, and the solution contains all of the flow configurations indicated above over different angular locations of the gap. Also shown is the pressure distribution (here $\Delta p = 0$ in view of the periodic geometry) and the plug speed $u_p(s)$, which varies in the pseudo-plugs, but is constant in configurations B and D when the plug is truly rigid and attached to

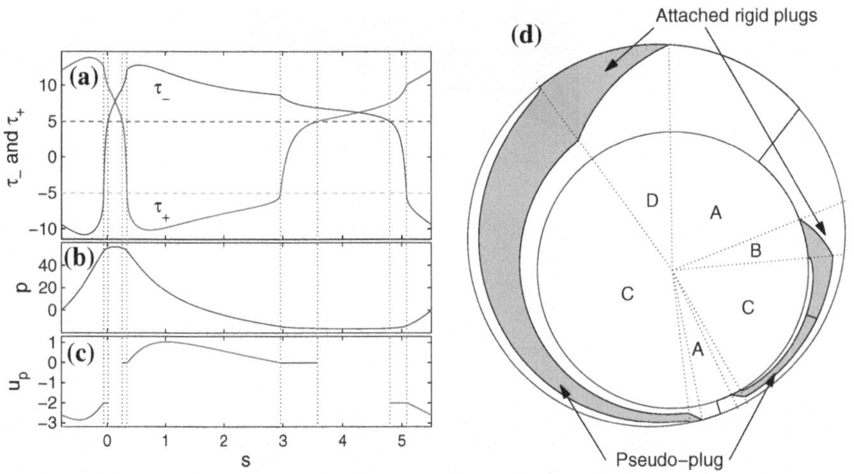

Fig. 6 A viscoplastic journal bearing solution, showing **a** $\tau_{\pm}(s)$, **b** $p(s)$, **c** $u_p(s)$ and **d** the fully yielded regions, true plugs and pseudo-plugs. The dotted lines show the borders of the regions with different flow configurations (as indicated in (**d**)). In (**a**), the dashed lines show $\pm B$. The origin of s (which corresponds to angle) is the location of the minimum gap. The outer cylinder (of radius 1.1) is fixed in place whilst the inner cylinder (of radius 0.8) rotates with angular speed 2 and its centre moves in the direction of the line of centres so as to close the minimum gap at speed 1. The fluid has a Bingham number of $B = 5$ and a power-law index of $n = \frac{1}{2}$

one of the cylinders. Figure 7 shows a slider bearing solution (with no net pressure drop underneath, $\Delta p = 0$). Again, the full flow solution contains all four possible configurations.

2.5 *Plastic Limit*

Returning to the Bingham squeeze flow problem, an interesting limit arises for $B \gg 1$. For such a parameter setting, the fake yield surfaces, $y = \pm Y$, must approach the bounding plates, $y = \pm\frac{1}{2}h$ (the solution of the cubic in (25) lies near $X = 0$ in Fig. 3). Thus $p_x \to -2B/h$ (or $p \to -2Bx/h$) and we arrive at

$$Y \sim \frac{1}{2}h - \sqrt{B^{-1}x}. \tag{40}$$

The fully yielded flow therefore becomes restricted to boundary layers of thickness $B^{-1/2}$. The pseudo-plug, on the other hand, has the stress solution,

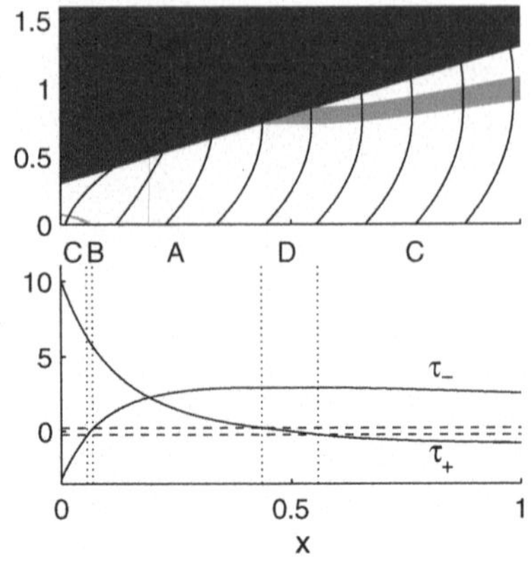

Fig. 7 A viscoplastic slider bearing solution, showing on top the geometry, true and pseudo plugs (shaded) and sample horizontal velocity profiles (solid lines), and on the bottom the surface shear stresses $\tau_{\pm}(x)$. The inclined slider (of length 1) moves to the right with speed 1. The dotted lines show the borders of the regions with different flow configurations (as indicated), and the dashed lines show $\pm B = 0.2$ (with power-law index $n = 1$). There are small C and B regions underneath the slider near the narrowest part of the gap

$$(\tau, \sigma) \sim B\left(-\frac{2y}{h}, \sqrt{1 - \frac{4y^2}{h^2}}\right), \tag{41}$$

and plug speed $u_p \sim x/h$.

It is informative to compare the solution (41) with a classical solution of Prandtl in ideal plasticity theory (Prager and Hodge 1968). This solution pertains to the squeeze flow of an ideal plastic material without any immediate assumption regarding the slenderness of the configuration. Indeed, one can check that (41), together with the pressure solution

$$p = -\frac{2Bx}{h} - B\sqrt{1 - \frac{4y^2}{h^2}}, \tag{42}$$

satisfies the full force-balance equations,

$$\frac{\partial}{\partial x}(\sigma - p) + \frac{\partial \tau}{\partial y} = 0 \quad \& \quad \frac{\partial \tau}{\partial x} - \frac{\partial}{\partial y}(\sigma + p) = 0, \tag{43}$$

along with the yield condition $\sigma^2 + \tau^2 = B^2$. In other words, our pseudo-plug solution is nothing more than the shallow limit of Prandtl's solution. Importantly, the presence of the fully yielded boundary layer adjacent to the plates is equivalent to Prandlt's assumption that those surfaces are "fully rough." That is, u must vanish on the plates whatever the stress exerted there, implying $u_x = 0$ and therefore $\sigma = 0$.

2.6 Sliplines

In ideal plasticity, an elegant way of solving (43) is using the method of characteristics (Prager and Hodge 1968). This method constructs two families of special curves along which certain quantities are conserved, all of which can be recognized by defining

$$(\sigma, \tau) = B(-\sin 2\vartheta, \cos 2\vartheta), \tag{44}$$

and then rewriting (43) in the two forms,

$$\left(\cos\vartheta\frac{\partial}{\partial x} + \sin\vartheta\frac{\partial}{\partial y}\right)(p + 2B\vartheta) = \left(\sin\vartheta\frac{\partial}{\partial x} - \cos\vartheta\frac{\partial}{\partial y}\right)(p - 2B\vartheta) = 0. \tag{45}$$

In plasticity theory, the two families of characteristic curves are called sliplines and have parametric equations,

$$\frac{\mathrm{d}y}{\mathrm{d}x} = \tan\vartheta \qquad \& \qquad \frac{\mathrm{d}y}{\mathrm{d}x} = -\cot\vartheta. \tag{46}$$

Evidently the two sets of sliplines are orthogonal to one another, and along these curves the (Riemann) invariants are $p \pm 2B\vartheta$ (respectively). The sliplines for Prandtl's squeeze flow solution, given in $x > 0$ by

$$\frac{x - x_0}{h} = \pm\tan^{-1}\sqrt{\frac{\frac{1}{2}h + y}{\frac{1}{2}h - y}} - \sqrt{1 - \frac{4y^2}{h^2}}, \tag{47}$$

where $x = x_0$ is the starting point of the curve on the bottom plate, are illustrated in Fig. 8. (The slipline pattern has reflection symmetry about both coordinate axes.)

Although it is not a slender approximation, Prandtl's solution requires correction at the edges and core of the squeeze flow (Smyrnaios and Tsamopoulos 2001; 2006): at the edges, the flow must be terminated somehow, perhaps by allowing the fluid to extrude into free space beyond the ends of the plate. At the centre, the slipline pattern must be adjusted to allow for the symmetry condition along $x = 0$. One way to do this is to assume that the sliplines that intersect the origin are yield surfaces and

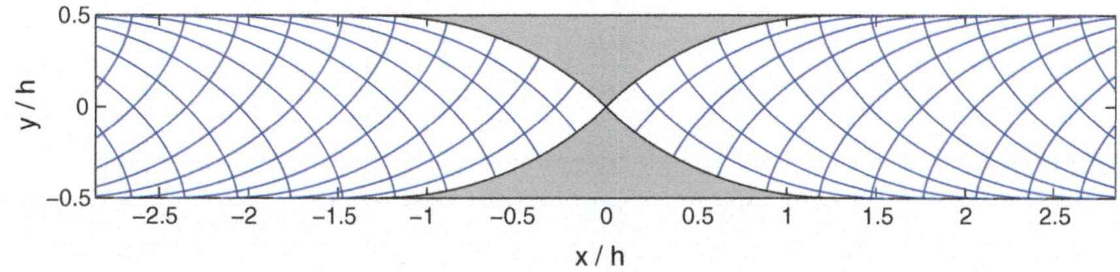

Fig. 8 Prandtl's slipline solution with the central plugs shaded

enclose a plug, as indicated in Fig. 8. Such a construction is allowed because further theory of the slipline field establishes that any of the sliplines can play the role of a yield surface provided the resulting stress field is consistent.

In more general geometries, the construction of the slipline network is more challenging. Practically, the most straghtforward situation is when p and ϑ are both known along a given curve. The sliplines can then be constructed by integrating (46) out from the known curve into the region of plastic deformation, exploiting the known invariants $p \pm 2B\vartheta$ and perhaps a simple finite-difference approximation (Prager and Hodge 1968). Unfortunately, only in a limited number of problems does one have the necessary known curve. More often, one must a boundary-value type problem because one only has incomplete information on more than one curve. Worse, problems in which boundary conditions on the velocity are required demand a simultaneous calculation of the velocity field. Despite this, slipline theory has proved useful in some specific plasticity problems that can be viewed as the plastic limits of viscoplastic flow problems (Nye 1967; Randolph and Houlsby 1984; Chamberlain et al. 2001). Alternatively, a useful test of the fidelity of a numerical scheme in the plastic limit can be provided by examining the slipline field predicted by a computed solution.

Note that the length of the adopted plugs in Fig. 8 is of the same order as the slot thickness and are therefore inaccessible in the lubrication theory even if the slot is narrow. Indeed, as discussed in Walton and Bittleston (1991) and Putz et al. (2009), one expects that true plugs could be hidden within the pseudo-plugs of slender flows around points of symmetry or stagnation; their detection and characterization requires a less controlled approximation strategy in view of their lack of scale separation. When the geometry of the conduit itself is more complicated, plugs can also clog wells and corners if their lengths are too short (Roustaei et al. 2015).

2.7 Extension to Three Dimensions

The lubrication analysis can be readily extended to three dimensions. We summarize the generalization for Herschel-Bulkley fluid flowing down a largely planar slot that is symmetrical about its midplane. The flow is pressure-driven and the walls are fixed in time. We align the midplane with the (x, z)−coordinate plane ($y = 0$; see Fig. 9). The dominant balance of forces is then

$$\frac{\partial p}{\partial x} = \frac{\partial \tau_{xy}}{\partial y}, \quad \frac{\partial p}{\partial y} = 0, \quad \frac{\partial p}{\partial z} = \frac{\partial \tau_{zy}}{\partial y}, \tag{48}$$

leading to

$$p = p(x, z), \quad (\tau_{xy}, \tau_{zy}) = z\,(p_x, p_z)\,. \tag{49}$$

The shear rates (u_y, w_y) across the slot again dominate the strain-rate tensor. Hence, the consitutive law, suitably non-dimensionalized using the characteristic

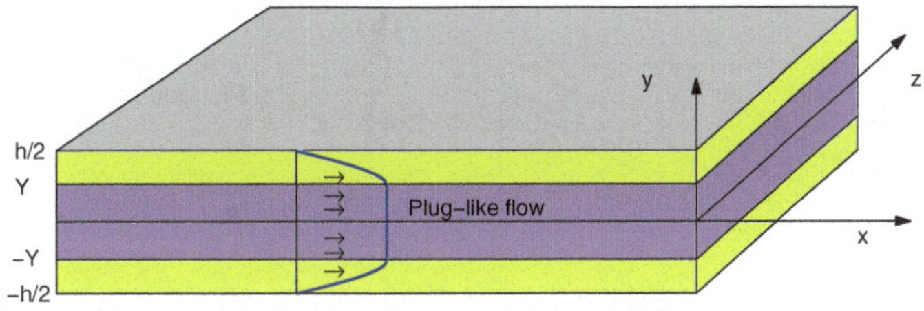

Fig. 9 Sketch of the geometry for a viscoplastic fluid in a Hele-Shaw cell

viscosity $\mu_* = K(\mathcal{U}/\mathcal{H})^{n-1}$ where K is the consistency and n is the power-law index, indicates that

$$\begin{pmatrix} \tau_{xy} \\ \tau_{zy} \end{pmatrix} \approx \left(\dot{\gamma}^{n-1} + \frac{B}{\dot{\gamma}}\right) \begin{pmatrix} u_y \\ w_y \end{pmatrix} \quad \text{if} \quad \tau \approx \sqrt{\tau_{xy}^2 + \tau_{zy}^2} > B, \tag{50}$$

where

$$\dot{\gamma} \approx \sqrt{u_y^2 + w_y^2}. \tag{51}$$

The (u, w) velocity field now follows (see Fig. 9):

$$\begin{pmatrix} u \\ w \end{pmatrix} = -\frac{n}{n+1} S^{(1-n)/n} \left[\max(\frac{1}{2}h - Y, 0)^{(n+1)/n} \right.$$
$$\left. - \max(|y| - Y, 0)^{(n+1)/n} \right] \begin{pmatrix} p_x \\ p_z \end{pmatrix}, \tag{52}$$

where the "yield surface" and pressure gradient are

$$Y = \frac{B}{S} \qquad \& \qquad S = \sqrt{p_x^2 + p_z^2}. \tag{53}$$

Note that, in the squeeze flow problem of Sect. 2.2, flow was immediately forced by the motion of the walls. For the current, pressure-driven flow, the pressure gradient may not always be sufficient to drive fluid down the slot, in which case the conduit must clog up and a true plug bridge the gap. Mathematically, this translates to Y exceeding $\frac{1}{2}h$ in the formulae above; the switch, $\max(\frac{1}{2}h - Y, 0)$, takes care of the implied (genuine) yield condition.

We now define a streamfunction based on the flux down the slot:

$$\begin{pmatrix} -\psi_z \\ \psi_x \end{pmatrix} = \int_{-\frac{1}{2}h}^{\frac{1}{2}h} \begin{pmatrix} u \\ w \end{pmatrix} \mathrm{d}y. \tag{54}$$

After a little algebra one then finds

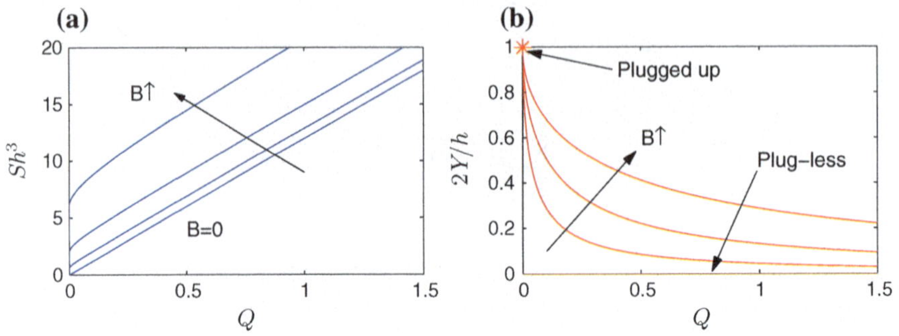

Fig. 10 **a** Flux-pressure-gradient relation for Bingham flow down a Hele-Shalw cell. **b** shows the corresponding fake yield surface. ($n = 1$)

$$\begin{pmatrix} \psi_z \\ -\psi_x \end{pmatrix} = \frac{Q}{S} \begin{pmatrix} p_x \\ p_z \end{pmatrix}, \tag{55}$$

involving the flux-pressure-gradient relation,

$$Q \equiv \sqrt{\psi_x^2 + \psi_z^2} = \frac{2n}{(n+1)(2n+1)} \left[\max\left(\frac{1}{2}h - Y), 0 \right) \right]^{1+1/n} \cdot [\frac{1}{2}(n+1)h + nY] S^{1/n}, \tag{56}$$

which is plotted in Fig. 10 (for $n = 1$). Finally, $p_{zx} = p_{xz}$ implies

$$(Q^{-1} S \psi_x)_x + (Q^{-1} S \psi_z)_z = 0, \tag{57}$$

a nonlinear elliptic equation that is similar to models proposed for viscoplastic displacement flows and nonlinear filtration (e.g. Bittleston et al. 2002, Pelipenko and Frigaard 2004).

For a Newtonian fluid, $Q \to (h^3/12)S$, (57) reduces to Laplace's equation and (56) is equivalent to Darcy's law, thus connecting flow in a Hele-Shaw cell with that through a porous medium. By analogy with this connection, and with a more general relation between Q and S than in (56), the nonlinear elliptic problem in (57) has been posed as a model for viscoplastic flow through a porous medium (e.g. Entov (1970)). Recent computations and experiments have even attempted to calibrate the $Q - S$ relation for particular idealized porous media (Chevalier et al. 2013a; Talon and Bauer 2013; Bleyer and Coussot 2014), following on from earlier Russian literature focussed on oil recovery problems (e.g. Bernadiner and Protopapas 1994). Importantly, as for the original viscoplastic constitutive law, the $Q - S$ relation features a threshold for flow to begin (Fig. 10), corresponding to the yield condition encoded in the switch, $\max(\frac{1}{2}h - Y, 0)$.

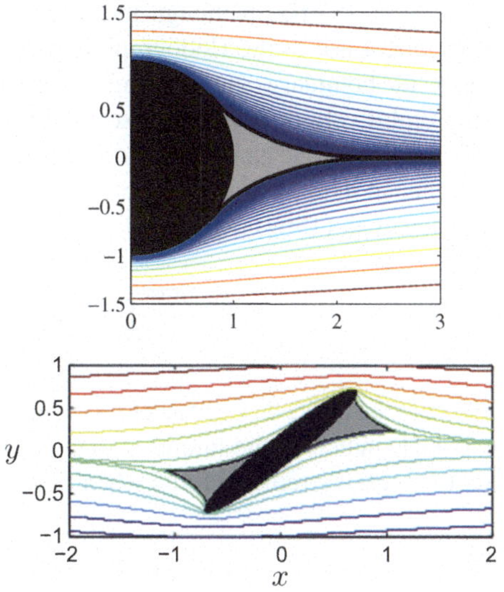

Fig. 11 Viscoplastic flow around a disk and an ellipse in a uniform Hele-Shaw cell with $B = 32$ and 20, respectively. Shown are a selection of streamlines. Lengths and velocities are scaled such that the mean flux in the x−direction is 2 well upstream or downstream of the obstacles, which have unit radius or semi-major axis, and the half-thickness of the cell is unity. Obstacles are shaded black; plugs are grey

The connection between Newtonian flow through a slot and in a porous medium originally motivated Henry Selby Hele-Shaw to visualize potential flow around obstacles. Along this vein, some numerical solutions to the model in (56)–(57) are shown in Fig. 11 for viscoplastic flow around a disk and an ellipse (obtained using an Augmented Lagrangian algorithm, see lectures by A. Wachs). These solutions contain genuine plug regions fore and aft of the obstacles, where Y reaches $\frac{1}{2}h$, $Q \to 0$ and flow grinds to a halt.

A clever way to deal with (57) (and some other viscoplastic flow problems; see Entov (1970)) is to use the Hodograph transformation, which can be applied when the slot is uniform (h is constant): this technique introduces the polar coordinate representations,

$$\begin{pmatrix} p_x \\ p_z \end{pmatrix} = S \begin{pmatrix} \cos\Theta \\ \sin\Theta \end{pmatrix} \quad \text{and} \quad \begin{pmatrix} \psi_x \\ \psi_z \end{pmatrix} = Q \begin{pmatrix} \sin\Theta \\ -\cos\Theta \end{pmatrix}, \tag{58}$$

and then makes the transformation, $(x, y) \to (Q, \Theta)$. The result is a linear elliptic problem for the streamfunction (Entov 1970):

$$\frac{Q^2}{S}\left(\frac{S^2}{QS'}\psi_Q\right)_Q + \psi_{\Theta\Theta} = 0, \tag{59}$$

along with

$$p_\Theta = -\frac{S^2}{QS'}\psi_Q, \qquad p_Q = \frac{S}{Q^2}\psi_\Theta, , \tag{60}$$

where $S' = \mathrm{d}S/\mathrm{d}Q$. Although the original elliptic equation is thereby linearized, the practical use of the Hodograph transformation is limited because one can only impose the boundary conditions after mapping the problem back to real space, which can prohibitively complicate those conditions.

Despite this, one can manufacture exact solutions for some special problems. For example, an interesting separable solution of (59) is given by $\psi = a(Q)\sin 2\Theta$ (cf. Alishaev et al. (1969)), where

$$\frac{Q^2}{S}\left(\frac{S^2}{QS'}a_Q\right)_Q - 4a = 0, \tag{61}$$

subject to the limits $a \propto Q^\alpha$ as $Q \to \infty$ and $a \propto Q^{(2n+1)/(n+1)}$ as $Q \to 0$ (which correspond to matching the solution to a far-field flow of power-law fluid and demanding that the pressure remains finite when flow halts), where

$$\alpha = \frac{1}{2}\sqrt{(n-1)^2 + 16n} - \frac{1}{2}(n-1). \tag{62}$$

The corresponding streamline pattern of the solution can be found by integrating

$$\left(\frac{\mathrm{d}x}{\mathrm{d}\Theta}, \frac{\mathrm{d}y}{\mathrm{d}\Theta}\right) = S^{-1}\frac{\mathrm{d}p}{\mathrm{d}\Theta}(\cos\Theta, \sin\Theta) \tag{63}$$

along curves of constant ψ, as illustrated in Fig. 12 for $n = 1$. Evidently, this solution corresponds to a stagnation-point flow with an embedded plug. The yield surfaces bordering this region can be constructed analytically (exploiting the fact that this curve is the limiting streamline for $Q \to 0$, along which $p \propto \cos 2\Theta$); in the first quadrant,

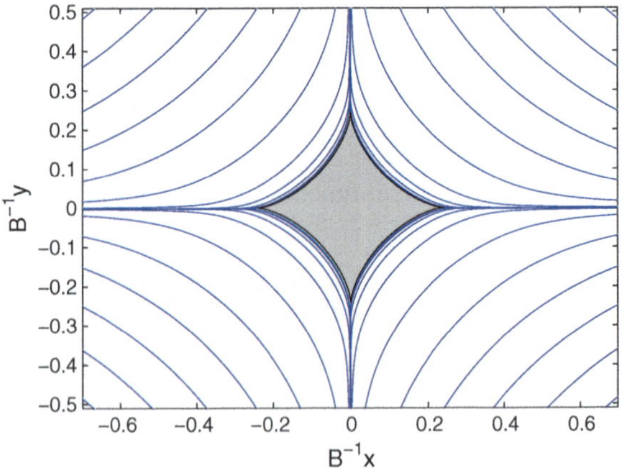

Fig. 12 Stagnation-point flow in a Hele-Shaw cell filled with Bingham fluid, matching to $\psi = 2xy$ in the far field

$$(x, y) = c(n)B^{(\alpha-1)/n}(\cos 3\Theta + 3\cos\Theta, 3\sin\Theta - \sin 3\Theta), \quad (64)$$

where $c(n)$ is an $n-$dependent constant that is determined from integrating the ODE in (61). Further details, as well as a discussion of other types of solutions can be found in Alishaev et al. (1969), Entov (1970), Balmforth et al. (1994) and Hewitt et al. (2016).

3 Viscoplastic Boundary Layers

The fully yield boundary layer arising in lubrication theory for $B \gg 1$ is an example of a viscoplastic boundary layer, in the sense that over this relatively thin region both viscous stresses and the yield stress complete the balance of forces. However, this is not the only situation in which viscoplastic boundary layers can arise. Indeed, Oldroyd (1947) in a classical paper suggested how such layers might arise in a variety of contexts and presented an analysis of the possible boundary-layer structure for Bingham fluid. A drawback of Oldroyd's theory is that it requires the solution of a rather challenging-looking PDE, and likely as a consequence, there have no attempts since to solve the problem in general settings. Oldroyd did, however, show that the equation had solutions with similarity form. Nevertheless, the relevance of these solutions has never been confirmed in any particular flow configuration. Moreover, the boundary layer structure (an in particular its scaling with Bingham number) appears to be different than in Oldroyd's theory in a number of other contexts (Tokpavi et al. 2008; Boujlel et al. 2012).

3.1 Oldroyd's Viscoplastic Shear Layer

Oldroyd's shear layer is the border between two plug flows which becomes thin in the limit that the Bingham number, $B = \mathcal{H}\tau_Y/(\mu\mathcal{U})$, is large, where $\mathcal{H}$ is a characteristic thickness of the layer and $\mathcal{U}$ is the velocity jump across it. For a straight boundary layer orientated along the horizontal line $y = y_0$ and with a characteristic aspect ratio of thickness to length of $\varepsilon \ll 1$, we resolve the boundary layer with a stretched coordinate and rescale for $B = \varepsilon^{-3} \gg 1$:

$$y = y_0 + \varepsilon\zeta, \quad u = u_m + U(x, \zeta), \quad v = \varepsilon V(x, \zeta), \quad (65)$$

where u_m is the mean horizontal velocity of the boundary layer, and

$$p = \frac{P(x, \zeta)}{\varepsilon^2}, \quad \tau = -\frac{1}{\varepsilon^3} + \frac{\check{\tau}(x, \zeta)}{\varepsilon}, \quad \sigma = \frac{\check{\sigma}(x, \zeta)}{\varepsilon^2}, \quad (66)$$

assuming $u_y < 0$, so that the leading-order shear stress is $\tau = -B$. These rescalings are designed so that, although τ dominates the stress state, the extensional stresses and pressure gradient still enter the force balance along with that dominant component.

The consequence of this design is that the thickness of the boundary layer has the scaling $\varepsilon = B^{-1/3}$.

The rescaled conservation equations are, to leading order,

$$U_x + V_\zeta = 0, \qquad P_x = \frac{\partial \check{\tau}}{\partial \zeta} + \frac{\partial \check{\sigma}}{\partial x} \qquad \& \qquad P_\zeta = -\frac{\partial \check{\sigma}}{\partial \zeta} + O(\varepsilon^2). \tag{67}$$

The expansion of the constitutive law for yielded fluid gives

$$\check{\tau} = U_\zeta + \frac{2U_x^2}{U_\zeta^2} \qquad \& \qquad \check{\sigma} = -\frac{2U_x}{U_\zeta} \tag{68}$$

Eliminating the pressure furnishes Oldroyd's boundary-layer equation,

$$\left[U_\zeta + \frac{2U_x^2}{U_\zeta^2}\right]_{\zeta\zeta} - 4\left(\frac{U_x}{U_\zeta}\right)_{x\zeta} = 0. \tag{69}$$

3.2 *Self-similar Solutions*

As indicated by Oldroyd, the boundary-layer equations have a solution with the self-similar form,

$$U = f(\chi) \qquad \& \qquad \chi = \frac{\zeta}{Y(x)}. \tag{70}$$

Assuming anti-symmetry about the centreline of the boundary-layer, and plugging (70) into (69), we find

$$f_{\chi\chi} = \lambda\chi \qquad \& \qquad Y_{xx} = -\frac{\lambda}{4Y^2}, \tag{71}$$

where λ is a separation constant. Imposing $f = \mp\frac{1}{2}$ and $f_\chi = 0$ at $\chi = \pm 1$ now gives

$$f = \frac{1}{4}\chi(\chi^2 - 3) \qquad \& \qquad \lambda = \frac{3}{2}. \tag{72}$$

One then has to solve (71a) for Y. There are solutions with $Y = Y_e$ and $Y_x = 0$ at the right-hand end of the boundary layer (giving $U_x = 0$) with

$$Y_e^{3/2}\left[\tan^{-1}\sqrt{\frac{s}{1-s}} - \sqrt{s(1-s)}\right]_{s=Y_0/Y_e}^{Y/Y_e} = \frac{\sqrt{3}}{2}(x - x_0), \tag{73}$$

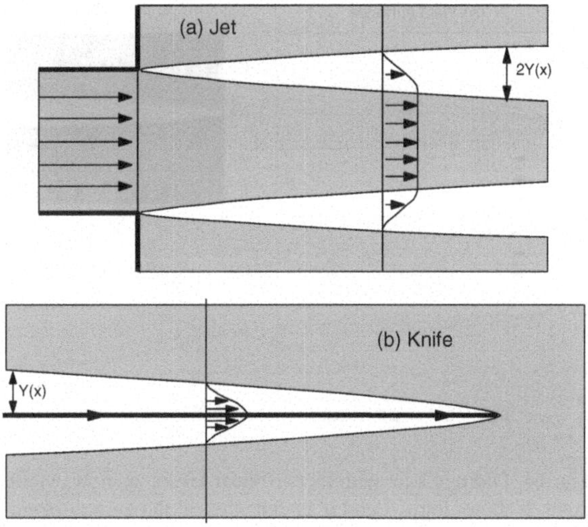

Fig. 13 Oldroyd's two canonical boundary layer problems: **a** A jet or finger of unyielded viscoplastic fluid is pushed out of a vent in a wall into a semi-infinite region filled with stagnant fluid; two thin shear layers buffer the plugged core of the jet from the rigid ambient. **b** The flow around a knife or plate pushed through an infinite viscoplastic fluid; a viscoplastic boundary layer coats the moving object

where $(x, Y) = (x_0, Y_0)$ denotes the start of the boundary layer. For $Y_e \to \infty$, we recover Oldroyd's power-law solution, $Y \propto x^{2/3}$.[2]

3.3 Sample Flow Solutions with Boundary Layers

Oldroyd suggested two canonical problems in which viscoplastic boundary layers might appear. The first consists of a finger or jet of Bingham fluid pushed out of a vent in a wall into semi-infinite region filled with stagnant fluid (see Fig. 13a). Oldroyd proposed that the borders between the finger and ambient would yield to form two thin viscoplastic shear layers with structure given by (72)–(73). Figure 14 shows a numerical solution of this problem, but set in a finite domain (with symmetry conditions imposed to the right and on the top and bottom) and computed using an augmented Lagrangian scheme. As predicted by Oldroyd, the sides of the finger form widening shear layer whose thickness scales with $B^{-1/3}$. The velocity profile and shape of the shear layer both agree with the self-similar solution.

Note that if one considers a narrower jet entering the same domain the solution is qualitatively different: the fluid yields almost immediately as it leaves the vent to form a wide plastically deforming region (Fig. 15); the plastic zones begin at the edges of the vent and merge together, leaving a small triangular tip to the incoming rigid jet. The plastic zone subsequently splits apart again further down the extrusion at a

[2]The determination of solutions of similarity form is a mathematical subject of itself. One option for the task is to introduce the rescalings, $(\hat{x}, \hat{\zeta}) = (kx, \ell\zeta)$, and then to establish what combinations (if any) of the parameters k and ℓ lead to the same equation as (69). One finds this to be so if $k^2 = \ell^3$. The combination $\chi = x/\zeta^{3/2} = \hat{x}/\hat{\zeta}^{3/2}$ is therefore invariant under the transformation and implies the existence of a self-similar solution with $U = U(\chi)$.

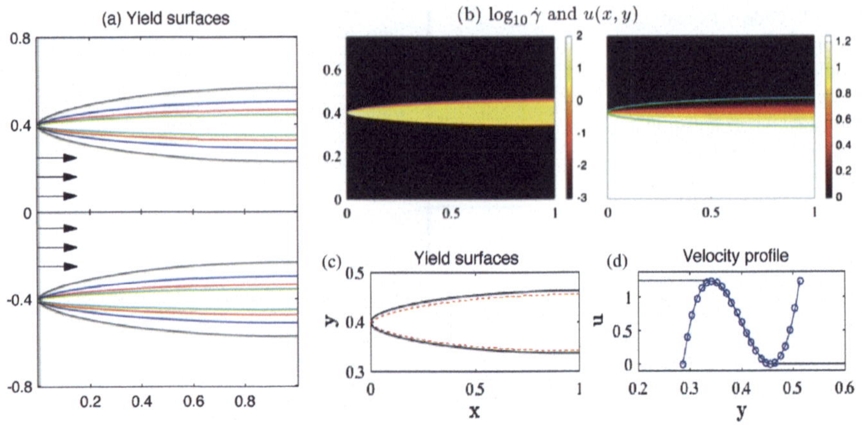

Fig. 14 Oldroyd's jet with its two shear layers. **a** shows yield surfaces for $B = 32$, 128, 512 and 2048. **b** shows $\log_{10}\dot{\gamma}$ and u as densities on the (x, y)−plane (half of the jet is shown), with the yield surfaces indicated, for $B = 20148$. **c** and **d** compare the yield surfaces and horizontal velocity profile of the upper shear layer with the prediction of the self-similar boundary layer solution (dotted red line and blue line with circles)

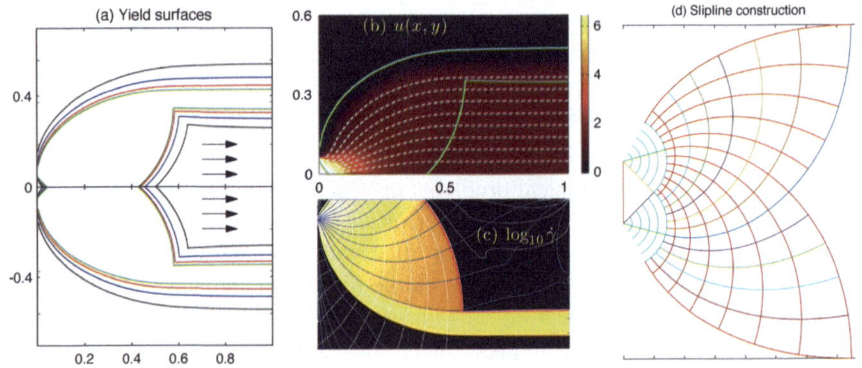

Fig. 15 A narrower jet. Again (**a**) shows the yield surfaces for $B = 32, 128, 512$ and 2048. **b** shows u, the yield surfaces and sample streamlines for $B = 2048$. **c** shows $\log_{10}\dot{\gamma}$ along with a selection of sliplines, as reconstructed from p and $\vartheta = -\frac{1}{2}\tan^{-1}(\sigma/\tau)$. **d** shows a slipline construction directly from the characteristic equations, beginning from two expansion fans and demanding that the sliplines meet at fortyfive degrees along the symmetry line

yield surface that eventually closes off the plastic region. Two horizontal viscoplastic boundary layers then remain that divide the surrounding rigid ambient from a moving plugged jet, and whose structure is again predicted by Oldroyd's self-similar solution. Thus, the extrusion adjusts to provide a minimum jet thickness if the vent is too narrow, and Oldroyd's expected flow configuration is recovered downstream.

As illustrated in Fig. 15c, d, the solution over the plastic region is described by a slipline field that begins with two expansion fans centred at the vent's edges and meets the symmetry line along the axis of the jet ($y = 0$) at $\pm 45°$ (a consequence of $\tau = 0$ there). The sliplines that border the triangular tip of the rigid jet form straight lines inclined at 45°. In turn, this implies that all the other sliplines of the same family in the fan form (straight) radial spokes of equal length and that the other family forms a set of concentric circular arcs (this follows from Hencky's rules, that dictate the geometry of the slipline field; see Prager and Hodge (1968)). Further from the centre of the fans, the outgoing sliplines begin to curve so that the other slipline family meets the symmetry line at the required angle. The slipline field must now be constructed by integrating the slipline equations. In principle, this direct construction of the slipline field offers a means to predict the emergent jet thickness.

Oldroyd's second problem studied the boundary layer around a knife or plate pushed through an infinite Bingham fluid (see Fig. 13b). This configuration has recently been the focus of an experimental study (Boujlel et al. 2012), which claims that one needs to supplement the viscoplastic boundary-layer solution with elastic-type deformation within the plug zones to generate a complete solution for the flow field. A computation of the steady flow around a finite plate is shown in Fig. 16. In this purely Bingham computation, the flow takes a boundary-layer form against the plate, as found experimentally and suggested by Oldroyd. The scaling of this boundary layer is not $B^{-1/3}$, however, but $B^{-1/2}$. In addition, fluid also circulates around to either side of the plate in two rigidly rotating vortices with diameter comparable to the length of the plate. The rigid rotation is bordered by a circular shear layer whose structure is actually of the Oldroyd $B^{-1/3}$–type. Besides the vortices and boundary layers, there are also regions of nearly plastic flow at the leading and trailing edges of the plate. The sliplines over those regions contain expansion fans centred at the plate's end and a network emerging from the viscoplastic boundary layer along the plate somewhat like Prandtl's construction (Fig. 8). The two are stitched together by a stress discontinuity, which is permissible in ideal plasticity if the tangential stress is discontinuous but the normal stress is continuous. Overall, the flow structure is quite different from Oldroyd's expectations, but matches solutions found for flow around elongated ellipses and rectangles (Chaparian and Frigaard 2017). Having said that, it is not so clear how one could extrapolate this solution to Oldroyd's semi-infinite knife.

Difficulties with the boundary-layer theory of Sects. 3.1–3.2 for the plate problem were already pointed out by Oldroyd: instead of imposing a symmetry condition at $\zeta = 0$, as for the shear layer, one must impose no-slip: $U(x, 0) = V(x, 0) = 0$. One must also match the boundary-layer solution to a plug or plastic flow outside the boundary layer, which corresponds to demanding that U, U_ζ and V should all become small at $\zeta = Y$. But the two conditions on $V(x, \zeta)$ cannot both be satisfied in Oldroyd's theory because V follows from integrating the continuity equation, $V_\zeta = -U_x$, which is only first order in ζ. Instead, Oldroyd proposed that one should impose $V(x, 0) = 0$ and omit the other boundary condition, suggesting that the nonvanishing transverse velocity at the edge of the boundary layer could be accounted for by allowing a modest elastic deformation over the region outside the boundary layer. Evidently, however, the core of the problem is that the boundary-layer scaling is not $B^{-1/3}$.

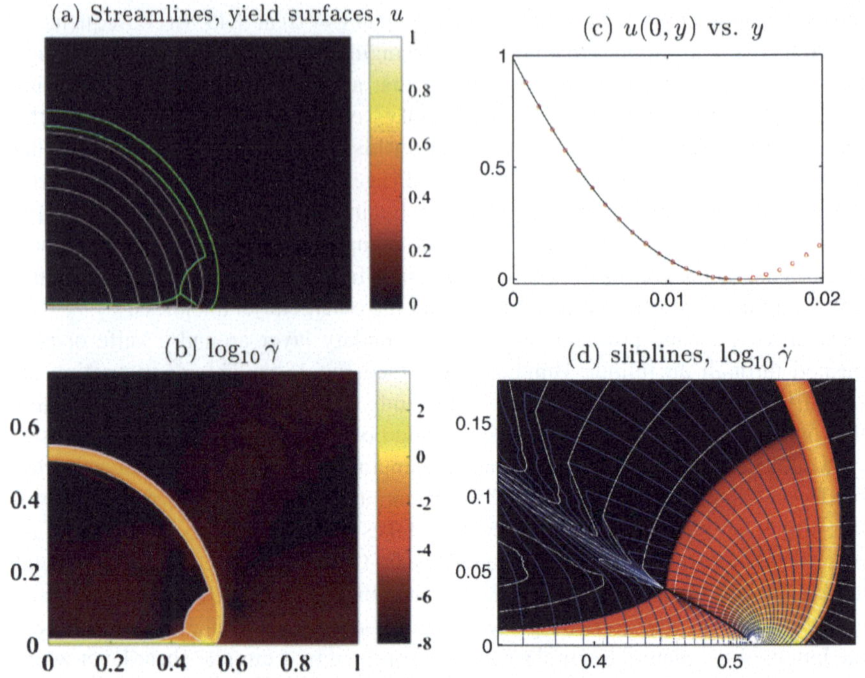

Fig. 16 Oldroyd's moving plate, showing the solution in the first quadrant; the plate occupies $-\frac{1}{2} < x < \frac{1}{2}$ and $y = 0$. **a** and **b** show u and $\log_{10}\dot{\gamma}$ as densities on the (x, y)−plane. **a** includes the yield surfaces and a selection of streamlines. **c** compares the boundary-layer velocity profile $u(0, y)$ with the asymptotic prediction in (75), taking Y from the numerical solution. **d** shows a magnification of the plastic region, with a reconstruction of the slipline field overlaid on a density map of $\log_{10}\dot{\gamma}$. $B = 2048$

The narrower $B^{-1/2}$ boundary layer is analogous to the large B limit of the fully yielded regions in lubrication theory (cf. (40)). For the flow around the plate, the appropriate scalings and main balances are

$$p = \frac{P(x, \eta)}{\varepsilon^2}, \quad \tau = -B + \frac{1}{\varepsilon}u_\zeta + \ldots, \quad P_x = u_{\zeta\zeta} \quad \& \quad P_\zeta = 0. \tag{74}$$

The link between the boundary layer thickness ε and B does not follow from the force balance equations here. Instead, one must argue that $\varepsilon = B^{-1/2}$ is demanded by matching the boundary-layer solution with the other regions in the flow. In particular, over the plastic region at the front and back of the plate, we must have $p = O(B)$ for a non-trivial slipline solution (see Sect. 2.6). Given $u(x, 0) = 1$ (the velocity of the plate), Eq. (74) predict the parabolic velocity profile,

$$u = \left[1 - \frac{\zeta}{Y(x)}\right]^2, \tag{75}$$

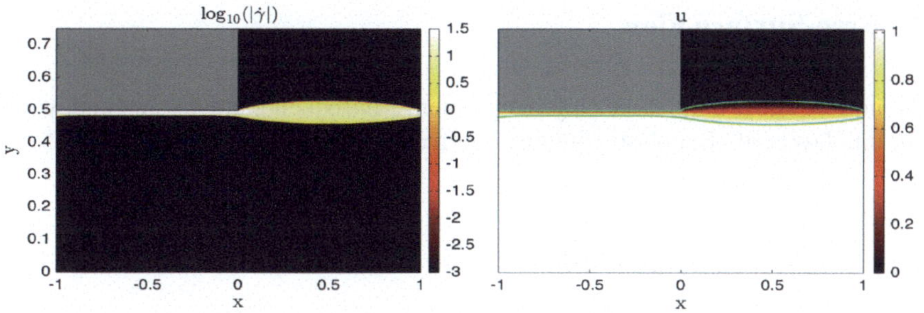

Fig. 17 Channel flow past a rectangular expansion at $B = 2048$, showing density maps of $\log_{10}\dot{\gamma}$ and u. The upper half of the solution is displayed. Plugs are shaded black; grey shows the bounding wall

where $\zeta = Y = \sqrt{2P_x}$ now denotes where the boundary layer meets either the rigid rotation or the plastic flow (where u and u_ζ become small; cf. Fig. 16).

Piau (2002) has also considered $B^{-1/2}$ boundary layers around the plate in a revision of Oldroyd's original analysis. Despite the different balances that this alternative scaling implicitly implies, Piau (somewhat obscurely) analyses the problem in a non-asymptotic fashion, keeping both the leading order and some of the correction terms. Instead of the simple equations outlined above (and their equally straightforward solution), he then arrives at Oldroyd's boundary-layer equations and writes down another self-similar solution. None of this elabouration appears necessary, although in some situations, retaining additional correction terms along with the leading order can improve an asymptotic solution (cf. Sect. 4.4).

For a third example, we consider channel flow through an expansion, as considered experimentally by Chevalier et al. (2013b), and who argued this to be a canonical viscoplastic analogue of the shear-banding of plastic materials in "frustrated flows" (Chevalier et al. considered pipe flow; we continue with the two-dimensional problem here). In the limit of large yield stress, the flow down the uniform part of the channel is largely plug-like, lubricated by thin boundary layers. When the plug meets the expansion, slightly less narrow shear layers develop to isolate the moving plug from rigid fluid in the clogged-up expansion. Figure 17 shows a computation of the configuration (assumed periodic, and again using the Augmented Lagrangian algorithm). The free shear layers have the $B^{-1/3}$–scaling and are well described by Oldroyd's boundary-layer analysis and similarity solution; the boundary layers against the channel walls, on the other hand, follow the $B^{-1/2}$ scaling.

4 Free-Surface Flow

We now consider a different range of problems, introducing gravity and endowing our thin layers of viscoplastic fluid with a free surface.

4.1 Sticky Viscoplastic Films

Returning to the lubrication model of Sect. 2, we consider the gravity-driven flow of a film of Bingham fluid over an inclined planar surface. We align an (x, z)–Cartesian coordinate system with the inclined plane, with x pointing down slope. The free surface lies at $z = \eta(x, t)$. See Fig. 18. The dimensional thin-film equations are

$$u_x + w_z = 0, \quad p_x = \tau_z + \rho g \sin\theta, \quad \& \quad p_z = -\rho g \cos\theta, \tag{76}$$

where g is gravity, ρ the fluid density and θ the angle of inclination with respect to the horizontal. With the scalings

$$x = \mathcal{L}\hat{x}, \quad (z, \eta) = \mathcal{H}(\hat{z}, \hat{\eta}), \quad (u, w) = \mathcal{U}(\hat{u}, \epsilon\hat{w}), \quad (p, \tau, \sigma) = \frac{\mu\mathcal{U}\mathcal{L}}{\mathcal{H}^2}(\hat{p}, \epsilon\hat{\tau}, \epsilon\hat{\sigma}), \tag{77}$$

and after dropping the hat decoration, we convert the force balance equations into the dimensionless form,

$$p_x = \tau_z + \mathrm{G_x}, \quad \& \quad p_z = -\mathrm{G_z}, \tag{78}$$

where the gravity parameters $\mathrm{G_x} = \rho g \mathcal{H}^3 \cos\theta / (\mu\mathcal{U}\mathcal{L})$ and $\mathrm{G_z} = \rho g \mathcal{H}^3 \cos\theta //$ $(\epsilon\mu\mathcal{U}\mathcal{L})$, both of which are assumed to be order one, which requires that the slope is relatively shallow. The Bingham law implies

$$\tau = u_z + B\,\mathrm{sgn}(u_z) \text{ if } |\tau| > B, \tag{79}$$

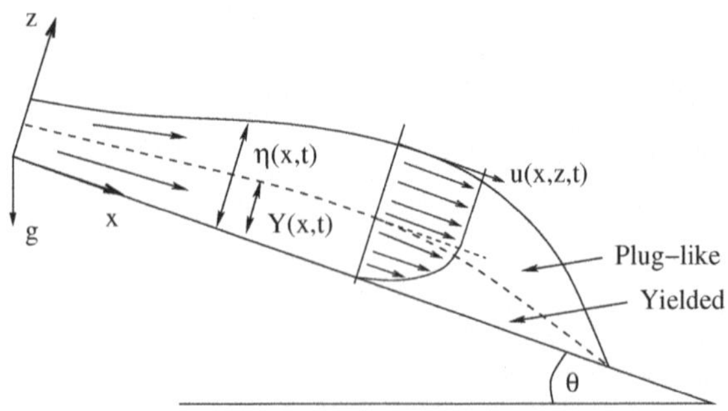

Fig. 18 Sketch of the geometry for a sticky viscoplastic film flow

and $u_z = 0$ otherwise, where $B = \tau_Y \mathcal{L}/(\rho g \mathcal{H}^2)$. At the free surface of the film, the thin-film scalings implies the stress conditions,

$$p(z, \eta, t) = \Pi(x, t) - \Gamma \eta_{xx} \qquad \& \qquad \tau(x, \eta, t) = 0, \tag{80}$$

where $\Gamma = \gamma \mathcal{H}^3/(\mu \mathcal{U} \mathcal{L}^3)$ is an inverse Capillary number and the normal surface force Π accounts for the ambient pressure of an overlying (inviscid) fluid such as air or water (*cf.* I. Wilson's lectures on cleaing and fouling). The kinematic condition demands

$$\eta_t + u(x, \eta, t)\eta_x = w(x, \eta, t). \tag{81}$$

It now follows that the pressure distribution is hydrostatic,

$$p = P - \mathrm{G_x} z \equiv \Pi - \Gamma \eta_{xx} + \mathrm{G_x}(\eta - z) \tag{82}$$

(P being the base pressure), and related to the shear stress by

$$\tau = (\mathrm{G_x} - P_x)(\eta - z) = (\mathrm{G_x} - \mathrm{G_z}\eta_x - \Pi_x + \Gamma \eta_{xxx})(\eta - z). \tag{83}$$

The velocity profile is

$$u = \frac{1}{2}(\mathrm{G_x} - P_x) \times \begin{cases} z(2Y - z), & 0 < z < Y, \\ Y^2, & Y < z < \eta, \end{cases}$$
$$Y = \mathrm{Max}\left(0, \eta - \frac{B}{|\mathrm{G_x} - P_x|}\right), \tag{84}$$

if there is no slip on the inclined plane ($u = w = 0$ on $z = 0$). The region $Y < z < \eta$ is occupied by a pseudo-plug (Fig. 18).

Finally, given the $z-$integrated mass conservation equation,

$$\frac{\partial \eta}{\partial t} + \frac{\partial}{\partial x} \int_0^\eta u \, \mathrm{d}z = 0, \tag{85}$$

we find the evolution equation for the fluid depth,

$$\frac{\partial \eta}{\partial t} = \frac{\partial}{\partial x} \left[\frac{1}{6}(3\eta - Y)Y^2(P_x - \mathrm{G_x}) \right]. \tag{86}$$

Without surface pressure variations and tension ($\Gamma = \Pi_x = 0$) this equation and some of its generalizations are reviewed in Balmforth et al. (2007b) (a first derivation being given by Liu and Mei (1989)).

The main feature of the model that we point out here is that fluid yields and spreads provided $Y > 0$. Hence, when $Y \to 0$, flow must come to a halt, furnishing an equation for a slumped deposit:

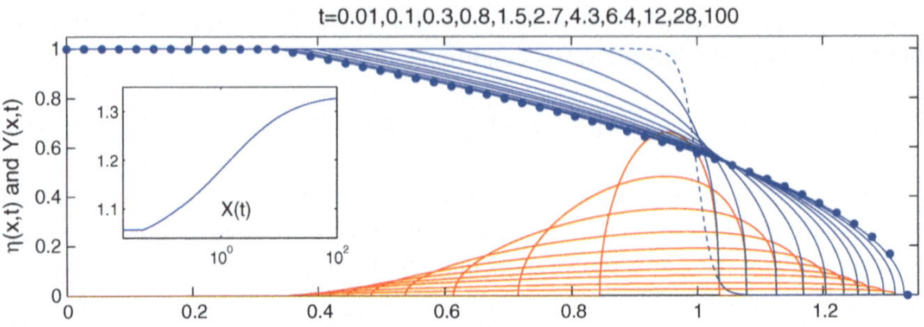

Fig. 19 Spreading viscoplastic current with $B = 0.5$, $\mathrm{G_z} = 1$, $\mathrm{G_x} = 0$, $\Gamma = 0$ and $\Pi = 0$, starting from a smoothed dambreak configuration that includes a pre-wetted film of thickness 10^{-3} to ease the computation and avoid a true contact line ($\eta(x, 0) = \frac{1}{2}[1 - \tanh 20(x^2 - 1)] + 10^{-3}$; dashed line). Shown are snapshots of $\eta(x, t)$ and $Y(x, t)$ at the times indicated; the dots show the final shape from (88). The solution is symmetrical about $x = 0$ and only the half in $x > 0$ is plotted. The inset shows the evolution of the front of the current $X(t)$

$$|\mathrm{G_x} - P_x|\eta = B. \tag{87}$$

On a flat surface and omitting surface tension and pressure variations ($\mathrm{G_x} = \Gamma = \Pi = 0$), we then find (cf. Nye (1952)).

$$\eta = \sqrt{1 - 2{\mathrm{G_z}}^{-1} B|x|}, \tag{88}$$

if we choose $\mathcal{H}$ as the maximum depth, and after shifting the position of the maximum to $x = 0$. The progress to this state is illustrated in Fig. 19, which shows a numerical solution to (86), starting from a dambreak-type configuration (an initial-value problem that is relevant to the "Bostwick consistometer," a practical rheometer used in food science; see Balmforth et al. (2007a)).

A numerical solution that includes surface tension ($\Gamma \neq 0$) is shown in Fig. 20. In this case, once the fluid is released, the high curvature of the initial state generates a rapid adjustment in which a prominent ridge and capillary wave train propagate towards the centre of the fluid, yielding the fluid layer en route. Flow eventually subsides, as in Fig. 19, but with the entire initial reservoir having collapsed. Again the final state is predicted by (87), which for the current parameter settings becomes

$$\Gamma\eta_{xxx} - \mathrm{G_x}\eta_x = \frac{B}{\eta}, \tag{89}$$

assuming $P_x < 0$ and the boundary conditions $\eta(X) = \eta'(X) = 0$ at the fluid edge $x = X$ (some care needs to be taken in arriving at this boundary condition; Jalaal (2016)).

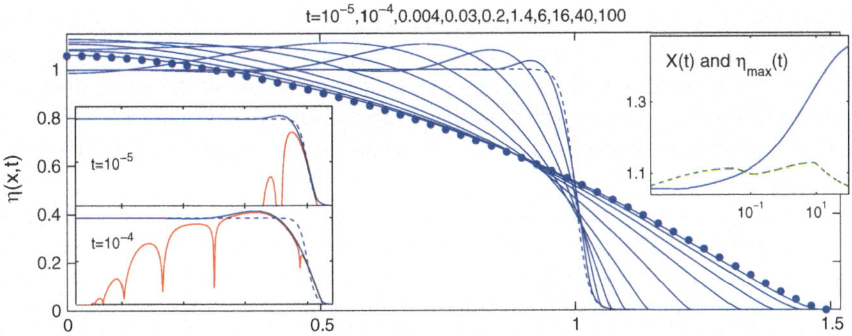

Fig. 20 Dambreak with $B = 0.5$ and $\Gamma = 0.1$ ($\mathrm{G_z} = 1$, $\mathrm{G_x} = \Pi = 0$; $\eta(x, 0) = \frac{1}{2}[1 - \tanh 20(x^2 - 1)] + 10^{-3}$). Shown are snapshots of $\eta(x, t)$ at the times indicated; the dots show the final shape expected from (89). The insets show the evolution of the front of the current $X(t)$ and its maximum depth $\eta_{max}(t)$, and the two earliest snapshots including $Y(x, t)$

The model in (86) is analogous to that used for viscous gravity currents (Craster and Matar 2009). As in that Newtonian problem, rather elabourate machinations are required to add inertia to the theory (Fernández-Nieto et al. 2010). This arises because the z−dependence of the velocity field forbids a simple approximation of the inertial terms and also changes with their introduction. Perhaps the most significant extension of the viscoplastic film theory for the future is to explore the inertial generalization of the model.

4.2 *Sliding Films*

What happens when the condition on the base of the film is not no-slip? In some geophysical problems, like glacier flow, this is a very real consideration in view of the physical condition of the interface between ice and the underlying terrain and the lubricating action of any water (Schoof and Hewitt 2013). For many non-Newtonian fluids in engineering this is also important because of the tendency for these materials to suffer apparent slip (e.g. (Barnes 1995); lectures by I. Wilson). Indeed, slip has been observed directly in drops of viscoplastic fluid spreading over glass surfaces Jalaal et al. (2015). Nevertheless, despite this widespread occurence, physical models of slip are rare and slip laws are mostly based on empiricism.

To consider the effect of slip on the film model above, we abandon the no slip condition in favour of a relation between the basal shear stress, $\tau_b \equiv \eta(\mathrm{G_x} - \eta_x)$, and slip velocity u_s. In particular, we take

$$u_s = k\ \mathrm{Max}(0, |\tau_b| - B_w)^m\ \mathrm{sgn}(\tau_b), \tag{90}$$

with (suitably scaled) dimensionless parameters k, B_w and m. A simple power-law of the form, $u_s = k|\tau_b|^m \,\mathrm{sgn}(\tau_b)$, for the slip law has the unappealing feature that material would slide over the bounding surface however small the basal stress, in contrast to the everyday observation that drops of viscoplastic fluid do adhere to shallowly inclined surfaces without flowing or sliding. The inclusion of B_w accounts for a sliding threshold.

With the sliding law in (90), spreading will either not occur at all if $|\tau_b| < (B_w, B)$, via sliding alone if $B_w < |\tau_b| < B$, through internal shear flow with no slip if $B < |\tau_b| < B_w$, or both shear and slip if $|\tau_b| > (B, B_w)$. In most situations one expects that $B_w < B$. Motion will then halt only after sliding ceases for $\tau_b = |\mathrm{G_x} - P_x|\eta \to B_w$, which predicts the same final shapes as above save that B is replaced by B_w. Hence in spreading tests, there is a danger that one might incorrectly diagnose the yield stress from the sliding threshold.

The final phase of spreading (with $B_w < |\tau_b| < B$) is controlled purely by the basal friction from the slip layer; the overlying flow is plug-like with $u = u_s$ and the analogue of (86) is

$$\frac{\partial \eta}{\partial t} = -\frac{\partial}{\partial x}(\eta u_s) = -k\frac{\partial}{\partial x}\left[\eta Y_w^m |P_x - \mathrm{G_x}|^{m-1}(P_x - \mathrm{G_x})\right],$$
$$Y_w = \eta - \frac{B_w}{|\mathrm{G_x} - P_x|}. \tag{91}$$

Evidently, the dynamics appears to be independent of the bulk rheology, although the spreading fluid cannot be rigid (u_s will in general depend on x). In fact, the whole layer is a pseudo-plug due to its yield stress.

4.3 Viscoplastic Membrane Models

A curious feature of the sliding film model is that the basal friction controls spreading and the only role of the bulk fluid rheology is to suppress shear and establish the pseudo-plug. If the underlying surface is very slippy, however, this is unrealistic as one expects that extensional stresses in the plane of the film ought to limit the spreading of the fluid. Above, such stresses have been omitted from the main balance of forces; to restore them, a different set of scalings is needed. The point is that the relatively free sliding of the fluid layer relieves the shear stress sufficiently to enhance the extensional stresses. We therefore introduce

$$(p, \tau, \sigma) = \frac{\mu \mathcal{U}}{\mathcal{L}}(\hat{p}, \epsilon\hat{\tau}, \hat{\sigma}), \tag{92}$$

to furnish the force balance equations

$$p_x = \sigma_x + \tau_z + \mathrm{G_x}, \quad \& \quad p_z + \sigma_z = -\mathrm{G_z}, \tag{93}$$

where we redefine the gravity parameters as

$$(\mathrm{G_x}, \mathrm{G_z}) = \rho g \mathcal{L}^3(\epsilon^{-1}\sin\theta, \cos\theta)/(\mu\mathcal{U}).$$

The stress conditions at the surface must also be revised:

$$p(z,\eta,t) + \sigma(x,\eta,t) = \Pi(x,t) - \Gamma\eta_{xx} \quad \& \quad \tau(x,\eta,t) - 2\eta_x\sigma(x,\eta,t) = 0, \tag{94}$$

with $\Gamma = \gamma\mathcal{H}/(\mu\mathcal{U}\mathcal{L})$. Thus, integrating both relations in (93) in z,

$$p + \sigma = \Pi + \eta - z \ \& \ 0 = \eta(\mathrm{G_x} - \mathrm{G_z}\eta_x - \Pi_x + \Gamma\eta_{xxx}) - \tau_b + 2\frac{\partial}{\partial x}\int_0^\eta \sigma \mathrm{d}z, \tag{95}$$

in which we see the emergence in importance of the extensional stress σ.

Despite the satisfying balances in (95), there is an apparent inconsistency with the constitutive law which indicates that $(\sigma, \epsilon\tau) \propto (2u_x, \epsilon^{-1}u_z)$ in view of (92). However, we now operate in the limit that sliding is relatively free, and so we cannot expect that much shear builds up in u. We therefore set

$$\dot{\gamma} = 2\left|\frac{\partial u_s}{\partial x}\right| + O(\epsilon^2), \quad u = u_s(x,t) + \epsilon^2 u_2(x,z,t), \quad w = -z\frac{\partial u_s}{\partial x} + O(\epsilon^2), \tag{96}$$

and

$$(\sigma, \tau) = \left(1 + \frac{B}{\dot{\gamma}}\right)\left(2\frac{\partial u_s}{\partial x}, \frac{\partial u_2}{\partial z}\right), \qquad \text{if } |\sigma| > B, \tag{97}$$

and $\partial u_s/\partial x = 0$ otherwise, where $B = \tau_Y\mathcal{L}/(\mu\mathcal{U})$.

At this stage, computing the correction u_2 is not needed and we arrive the paired evolution equations,

$$\eta_t + (\eta u)_x = 0 \ \& \ \mathrm{Re}\,(u_t + uu_x) = \mathrm{G_x} - \mathrm{G_z}\eta_x - \Pi_x + \Gamma\eta_{xxx} - \frac{1}{\eta}\tau_b + \frac{2}{\eta}\frac{\partial}{\partial x}(\eta\sigma), \tag{98}$$

given $u \approx u_s$ and where we have also restored the leading-order inertial terms with Reynolds number $\mathrm{Re} = \rho\mathcal{U}\mathcal{L}/\mu$, which is acceptable in our asymptotic scheme in view of the plug-like character of the flow (unlike in Sect. 4.1). We may also immediately generalize to the Herschel-Bulkley version of the model by writing

$$\sigma = 2u_x\left(|2u_x|^{n-1} + \frac{B}{2|u_x|}\right) \quad \text{for } |\sigma| > 0 \ \ \& \ \ u_x = 0 \text{ otherwise} \tag{99}$$

(though we consider only the Bingham case, $n = 1$, in all examples).

The model in (98)–(99) is what one might call a membrane model in solid mechanics (Mansfield 2005). In fluid dynamics, the model is equivalent to that for free viscous films and jets (Craster and Matar 2009), Savage-Hutter-like models for granular flows (Balmforth and Kerswell, 2005), fast-moving ice streams and floating ice shelves (with $\mathrm{Re} \to 0$; Schoof and Hewitt 2013), and viscoplastic threads (Balmforth et al. 2010). Indeed, by dropping the basal drag τ_b, the model becomes identical to that for the evolution of a viscoplastic jet under varicose disturbances (perturbations symmetric about the midline), with $\eta(x,t)$ re-interpreted as the half-thickness.

To reconsider the gravitational spreading problem of Sects. 4.1–4.2, we set $\mathrm{Re} = \Gamma = \Pi_x = 0$ and take (90) for τ_b. Assuming that the fluid approaches rest with $(u, u_x) \to 0$ and $\mathrm{sgn}(u_x) = s = \pm 1$, the final shape is dictated by balancing the remaining terms the momentum equation in (98): given that $\sigma \to sB$ and $|\tau_b| \to B_w$ these terms combine into the relation,

$$|(2sB - \eta \mathrm{G_z})\eta_x - \eta \mathrm{G_x}| = B_w. \tag{100}$$

To examine the shape this predicts in more detail, we consider a flat plane ($\mathrm{G_x} = 0$). If the centre of the drop, with $\eta = \eta_{max}$, is at $x = 0$, $\eta_x < 0$ and $\eta - sB > 0$ just to the right. Hence

$$\eta = \sqrt{(\eta_{max} - 2s\mathrm{G_z}^{-1}B)^2 - 2\mathrm{G_z}^{-1}B_w x} + 2s\mathrm{G_z}^{-1}B. \tag{101}$$

For a flow in compression ($s < 0$), this profile can decrease to zero, with finite slope at the edge. On the other hand, when the flow comes to rest in expansion ($s > 0$), the solution must terminate at a finite height to avoid becoming multi-valued. This suggests either that the edge is vertical (in violation of the slender asymptotics) or that a rim of unyielded fluid borders the main body of the slumped current.

Sample numerical solutions to (98) are shown in Fig. 21, and again correspond to dam-break-like initial-value problems (with neither surface pressure variation nor tension, $\Pi = \Gamma = 0$). The first example includes both an extensional yield stress and bottom drag with a finite threshold; no drag is included for the second case. Both slide to a halt with a shape given by (101). The flows are in extension, and the final shapes terminate in abrupt cliffs of unyielded fluid that were pushed ahead of the extending current.

The second example of Fig. 21 spreads out with a distinctive flat-topped profile, revealing how an analytical solution of the problem is feasible in this case: although the Reynolds number used for the computation is $\mathrm{Re} = 1$, the velocity u remains small throughout. In this situation, the inertial terms are small and over the bulk of the current one may take the solution,

$$\eta = N(t) \qquad \& \qquad u = x\Upsilon(t), \tag{102}$$

with (from the first relation in (98))

$$\frac{\mathrm{d}N}{\mathrm{d}t} + \Upsilon N = 0. \tag{103}$$

This spatial profile must be terminated by a jump or shock at the flow front, $x = X(t)$. Two further relations then follow from imposing the jump conditions,[3]

$$\frac{\mathrm{d}X}{\mathrm{d}t} = \Upsilon X \quad \& \quad \frac{1}{2}\mathrm{G_z}N = 2\sigma \equiv 4\Upsilon + 2B. \tag{104}$$

Thus, $NX = 1$ if $N(0) = X(0) = 1$ (an obvious demand of mass conservation) and

$$\frac{\mathrm{d}N}{\mathrm{d}t} = -\frac{1}{8}N(N-4B), \quad \text{or} \quad N = 4B\left[1-(1-4B)e^{-Bt/2}\right]^{-1}, \tag{105}$$

if $\mathrm{G_z} = 1$. This solution, which is compared to the numerical computation in Fig. 21, exposes the final state, $\eta \to 4B$, and yield condition, $4B < 1$. The latter is characteristic of the initial condition and arises because the jump in the net hydrostatic pressure at the edge of the fluid can be balanced by the extensional stress 2σ without failure provided σ does not exceed the yield stress. When the initial extensional stress exceeds B, sliding ensues, until the extension of the layer reduces the net hydrostatic pressure on the side face to the equilibrium level $4B$. Much that same happens when $B_s \neq 0$ although that threshold affects both the failure condition and final rest state.

4.4 Jet Instability

To explore the (varicose) instability of a high-speed viscoplastic jet in air, we need a prescription for the dynamic pressure due to the irrotational flow of ambient air. For this, we need to solve Laplace's equation for the velocity potential $\mathcal{UH}\phi(x, y, t)$ (with y scaled with $\mathcal{L}$ rather than the film thickness) subject to the boundary condi-

[3] The first shock condition (Rankine-Hugoniot relation; see Whitham 1974) follows from first writing the mass-conservation equation in (98) as a integral conservation law,

$$0 = \frac{\mathrm{d}}{\mathrm{d}t}\int_{x_1}^{x_2} \eta(x,t)\mathrm{d}x + [J(x,t)]_{x=x_1}^{x=x_2} = \int_{x_1}^{X} \eta_t(x,t)\mathrm{d}x$$
$$+ \int_{X}^{x_2} \eta_t(x,t)\mathrm{d}x + \left[J(x,t) - \dot{X}\eta(x,t)\right]_{x=x_1}^{x=x_2},$$

where $J = hu$ is the mass flux, x_1 and x_2 are arbitrary locations straddling the shock, and $X^{\pm}$ implies the limit from either the left or right. By taking $x_1 \to X^-$ and $x_2 \to X^+$ and arguing that the remaining integrals then become negligible, we then find

$$\dot{X}\,[\eta(x,t)]_{x=X^-}^{x=X^+} = [J(x,t)]_{x=X^-}^{x=X^+}.$$

The second relation follows from applying the same argument to the momentum equation in (98) ($\Gamma = 0$):

$$\dot{X}\,\mathrm{Re}\,[\eta u]_{x=X^-}^{x=X^+} = \left[2\eta\sigma - \frac{1}{2}\mathrm{G_z}\eta^2\right]_{x=X^-}^{x=X^+}.$$

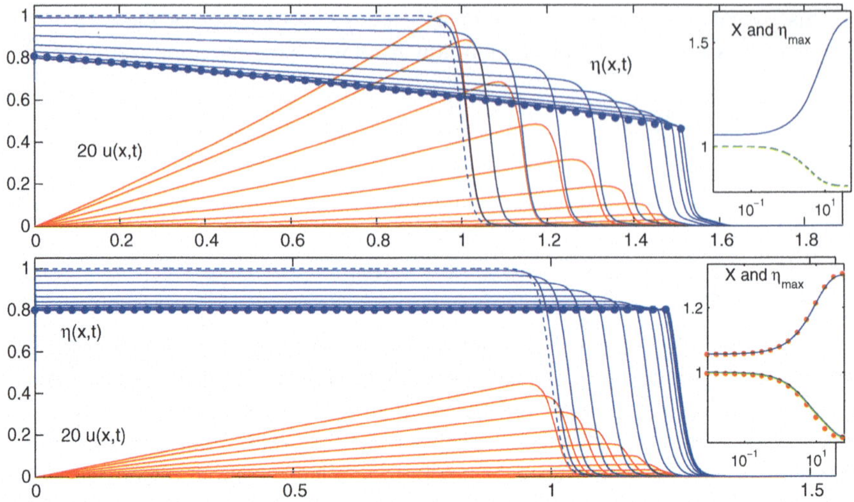

Fig. 21 Dambreak of a sliding viscoplastic current with **a** $B = 0.1$, $B_s = 0.1$, $m = 1$ and $k = 1$, and **b** $B = 0.2$ and $B_s = k = 0$. Both have $\mathrm{Re} = \mathrm{G_Z} = 1$, $\mathrm{G_X} = \Gamma = \Pi = 0$, $u(x, 0) = 0$ and $\eta(x, 0) = \frac{1}{2}[1 - \tanh 20(x^2 - 1)] + 10^{-3}$ (dashed line). Shown are snapshots of $\eta(x, t)$ and $u(x, t)$ at $t = j^2/5$, $j = 1, 2, ..., 12$. The dots show the final shape from (101) given the final maximum depth η_{max}. The insets show the evolution of the front of the current $X(t)$ and $\eta_{max}(t)$; in (**b**), the dots show the prediction in (105), with the starting value of X adjusted to account for the finite thickness of the front (i.e. $X = 1/N + X(0) - 1$)

tions on the fluid-air interface. Those conditions are

$$\phi_y(x, 0, t) = \eta_t + O(\epsilon) \quad \& \quad \frac{\partial}{\partial x}[\epsilon \varrho \,\mathrm{Re}\, \phi_t(x, 0, t) + p_a] = O(\epsilon^2), \qquad (106)$$

where ϱ is the ratio of air to fluid density, p_a is the interfacial air pressure (scaled by $\mu\mathcal{U}/\mathcal{L}^2$), which follow from the leading order kinematic condition and Bernoulli's law. The last of these relations indicates that the air pressure contribution to Π if $O(\epsilon)$ if $\varrho\,\mathrm{Re} = O(1)$; we keep this correction term along with the leading order in view of the fact that it is the source of instability. Further justification can be given from the observation that all the corrections to (98) are $O(\epsilon^2)$. It is therefore consistent to include an $O(\epsilon)$ approximation to p_a.

Laplace's equation can be solved using the Fourier transform (Sneddon 1957): if we denote the transform of $f(x)$ and its inverse by

$$\check{f}(k) = \mathcal{F}\{f(x)\} = \int_{-\infty}^{\infty} f(x) e^{-ikx} \mathrm{d}x \quad \&$$

$$f(x) = \mathcal{F}^{-1}\{\check{f}(k)\} = \int_{-\infty}^{\infty} \check{f}(k) e^{ikx} \frac{\mathrm{d}k}{2\pi}, \qquad (107)$$

Fig. 22 Sketch of the geometry for a viscoplastic beam

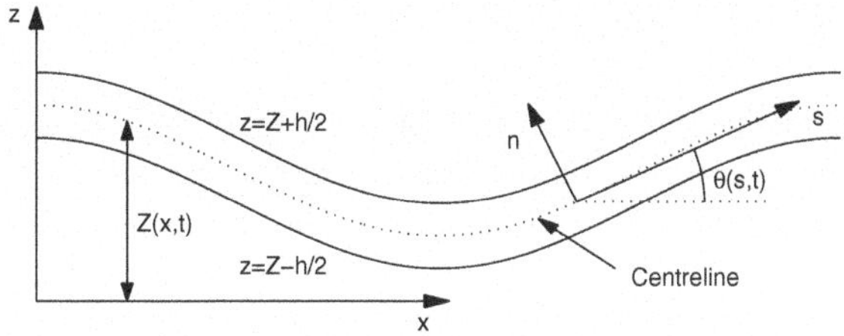

then one can verify that $\check{\phi}(k, y, t) = e^{-|k|y}\check{\phi}(k, 0, t)$. The conditions in (106) now imply

$$-|k|\check{\phi}(k, 0, t) = \mathcal{F}\{\eta_t\} \quad \& \quad ik\epsilon\varrho\mathrm{Re}\,\check{\phi}_t(k, 0, t) + \mathcal{F}\{\partial p_a/\partial x\} = 0. \quad (108)$$

Hence

$$\frac{\partial}{\partial x}p_a = \epsilon\varrho\mathrm{Re}\ \ y\mathcal{F}^{-1}\{i\ \mathrm{sgn}(k)\mathcal{F}\{\eta_{tt}\}\} \equiv -\epsilon\varrho\mathrm{Re}\ \mathrm{H}\{\eta_{tt}\}, \quad (109)$$

where

$$\mathrm{H}\{f(x)\} = \frac{1}{\pi}\fint \frac{f(z)\,\mathrm{d}z}{x - z} \quad (110)$$

is the Hilbert transform (Sneddon 1957) (and the extra decoration on the integral symbol indicates that the principal value must be taken to render the integral non-singular).

Now consider the stability of a viscoplastic jet, ignoring gravity $\mathrm{G_x} = \mathrm{G_z} = 0$ and basal drag $\tau_b = 0$. The evolution equations become

$$\eta_t + (\eta u)_x = 0 \quad \& \quad \mathrm{Re}\,(u_t + uu_x) = \epsilon\varrho\mathrm{Re}\ \mathrm{H}\{\eta_{tt}\} + \Gamma\eta_{xxx} + \frac{2}{\eta}\frac{\partial}{\partial x}(\eta\sigma). \quad (111)$$

For a Newtonian jet, it is straightforward to consider the stability of small perturbations about the uniform equilibrium $u = \eta = 1$ with dependence $\exp ik(x - ct)$, where k is the wavenumber and c the (complex) wavespeed. One can thereby establish that perturbations are unstable for $k < \epsilon\varrho\mathrm{Re}\,/\,\Gamma$, and one expects these modes to grow to break up the jet.[4] For a Bingham fluid, on the other hand, the equilibrium is unconditionally stable because the stress of the basic state, $\sigma = 0$, is finitely below the yield stress B. Thus, the jet can only break up given a finite-amplitude initial perturbation. The situation is closely related to the removal of the Rayleigh instability in viscoplastic threads with surface tension (Balmforth et al. 2010) and mirrors many other viscoplastic problems of hydrodynamic stability (see lectures by I. Frigaard and S. Hormozi).

---

[4]The restriction to varicose perturbations ignores the (very real) possibility that sinuous perturbations may be more unstable and dominate the break-up dynamics. Sinuous perturbations require a consideration of the bending of the midline of the fluid layer, along the lines considered in Sect. 4.5.

4.5 Viscoplastic Beams

When the fluid is not stuck down to an underlying surface, it can become easier for the viscoplastic film to buckle out of plane rather than deform symmetrically with respect to the midline (Fig. 22). This calls for a different development of the thin-layer equations (Balmforth and Hewitt 2013). We begin with the scaling of the stress components:

$$(p, \tau, \sigma) = \frac{\mu \mathcal{U} \mathcal{H}}{\mathcal{L}^2}(\hat{p}, \epsilon\hat{\tau}, \hat{\sigma}), \tag{112}$$

but add the further constraint $-p - \sigma = \epsilon^2 s_{zz}$. That is, $p = -\sigma + O(\epsilon^2)$, which reduces the normal stress in the z−direction to $O(\epsilon^2)$, thereby achieving the force balances,

$$p_x = \sigma_x + \tau_z + \mathrm{G_x}, \quad \& \quad 0 = \tau_x + \frac{\partial}{\partial z} s_{zz} - \mathrm{G_z}, \tag{113}$$

with the gravity parameters, $(\mathrm{G_x}, \mathrm{G_z}) = \rho g \mathcal{L}^2(\sin\theta, \epsilon^{-1}\cos\theta)/(\mu\mathcal{U}\mathcal{H})$, so the film is now either nearly vertical ($\theta \approx \frac{1}{2}\pi$; $\epsilon^{-1}\cos\theta) = O(1)$) or horizontal ($\mathrm{G_z} = O(1)$ and $\mathrm{G_x} \to 0$).

We place the surfaces at $z = \eta_\pm(x, t) = Z(x, t) \pm \frac{1}{2}h(x, t)$, where $z = Z(x, t)$ is the midline of the sheet, and ignore surface pressure variations and tension. The surface stress conditions are then

$$s_{zz} - \eta_x \tau = 0 \quad \& \quad \tau + \eta_x(p - \sigma) = 0, \tag{114}$$

at $z = \eta(x, t) = Z \pm \frac{1}{2}h$.

We now formulate integral expressions of force and torque balance by integrating over z the equations in (113) and then the first of these expressions multiplied by $z - Z(x, t)$. Given (114) we then arrive at[5]

$$\Sigma_x = h\mathrm{G_x}, \qquad Q_x = h\mathrm{G_z}, \qquad \& \qquad M_x + \Sigma Z_x - Q = 0, \tag{115}$$

where the "stress resultants" and effective moment are

$$\begin{aligned}[\Sigma, Q, M] &= \int_{Z-\frac{1}{2}h}^{Z+\frac{1}{2}h} [\sigma - p, \tau, (z - Z)(\sigma - p)]\, \mathrm{d}z \\ &= \int_{Z-\frac{1}{2}h}^{Z+\frac{1}{2}h} [2\sigma, \tau, 2(z - Z)\sigma]\, \mathrm{d}z + O(\epsilon^2).\end{aligned} \tag{116}$$

The specific ordering of the stress components in (112) demands a particular form for the velocity field in order to achieve the same ordering of the strain rates (with

[5]Note that there are typographical errors in the corresponding formulae for order-one curvature presented by Balmforth and Hewitt (2013). Specifically, the moment terms in their (95) and (116) should be $-\epsilon\kappa M_s$, and that in (98) should be $-\kappa M_s$.

(u, w) scaled by $\mathcal{U}$):

$$u = \epsilon[U - (z - Z)W_x] + O(\epsilon^2), \qquad w = W + O(\epsilon^2), \tag{117}$$

where $[U(x,t), W(x,t)]$ is the leading-order velocity of the centreline. Hence (with units of $\mathcal{U}/\mathcal{L}$ for strain rate), $\dot{\gamma}_{xz} = O(\epsilon)$ and

$$\dot{\gamma}_{xx} = 2\Delta - 2(z - Z)W_{xx}, \qquad \text{where} \qquad \Delta = U_x + Z_x W_x. \tag{118}$$

Given that $\dot{\gamma} \sim |\dot{\gamma}_{xx}|$, for the fluid to be rigid over a section of the film at a given position in x (i.e. for a finite range of z), we must have that $\Delta = W_{zz} = 0$. But if both these conditions hold, then $u_x = 0$ across this entire cross-section. In other words, the fluid is either fully yielded or rigid across its thickness. Where yielded, we may now integrate

$$2\sigma = 2\dot{\gamma}_{xx} + B\,\mathrm{sgn}(\dot{\gamma}_{xx}) \tag{119}$$

and $2(z - Z)\sigma$ in z to determine constitutive relations for Σ and M (the resultant Q is related to higher order corrections to the velocity field and is determined regardless by the second force balance in (115)). Before quoting the result, we note an additional simplification implied by the velocity field in (117): the kinematic conditions at the two surfaces, $z = Z \pm \frac{1}{2}h$, reduce to $W = (Z \pm \frac{1}{2}h)_t$. Hence

$$W \equiv Z_t \qquad \& \qquad h = 1 \tag{120}$$

(in view of our scaling of the thickness). The constitutive relations are then

$$\Sigma = 4\Delta + 2B\Xi\,\mathrm{sgn}(\Delta), \qquad \Xi = \mathrm{Min}\left(1, \left|\frac{2\Delta}{W_{xx}}\right|\right), \tag{121}$$

and

$$M = -\frac{1}{3}W_{xx} - \frac{1}{2}B(1 - \Xi^2)\,\mathrm{sgn}(W_{xx}). \tag{122}$$

If the fluid is rigid, on the other hand, we must have that

$$|\Sigma| < 2B\Xi \qquad \& \qquad |M| < \frac{1}{2}B(1 - \Xi^2), \tag{123}$$

or

$$|M| < \frac{1}{2}B\,\mathrm{Max}\left(0, 1 - \frac{\Sigma^2}{4B^2}\right), \tag{124}$$

Equations (115) and (120)–(122) provide a theory for viscoplastic beams or columns suffering relatively small deflections (analogous to the classical Euler beam theory of elasticity (Mansfield 2005). With more effort, this theory can be generalized to accommodate $O(1)$ deflections (curvature) using the arc-length based coordinate system for the centreline of the beam (cf. Sect. 2.1 and Fig. 22; Balmforth and Hewitt (2013)).

4.6 *Bending Viscoplastic Beams; Toppling Viscoplastic columns*

Assuming that the end of the column at $x = 0$ is free (so that $\Sigma(0, t) = Q(0, t) = 0$) but that at $x = 1$ is clamped ($U(1, t) = W(1, t) = W_x(1, t) = 0$), we may write

$$(\Sigma, Q) = x(\mathrm{G_x}, \mathrm{G_z}) \qquad \& \qquad M_x = \frac{1}{2}\mathrm{G_z}x^2 - \Sigma Z. \tag{125}$$

If the column is initially straight, then $Z(x, 0) = 0$ and

$$M(x, 0) = \frac{1}{2}\mathrm{G_z}x^2. \tag{126}$$

The yield condition is therefore

$$\mathrm{G_z}x^2 \geq B \,\mathrm{Max}\left(0, 1 - \frac{\mathrm{G_x}^2 x^2}{4B^2}\right). \tag{127}$$

Hence, the fixed end $x = 1$ is the most dangerous position for failure (yielding). The criterion (127) combines two effects: the viscoplastic beam can fail by bending under an imposed torque $M \to \frac{1}{2}\mathrm{G_z}$, or under the action of a compressive load $\Sigma \to \mathrm{G_x}$. The two evidently conspire, with the compressive load lowering the threshold for bending on the right of (127).

The horizontal beam, with $\mathrm{G_x} = 0$ and therefore $\Sigma(x, t) = 0$, fails purely by bending under the gravitational torque if $\mathrm{G_z} < B$. This is the viscoplastic version of Galileo's problem for the bending of a cantilever (Balmforth and Hewitt 2013). When failure occurs, the beam bends near its fixed end over the section $\sqrt{B/\mathrm{G_z}} < x < 1$, rotating the free end down to reduce the gravitational torque. In the small deflection theory, $Z(x, t)$ increases linearly with time because $M(x, t) = \frac{1}{2}x^2$, which prescribes a time-independent vertical speed $W(x)$ through (122) (with $\Delta = \Xi = 0$). Despite this, the reduction of the gravitational torque as the beam bends down eventually allows the yield stress to terminate the beam's fall; this is illustrated in Fig. 23, which shows a solution of the corresponding theory for $O(1)$ deflections (Balmforth and Hewitt 2013).

The vertical column with $\mathrm{G_z} = 0$ cannot fail by bending as there is no imposed torque at $t = 0$. Instead, it collapses when $\mathrm{G_x} > 2B$, which is the criterion for a slender vertical filament to yield symmetrically at its base (Balmforth et al. 2010). Once the column fails in that fashion, any deviation of its orientation from the vertical will allow it to topple over sideways. The situation is rather different to the Euler

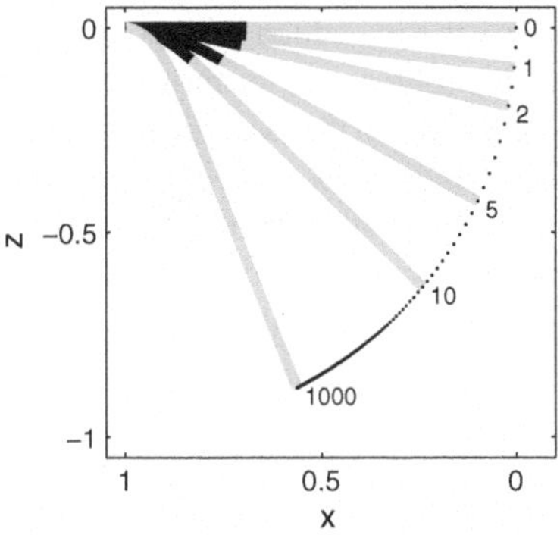

Fig. 23 Sagging viscoplastic beam with $B = 0.5$ and $G_z = 1$ at the times indicated. The dots show the position of the end of the beam at equally spaced times. The darker region is yielded; grey shading indicates the plug. The solution at $t = 1000$ is close to the final resting state. Note that the $x-$axis is drawn right to left in this plot, the convention in Balmforth and Hewitt (2013) (from which the solution is taken) being opposite to that used here

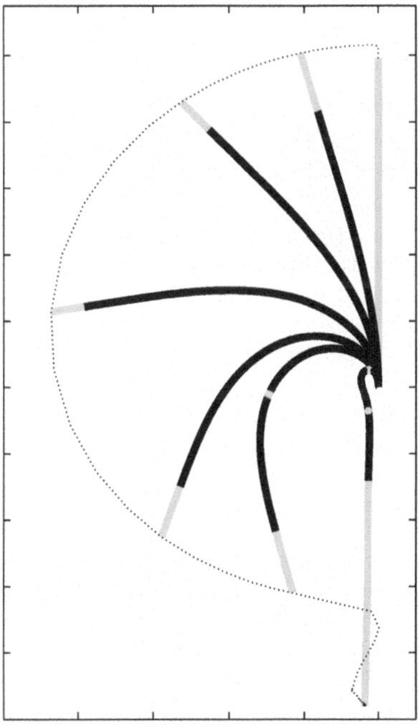

Fig. 24 A viscoplastic column buckling under gravity with $B = G_x = 1$ at the times $t = 0$, 0.41, 0.4125, 0.415, 0.4175, 0.42 and 1. To initiate collapse, the column is extruded almost upwards at fixed speed such that it reaches the critical length for collapse at $t = 0$; the direction of extrusion is offset from the vertical by an angle 10^{-6} to allow sideways toppling. The dotted curve shows the position of the free end of the column. The darker region is yielded; grey shading indicates the plug. At later times, the column is hanging vertically and begins to stretch and thin under gravity (i.e. $h(x,t) \neq 1$; another ingredient of the generalized theory)

buckling problem of elasticity theory, which is classically posed as a linear instability problem. For the viscoplastic problem, as one moves beyond the threshold for failure $G_x = 2B$, a narrow region yields at the base of the column. This region expands linearly with time as deflections grow explosively, corresponding to sudden toppling of the column; see the illustrative solution for $O(1)$ curvature in Fig. 24 and Balmforth and Hewitt (2013) for further details. An interesting possible application of this problem is in 3D printing (Hansen et al. 2009).

5 Concluding Remarks

These notes have presented an array of asymptotic analyses of viscoplastic flow problems, together with a sprinkling of other mathematical techniques. In almost all the problems in question, and as illustrated in several instances, the more analytical methodologies work hand-in-hand with numerical approaches (discussed by A. Wachs) to acquire a detailed understanding of the viscoplastic flow dynamics. We enjoyed slot flows, boundary layers, sticky and slippy films, and bending columns, highlighting the important effects introduced by the yield stress in each case. The emphasis has been on asymptotics and therefore on building approximate solutions, and complements approaches seeking to constrain solutions with variational and bounding techniques, as reviewed by I. Frigaard.

Acknowledgements I thank two Hewitts for their contributions to the work summarized in these notes: Ian Hewitt computed and prepared Figs. 5, 20 and 21. Duncan Hewitt computed and prepared Figs. 8, 11, 12, 13 and 14. Section 3 is a prelude to a more thorough discussion by Balmforth et al. (2017). I thank Richard Craster for the construction of the slipline field in Fig. 12. Lujia Liu contributed to the developments of Sect. 4.4.

References

Alishaev, M. G., Entov, V. M., & Segalov, A. E. (1969). Elementary solutions of plane nonlinear filtration problems. *Fluid Dynamics*, *4*(3), 77–84.

Balmforth, N. J., & Craster, R. V. (1999). A consistent thin-layer theory for Bingham plastics. *Journal of Non-Newtonian Fluid Mechanics*, *84*, 65–81.

Balmforth, N. J., & Hewitt, I. J. (2013). Viscoplastic sheets and threads. *Journal of Non-Newtonian Fluid Mechanics*, *193*, 28–42.

Balmforth, N. J., & Kerswell, R. R. (2005). Granular collapse in two dimensions. *Journal of Fluid Mechanics*, *538*, 399–428.

Balmforth, N. J., Craster, R. V., Hewitt, D. R., Hormozi, S., & Maleki, A. (2017). Viscoplastic boundary layers. *Journal of Fluid Mechanics*, *813*, 929–954.

Balmforth, N. J., Craster, R. V., Perona, P., Rust, A. C., & Sassi, R. (2007a). Viscoplastic dam breaks and the Bostwick consistometer. *Journal of Non-Newtonian Fluid Mechanics*, *142*, 63–78.

Balmforth, N. J., Craster, R. V., Rust, A. C., & Sassi, R. (2007b). Viscoplastic flow over an inclined surface. *Journal of Non-Newtonian Fluid Mechanics*, *142*, 219–243.

Balmforth, N. J., Dubash, N., & Slim, A. C. (2010) Extensional dynamics of viscoplastic filaments: I and II. *Journal of Non-Newtonian Fluid Mechanics*, *165*, 1139–1146 and 1147–1160.
Barnes, H. A. (1995). A review of the slip (wall depletion) of polymer solutions, emulsions and particle suspensions in viscometers: its cause, character, and cure. *Journal of Non-Newtonian Fluid Mechanics*, *56*, 221–251.
Bender, C. M. & Orszag, S. A. (1978) *Advanced mathematical methods for scientists and engineers*. McGraw-Hill.
Bernadiner, M. G. & Protopapas, A. L. (1994) Progress on the theory of flow in geologic media with threshold gradient. *Journal of Environmental Science & Health Part A*, *29*(1), 249–275.
Bittleston, S. H., Ferguson, J., & Frigaard, I. A. (2002). Mud removal and cement placement during primary cementing of an oil well: Laminar non-Newtonian displacements in an eccentric annular Hele-Shaw cell. *Journal of Engineering Mathematics*, *43*(2–4), 229–253.
Bleyer, J., & Coussot, P. (2014). Breakage of non-Newtonian character in flow through a porous medium: Evidence from numerical simulation. *Physical Review E*, *89*(6), 063018.
Boujlel, J., Maillard, M., Lindner, A., Ovarlez, G., Chateau, X., & Coussot, P. (2012). Boundary layer in pastes: Displacement of a long object through a yield stress fluid. *Journal of Rheology*, *56*, 1083–1108.
Chamberlain, J. A., Sader, J. E., Landman, K. A., & White, L. R. (2001). Incipient plane-strain failure of a rectangular block under gravity. *International Journal of Mechanical Sciences*, *43*, 793–815.
Chaparian, E., & Frigaard, I. A. (2017). Yield limit analysis of particle motion in a yield-stress fluid. *Journal of Fluid Mechanics*, *819*, 311–351.
Chevalier, T., Chevalier, C., Clain, X., Dupla, J. C., Canou, J., Rodts, S., et al. (2013a). Darcy's law for yield stress fluid flowing through a porous medium. *Journal of Non-Newtonian Fluid Mechanics*, *195*, 57–66.
Chevalier, T., Rodts, S., Chateau, X., Boujlel, J., Maillard, M., & Coussot, P. (2013b). Boundary layer (shear-band) in frustrated viscoplastic flows. *Europhysics Letters*, *102*, 48002.
Covey, G. H., & Stanmore, B. R. (1981). Use of the parallel-plate plastometer for the characterisation of viscous fluids with a yield stress. *Journal of Non-Newtonian Fluid Mechanics*, *8*, 249–260.
Craster, R. V., & Matar, O. K. (2009). Dynamics and stability of thin liquid films. *Reviews of Modern Physics*, *81*, 1131.
Entov, V. M. (1970). Analogy between equations of plane filtration and equations of longitudinal shear of nonlinearly elastic and plastic solids. *Journal of Applied Mathematics and Mechanics*, *34*(1), 153–164.
Fernández-Nieto, E. D., Noble, P., & Vila, J.-P. (2010). Shallow water equations for non-newtonian fluids. *Journal of Non-Newtonian Fluid Mechanics*, *165*, 712–732.
Fusi, L., Farina, A., & Rosso, F. (2012). Flow of a Bingham-like fluid in a finite channel of varying width: a two-scale approach. *Journal of Non-Newtonian Fluid Mechanics*, *177*, 76–88.
Hansen, C. J., Wu, W., Toohey, K. S., Sottos, N. R., White, S. R., & Lewis, J. A. (2009). Self-healing materials with interpenetrating microvascular networks. *Advanced Materials*, *21*, 4143–4147.
Hewitt, D. R., Daneshi, M., Balmforth, N. J., & Martinez, D. M. (2016). Obstructed and channelized viscoplastic flow in a hele-shaw cell. *Journal of Fluid Mechanics*, *790*, 173–204.
Hewitt, I. J., & Balmforth, N. J. (2012). Viscoplastic lubrication theory with application to bearings and the washboard instability of a planing plate. *Journal of Non-Newtonian Fluid Mechanics*, *169*, 74–90.
Jalaal, M. (2016). *Controlled spreading of complex droplets* (Doctoral dissertation, University of British Columbia).
Hinch, E. J. (1991). *Perturbation methods*. Cambridge University Press.
Jalaal, M., Balmforth, N. J., & Stoeber, B. (2015). Slip of spreading viscoplastic droplets. *Langmuir*, *31*(44), 12071–12075.
Lipscomb, G. G., & Denn, M. M. (1984). Flow of Bingham fluids in complex geometries. *Journal of Non-Newtonian Fluid Mechanics*, *14*, 337–346.

Liu, K. F., & Mei, C. C. (1989). Slow spreading of a sheet of Bingham fluid on an inclined plane. *Journal of Fluid Mechanics*, *207*, 505–529.
Mansfield, E. H. (2005). *The bending and stretching of plates*. Cambridge University Press.
Nye, J. F. (1952). The mechanics of glacier flow. *Journal of Glaciology*, *2*, 82–93.
Nye, J. F. (1967). Plasticity solution for a glacier snout. *Journal of Glaciology*, *6*(47), 695–715.
Oldroyd, J. G. (1947). Two-dimensional plastic flow of a Bingham solid. *Mathematical Proceedings of the Cambridge Philosophical Society*, *43*, 383–395.
Pelipenko, S., & Frigaard, I. A. (2004). Two-dimensional computational simulation of eccentric annular cementing displacements. *IMA journal of applied mathematics*, *69*(6), 557–583.
Piau, J.-M. (2002). Viscoplastic boundary layer. *Journal of Non-Newtonian Fluid Mechanics*, *102*, 193–218.
Pinkus, O., & Sternlicht, B. (1961). *Theory of hydrodynamic lubrication*. McGraw-Hill.
Prager, W., & Hodge, P.G. (1968). *Theory of perfectly plastic solids*. Dover.
Putz, A., Frigaard, I. A., & Martinez, D. M. (2009). The lubrication paradox & use of regularisation methods for lubrication flows. *Journal of Non-Newtonian Fluid Mechanics*, *163*, 62–77.
Randolph, M. F., & Houlsby, G. T. (1984). The limiting pressure on a circular pile loaded laterally in cohesive soil. *Géotechnique*, *34*(4), 613–623.
Roustaei, A., Gosselin, A., & Frigaard, I. A. (2015). Residual drilling mud during conditioning of uneven boreholes in primary cementing. part 1: Rheology and geometry effects in non-inertial flows. *Journal of Non-Newtonian Fluid Mechanics*, *220*, 87–98.
Schoof, C., & Hewitt, I. A. (2013). Ice-sheet dynamics. *Annual Review of Fluid Mechanics*, *45*, 217–239.
Smyrnaios, D. N., & Tsamopoulos, J. A. (2001). Squeeze flow of Bingham plastics. *Journal of Non-Newtonian Fluid Mechanics*, *100*, 165–190.
Smyrnaios, D. N., & Tsamopoulos, J. A. (2006). Transient squeeze flow of viscoplastic materials. *Journal of Non-Newtonian Fluid Mechanics*, *133*, 35–56.
Sneddon, I. N. (1957). *Elements of partial differential equations*. McGraw-Hill.
Talon, L., & Bauer, D. (2013). On the determination of a generalized Darcy equation for yield-stress fluid in porous media using a Lattice-Boltzmann TRT scheme. *The European Physical Journal E*, *36*, 139.
Tokpavi, D. L., Magnin, A., & Jay, P. (2008). Very slow flow of Bingham viscoplastic fluid around a circular cylinder. *Journal of Non-Newtonian Fluid Mechanics*, *154*, 65–76.
Walton, I. C., & Bittleston, S. H. (1991). The axial flow of a Bingham plastic in a narrow eccentric annulus. *Journal of Fluid Mechanics*, *222*, 39–60.
Whitham, G. B. (1974). *Linear and nonlinear waves*. Wiley.
Wilson, S. D. R. (1999). A note on thin-layer theory for Bingham plastics. *Journal of Non-Newtonian Fluid Mechanics*, *1*, 29–33.

Computational Methods for Viscoplastic Fluid Flows

Anthony Wachs

This document constitutes the lecture notes that I used over the summer school *Viscoplastic fluids: From theory to application* organized in the International Centre for Mechanical Sciences of Udine, Italy, from May 30th to June 3rd, 2016. It contains five sections corresponding to the five lectures I gave.

Understanding the flow of a viscoplastic material is a challenging scientific topic, at the experimental level as much as at the analytical and computational levels. Computing the flow of a given material, regardless of its rheological nature, requires to address various questions as, e.g., spatial discretization scheme, time algorithm, linear system solver, software implementation, etc. What makes viscoplastic material specific is the mathematical form of the constitutive equation and in particular its non-differentiability property. Over this series of 5 lectures, I will suggest different numerical approaches to properly compute a viscoplastic fluid flow. We will try to cover the body of knowledge existing in the literature and show how the different solution methods perform in assorted flow configurations. We will emphasize their assets and drawbacks and discuss the remaining challenges to compute faster solutions of satisfactory accuracy.

1 General Introduction, Concepts, Difficulties, Various Algorithms

Computing the flow of a fluid involves the following steps:

- properly formulating the problem: unknown fields, conservation equations, boundary conditions and, in case of an unsteady problem (initial value problem), initial conditions,

A. Wachs (✉)
Department of Mathematics, Department of Biological and Chemical Engineering,
University of British Columbia, Vancouver, Canada
e-mail: wachs@mail.ubc.ca

G. Ovarlez and S. Hormozi (eds.), *Lectures on Visco-Plastic Fluid Mechanics*,
CISM International Centre for Mechanical Sciences 583,
https://doi.org/10.1007/978-3-319-89438-6_3

- discretizing the problem: discrete representation of the continuous fields and discrete form of conservation equations that can generally be represented in a matrix-vector form,
- choosing a solution algorithm: time-stepping method, linear or non-linear solvers, direct or iterative method,
- linear system solution method: direct or iterative,
- implementation: serial or parallel.

Let's start by formulating the problem. Solving the problem of the isothermal flow of a material implies to solve the conservation of mass and the conservation of momentum. Additional constitutive equations are necessary to relate (i) the density ρ to other physical variables as, e.g., pressure p or temperature T, and (ii) the stress tensor $\boldsymbol{\tau}$ to the rate-of-strain tensor $\boldsymbol{D}$. Let's consider a domain Ω smoothly bounded by $\Gamma = \partial\Omega$ and a time interval $[0, T]$. Hence the general isothermal flow problem reads:

Find $(\boldsymbol{u}, p, \rho, \boldsymbol{\tau})$ such that:

$$\rho\left(\frac{\partial \boldsymbol{u}}{\partial t} + \boldsymbol{u}\cdot\nabla\boldsymbol{u}\right) - \nabla\cdot\boldsymbol{\tau} + \nabla p = \boldsymbol{f} \quad \text{in } \Omega \times [0, T] \tag{1}$$

$$\frac{\partial \rho}{\partial t} + \nabla\cdot\rho\boldsymbol{u} = 0 \quad \text{in } \Omega \times [0, T] \tag{2}$$

$$\rho = F(p) \quad \text{in } \Omega \times [0, T] \tag{3}$$

$$\boldsymbol{\tau} = G(\boldsymbol{D}) \quad \text{in } \Omega \times [0, T] \tag{4}$$

$$\text{Boundary conditions on } \Gamma \times [0, T] \tag{5}$$

$$\text{Initial conditions in } \Omega \tag{6}$$

where $\boldsymbol{u}$ is the velocity field, F and G are two functions relevant of the specific behavior of the material, t is the time variable, $\boldsymbol{f}$ is a body force (as, e.g., gravity), $\boldsymbol{D} = \frac{1}{2}(\nabla\boldsymbol{u} + \nabla\boldsymbol{u}^t)$ and $\nabla\boldsymbol{u}$ is the velocity gradient tensor. For now, we investigate the case of an incompressible flow in which the density is constant, i.e., $F(p) = constant$, (we shall look at the weakly compressible case in Sect. 4) such that (2) simplifies to:

$$\nabla\cdot\boldsymbol{u} = 0 \tag{7}$$

If we further assume that the material is Newtonian, i.e., its behavior satisfies the following constitutive equation $\boldsymbol{\tau} = 2\mu\boldsymbol{D}$ (here we have $G(\boldsymbol{D}) = 2\mu\boldsymbol{D}$ and $\mu = constant$ is the viscosity), we can directly plug (4) into (1) to get the well known Navier & Stokes equations for an incompressible Newtonian material. Assuming Dirichlet boundary conditions for $\boldsymbol{u}$, the whole flow problem reads:

Find $(\boldsymbol{u}, p)$ such that:

$$\rho\left(\frac{\partial \boldsymbol{u}}{\partial t} + \boldsymbol{u}\cdot\nabla\boldsymbol{u}\right) - \mu\nabla^2\boldsymbol{u} + \nabla p = \boldsymbol{f} \quad \text{in } \Omega\times[0,T] \tag{8}$$

$$\nabla\cdot\boldsymbol{u} = 0 \quad \text{in } \Omega\times[0,T] \tag{9}$$

$$\boldsymbol{u} = \boldsymbol{g} \quad \text{on } \Gamma\times[0,T] \tag{10}$$

$$\boldsymbol{u}(t=0) = \boldsymbol{u}_0 \quad \text{in } \Omega \tag{11}$$

where $\boldsymbol{g}$ is a vectorial function of space and time and $\boldsymbol{u}_0$ a vectorial function of space.

Now if the material is purely viscous but its viscosity is not constant and depends on, e.g., the rate of strain, the formulation of the problem does not change much. In fact, we have:

Find $(\boldsymbol{u}, p)$ such that:

$$\rho\left(\frac{\partial \boldsymbol{u}}{\partial t} + \boldsymbol{u}\cdot\nabla\boldsymbol{u}\right) - \nabla\cdot 2\mu(\dot{\gamma})\boldsymbol{D} + \nabla p = \boldsymbol{f} \quad \text{in } \Omega\times[0,T] \tag{12}$$

$$\nabla\cdot\boldsymbol{u} = 0 \quad \text{in } \Omega\times[0,T] \tag{13}$$

$$\boldsymbol{u} = \boldsymbol{g} \quad \text{on } \Gamma\times[0,T] \tag{14}$$

$$\boldsymbol{u}(t=0) = \boldsymbol{u}_0 \quad \text{in } \Omega \tag{15}$$

where $\dot{\gamma} = 2\|\boldsymbol{D}\|$ denotes the generalized shear rate and $\|.\|$ corresponds to the *Euclidian norm*, defined as :

$$\|\boldsymbol{\mathcal{X}}\| = \sqrt{\frac{1}{2}\sum_{1\le i,j\le d}|\boldsymbol{\mathcal{X}}_{ij}|^2}\ , \quad \forall\boldsymbol{\mathcal{X}}\in\mathbb{R}^{d\times d}\ , \ d = 1, 2 \text{ or } 3 \tag{16}$$

At that stage, it is already worthwhile to make a few comments:

- for purely viscous materials, we can directly plug the rheological constitutive law (4) into the momentum equation (1) because $G(\boldsymbol{D}) = 2\mu\boldsymbol{D}$ and $G(\boldsymbol{D}) = 2\mu(2\|\boldsymbol{D}\|)\boldsymbol{D}$ are continuous and differentiable functions of $\boldsymbol{D}$ and thus of $\boldsymbol{u}$,
- although the generalized viscous material formulation of the flow problem (12)–(15) is not very different from the Newtonian formulation (8)–(11), it is however worth noticing that the viscous term in the momentum conservation equation becomes non-linear. This has important implications on the solution method in the case of creeping flows (low Reynolds number). In fact, even for a Newtonian flow, the momentum conservation equation is non-linear in $\boldsymbol{u}$ due to the convective (inertial) term $\rho\boldsymbol{u}\cdot\nabla\boldsymbol{u}$. However, if inertia is negligible, the remaining non-linearities in the problem come from the viscous term. For a Newtonian material, the creeping flow problem is linear while it is not for a generalized viscous material,
- non-linearities in the momentum conservation equation can be handled in different ways:

- for the inertial term $\rho \boldsymbol{u} \cdot \nabla \boldsymbol{u}$: fully explicitation of this term, semi-explicitation through linearization $\rho \tilde{\boldsymbol{u}} \cdot \nabla \boldsymbol{u}$ where $\tilde{\boldsymbol{u}}$ is a known velocity field, or fixed point iteration/Newton-Raphson method for full implicitation,
- for the viscous term $\nabla \cdot 2\mu(2\|\boldsymbol{D}\|)\boldsymbol{D}$: semi-explicitation through linearization $\nabla \cdot 2\mu(2\|\boldsymbol{D}(\tilde{\boldsymbol{u}})\|)\boldsymbol{D}(\boldsymbol{u})$ where $\tilde{\boldsymbol{u}}$ is a known velocity field, or fixed point iteration/Newton-Raphson method for full implicitation.

Most of the flows involving a viscoplastic material generally exhibit a low to moderate level of inertia. In a time-dependent problem, the fully explicit treatment of $\rho \boldsymbol{u} \cdot \nabla \boldsymbol{u}$ is deemed to be a good approximation and an easy to implement solution method. In a steady-state problem, either the solution is computed as the steady-state solution of a time evolution problem or the inertial term is computed by a fixed-point iteration/Newton-Raphson method. The fixed-point iterative method is conceptually easy to implement and avoid all the complications related to computing the Jacobian matrix in Newton-Raphson methods. However, it is generally slower to converge, as we shall see over the next lectures. To simplify the problem, we will assume at that stage that inertia is negligible.

So let's turn our attention to the solution of a viscoplastic flow problem. The specificity of viscoplastic fluid flows is the non-differentiability of the shear stress-shear rate constitutive law and indeterminacy of the stress field below the yield stress. We shall proceed with the simplest viscoplastic model: the Bingham model, that reads as follows:

$$\begin{cases} \boldsymbol{\tau} = 2\mu \boldsymbol{D} + \dfrac{\boldsymbol{D}}{\|\boldsymbol{D}\|}\tau_y & \text{if } \|\boldsymbol{\tau}\| > \tau_y \\ \boldsymbol{D} = 0 & \text{if } \|\boldsymbol{\tau}\| \le \tau_y \end{cases} \tag{17}$$

where τ_y denotes the yield stress and μ the constant plastic viscosity. Introducing the appropriate functional spaces for $\boldsymbol{u}$ and p, the whole Bingham flow problem reads:

Find $(\boldsymbol{u}, p) \in H^1(\Omega)^d \times L^2(\Omega)$ such that:

$$\rho \frac{\partial \boldsymbol{u}}{\partial t} - \nabla \cdot \boldsymbol{\tau} + \nabla p = \boldsymbol{f} \quad \text{in } \Omega \times [0, T] \tag{18}$$

$$\nabla \cdot \boldsymbol{u} = 0 \quad \text{in } \Omega \times [0, T] \tag{19}$$

$$\begin{cases} \boldsymbol{\tau} = 2\mu \boldsymbol{D}(\boldsymbol{u}) + \dfrac{\boldsymbol{D}(\boldsymbol{u})}{\|\boldsymbol{D}(\boldsymbol{u})\|}\tau_y & \text{if } \|\boldsymbol{\tau}\| > \tau_y \\ \boldsymbol{D}(\boldsymbol{u}) = 0 & \text{if } \|\boldsymbol{\tau}\| \le \tau_y \end{cases} \tag{20}$$

$$\boldsymbol{u} = \boldsymbol{g} \quad \text{on } \Gamma \times [0, T] \tag{21}$$

$$\boldsymbol{u}(t = 0) = \boldsymbol{u}_0 \quad \text{in } \Omega \tag{22}$$

where $L^2(\Omega)$ denotes the space of integrable functions on Ω and the Sobolev space $H^1(\Omega)$ is defined as:

$$H^1(\Omega) = \{v | v \in L^2(\Omega)\,,\ \frac{\partial v}{\partial x_i} \in L^2(\Omega)\,,\ \forall i = 1, \ldots, d\} \tag{23}$$

Please note that contrary to generalized viscous materials, the form of the constitutive Bingham law does not allow us to plug it into the momentum conservation equation.

Apart from the special method of Szabo and Hassager (1992) who computed the flow only in yielded regions and tracked the yield surfaces with a re-meshing approach, there are mainly two families of solution method. The first one is based on variational inequalities, dates from the early seventies and is due to Duvaut and Lions (1976), together with Glowinski (1974), Glowinski et al. (1976), Bristeau and Glowinski (1974). The method also involves using Lagrange multipliers and the corresponding solution algorithm is of the Uzawa type. The underlying mathematical developments might appear difficult to follow for some, but the practical implementation is fairly easy because of the modularity of the methodology. The second family of methods stems from the recognition that the flow could not be computed directly and thus requires a mean to circumvent the non-differentiability of the constitutive law. The idea due to Bercovier and Engleman (1980) and Papanastasiou (1987) involves approaching the true (discontinuous) constitutive law by an approximated continuous and differentiable one. The resulting approximated problem is then continuous and differentiable, but only at the expense of an additional numerical parameter. These models are usually called regularized models in the literature. The magnitude of the regularization parameter can be tuned such that the approximated model is as close as possible to the exact one. In practice, the smaller/larger is the regularization parameter, the closer the approximated model is to its exact form. Both methods have been widely used in the literature and have supplied valuable results. The salient features of both these two methods are detailed below.

1.1 Multipliers Techniques

This type of method relies on the ability to write a variational formulation representative of the problem at hand. The first work in this area is due to Duvaut and Lions (1976) who derived the corresponding variational inequality for the flow of a Bingham fluid. Once the basic inequality is established, the way the Lagrange multiplier is introduced leads to slightly different formulations and hence different solution algorithms:

- the Augmented Lagrangian (AL) algorithm is based on the introduction of a Lagrange multiplier field as well as an additional rate-of-strain tensor field,
- the projection-like algorithm based on the introduction of a single (but different than that in the case of the AL algorithm) Lagrange multiplier field only.

As a starting point, let us show briefly how to derive the variational inequality of Duvaut and Lions. There are various ways to achieve this objective (see Duvaut and Lions (1976) or Glowinski et al. (1976)). Here we suggest to start by applying the lemma of Ionescu and Sofonea (1986) to the Bingham model (17) with $(\boldsymbol{\tau}, \boldsymbol{D}) \in L^2(\Omega)^{d\times d} \times L^2(\Omega)^{d\times d}$ to get:

$$\int_\Omega 2\mu \boldsymbol{D} : (\boldsymbol{q} - \boldsymbol{D}) dx + \int_\Omega \sqrt{2}\tau_y (|\boldsymbol{q}| - |\boldsymbol{D}|) dx \geq \int_\Omega \boldsymbol{\tau} : (\boldsymbol{q} - \boldsymbol{D}) dx \ , \quad \forall \boldsymbol{q} \in L^2(\Omega)^{d\times d} \tag{24}$$

where the Frobenius norm and scalar product in $L^2(\Omega)^{d\times d}$ are respectively defined as:

$$|\boldsymbol{\mathcal{X}}| = \left(\sum_{1\leq i,j\leq d} |\mathcal{X}_{ij}|^2 \right)^{\frac{1}{2}} = \sqrt{2}\|\boldsymbol{\mathcal{X}}\| \ , \quad \boldsymbol{\mathcal{X}} : \boldsymbol{\mathcal{Y}} = \sum_{1\leq i,j\leq d} \mathcal{X}_{ij}\mathcal{Y}_{ij} \ , \quad \forall \boldsymbol{\mathcal{X}} \in L^2(\Omega)^{d\times d} \tag{25}$$

The variational formulation corresponding to the conservation equations (18)–(19) reads as follows:

$$-\int_\Omega q \nabla \cdot \boldsymbol{u} dx = 0 \ , \quad \forall q \in L^2(\Omega) \tag{26}$$

$$\int_\Omega \rho \frac{\partial \boldsymbol{u}}{\partial t} \cdot \boldsymbol{v} dx + \int_\Omega \boldsymbol{\tau} : \boldsymbol{D}(\boldsymbol{v}) dx - \int_\Omega p \nabla \cdot \boldsymbol{v} dx = \int_\Omega \boldsymbol{f} \cdot \boldsymbol{v} dx \ , \quad \forall \boldsymbol{v} \in H_0^1(\Omega)^d \tag{27}$$

where $H_0^1(\Omega) = \{v \in H^1(\Omega) | v = 0 \ on \ \Gamma\}$. Combining (24), (26) and (27), we get the mathematical formulation of the time-dependent flow of an incompressible Bingham fluid written in terms of a variational inequality as:

Find $\boldsymbol{u} \in H^1(\Omega)^d$, $p \in L^2(\Omega)$ such that:

$$-\int_\Omega q \nabla \cdot \boldsymbol{u} dx = 0 \ , \quad \forall q \in L^2(\Omega) \tag{28}$$

$$\int_\Omega \rho \frac{\partial \boldsymbol{u}}{\partial t} \cdot (\boldsymbol{v} - \boldsymbol{u}) dx - \int_\Omega p \nabla \cdot (\boldsymbol{v} - \boldsymbol{u}) dx + \int_\Omega 2\mu \boldsymbol{D}(\boldsymbol{u}) : (\boldsymbol{D}(\boldsymbol{v}) - \boldsymbol{D}(\boldsymbol{u})) dx + \int_\Omega \sqrt{2}\tau_y (|\boldsymbol{D}(\boldsymbol{v})| - |\boldsymbol{D}(\boldsymbol{u})|) dx \geq \int_\Omega \boldsymbol{f} \cdot (\boldsymbol{v} - \boldsymbol{u}) dx \ , \quad \forall \boldsymbol{v} \in H_0^1(\Omega)^d \tag{29}$$

The above problem can be written equivalently in terms of a minimization problem introducing the following functional $\Gamma(\boldsymbol{v}, q) : H^1(\Omega)^d \times L^2(\Omega) \to \mathbb{R}$ as:

$$\Gamma(\boldsymbol{v}, q) = \frac{1}{2} \int_\Omega \frac{\rho}{\Delta t} \boldsymbol{v}^2 dx + \int_\Omega \mu |\boldsymbol{D}(\boldsymbol{v})|^2 dx + \int_\Omega \sqrt{2}\tau_y |\boldsymbol{D}(\boldsymbol{v})| dx - \int_\Omega q \nabla \cdot \boldsymbol{v} dx - \int_\Omega \boldsymbol{f} \cdot \boldsymbol{v} dx \tag{30}$$

provided the unsteady term has been discretized with a first order backward Euler scheme. Then, $(\boldsymbol{u}, p) \in H^1(\Omega)^d \times L^2(\Omega)$ is the solution of the following minimization problem

$$\Gamma(\boldsymbol{u}, p) = \min_{\boldsymbol{v} \in H^1(\Omega)^d} \max_{q \in L^2(\Omega)} \Gamma(\boldsymbol{v}, q) \tag{31}$$

Both these formulations (either as a variational inequality or a minimization problem) are obviously an improvement compared to the original formulation, though they are still partly formal. The rigorous formulation is even more complicated and was thoroughly discussed by Duvaut and Lions (1976). The next step is to derive a tractable computational method with suitable convergence properties. At this step, there are mainly two variants:

1. Decomposition-coordination method
Here an auxiliary additional variable $\boldsymbol{d}$ representing the rate-of-strain tensor $\boldsymbol{D}(\boldsymbol{u})$ is introduced (decomposition step). Now the solution may be written as:

Find $\boldsymbol{u} \in H^1(\Omega)^d$, $p \in L^2(\Omega)$, $\boldsymbol{d} \in \mathcal{T}(\Omega)$ such that:

$$\Gamma^*(\boldsymbol{u}, p, \boldsymbol{d}) = \min_{\boldsymbol{v} \in H^1(\Omega)^d} \min_{\boldsymbol{q} \in \mathcal{T}(\Omega)} \max_{q \in L^2(\Omega)} \Gamma^*(\boldsymbol{v}, q, \boldsymbol{d}) \tag{32}$$

where:

$$\mathcal{T}(\Omega) = \{\boldsymbol{q} \in L^2(\Omega)^{d \times d} | \boldsymbol{q} = \boldsymbol{D}(\boldsymbol{v})\} \tag{33}$$

$$\Gamma^*(\boldsymbol{v}, q, \boldsymbol{q}) = \frac{1}{2} \int_\Omega \frac{\rho}{\Delta t} \boldsymbol{v}^2 dx + \int_\Omega \mu |\boldsymbol{D}(\boldsymbol{v})|^2 dx + \int_\Omega \sqrt{2} \tau_y |\boldsymbol{q}| dx - \int_\Omega q \nabla \cdot \boldsymbol{v} dx - \int_\Omega \boldsymbol{f} \cdot \boldsymbol{v} dx \tag{34}$$

The coordination step implies to introduce a Lagrange multiplier field $\boldsymbol{\lambda} \in L^2(\Omega)^{d \times d}$, homogeneous to a plastic stress, to relax the constraint $\boldsymbol{d} = \boldsymbol{D}(\boldsymbol{u})$. This leads to the following saddle point problem:

Find $\boldsymbol{u} \in H^1(\Omega)^d$, $p \in L^2(\Omega)$, $\boldsymbol{d} \in L^2(\Omega)^{d \times d}$, $\boldsymbol{\lambda} \in L^2(\Omega)^{d \times d}$ such that:

$$\tilde{\mathcal{L}}(\boldsymbol{u}, p, \boldsymbol{d}, \boldsymbol{\lambda}) = \min_{\boldsymbol{v} \in H^1(\Omega)^d} \max_{q \in L^2(\Omega)} \min_{\boldsymbol{q} \in L^2(\Omega)^{d \times d}} \max_{\alpha \in L^2(\Omega)^{d \times d}} \tilde{\mathcal{L}}(\boldsymbol{v}, q, \boldsymbol{q}, \boldsymbol{\alpha}) \tag{35}$$

where:

$$\tilde{\mathcal{L}}(\boldsymbol{v}, q, \boldsymbol{q}, \boldsymbol{\alpha}) = \frac{1}{2} \int_\Omega \frac{\rho}{\Delta t} \boldsymbol{v}^2 dx + \int_\Omega \mu |\boldsymbol{D}(\boldsymbol{v})|^2 dx + \int_\Omega \sqrt{2} \tau_y |\boldsymbol{q}| dx - \int_\Omega q . \nabla \cdot \boldsymbol{v} dx - \int_\Omega \boldsymbol{f} . \boldsymbol{v} dx + \int_\Omega \boldsymbol{\alpha} . (\boldsymbol{D}(\boldsymbol{v}) - \boldsymbol{q}) dx \tag{36}$$

The functional can be further augmented in the following way:

$$
\begin{aligned}
\tilde{\mathcal{L}}_r(\boldsymbol{v}, q, \boldsymbol{q}, \boldsymbol{\alpha}) &= \tilde{\mathcal{L}}(\boldsymbol{v}, q, \boldsymbol{q}, \boldsymbol{\alpha}) + \frac{r}{2}\int_\Omega |\boldsymbol{D}(\boldsymbol{u}) - \boldsymbol{q}|^2 dx \\
&= \frac{1}{2}\int_\Omega \frac{\rho}{\Delta t}\boldsymbol{v}^2 dx + \int_\Omega \mu|\boldsymbol{D}(\boldsymbol{v})|^2 dx + \int_\Omega \sqrt{2}\tau_y|\boldsymbol{q}|dx \\
- \int_\Omega q.\nabla\cdot\boldsymbol{v}dx - &\int_\Omega \boldsymbol{f}.\boldsymbol{v}dx + \int_\Omega \boldsymbol{\alpha}.(\boldsymbol{D}(\boldsymbol{v}) - \boldsymbol{q})dx + \frac{r}{2}\int_\Omega |\boldsymbol{D}(\boldsymbol{u}) - \boldsymbol{q}|^2 dx
\end{aligned} \tag{37}
$$

where $r \geq 0$ is the augmentation parameter.
The saddle point problem is then solved by a variant of the fully decoupled Uzawa algorithm ALG2, see Fortin and Glowinski (1982, 1983); Glowinski and Tallec (1989). The advantage of this approach is that the optimality condition associated with $\boldsymbol{d}$ can be written as a minimization problem, which is, in turn, equivalent to a slightly modified Bingham equation. Therefore we end up with an explicit scheme to compute the rate-of-strain tensor $\boldsymbol{d}$ using the Lagrange multiplier $\boldsymbol{\lambda}$ and the rate-of-strain tensor $\boldsymbol{D}(\boldsymbol{u})$ computed with the velocity field, thus circumventing the singularity at vanishing shear rates.

2. Orthogonal-projection operator
This latter variant is a direct consequence of Theorem 9.1 in Duvaut and Lions (1976), Glowinski et al. (1976) which provides the existence of a Lagrange multiplier $\boldsymbol{\lambda}$ satisfying some specific properties presented here. When applied to problem (28)–(29), the results of Duvaut and Lions can be formulated as:
If $(\boldsymbol{u}, p)$ is a solution of (28)–(29), there exists a tensorial field $\boldsymbol{\lambda}$ s.t.

$$\boldsymbol{\lambda} \in L^2(\Omega)^{d\times d} \quad , \quad \boldsymbol{\lambda} = \boldsymbol{\lambda}^t \tag{38}$$

$$|\boldsymbol{\lambda}| \leq 1 \text{ in } \Omega \tag{39}$$

$$\boldsymbol{\lambda} : \boldsymbol{D}(\boldsymbol{u}) = |\boldsymbol{D}(\boldsymbol{u})| \text{ in } \Omega \tag{40}$$

$$\rho\frac{\partial \boldsymbol{u}}{\partial t} - \nabla\cdot 2\mu\boldsymbol{D}(\boldsymbol{u}) - \sqrt{2}\tau_y\nabla\cdot\boldsymbol{\lambda} + \nabla p = \boldsymbol{f} \quad \text{in } \Omega \tag{41}$$

$$\nabla\cdot\boldsymbol{u} = 0 \text{ in } \Omega \tag{42}$$

This Lagrange multiplier can be constructed by approximating the functional by regularization (i.e., replacing it by a differentiable functional) and by making the regularization parameter tends to zero (see for instance Glowinski et al. (1976); Glowinski 2003). Finally, a nice consequence at the computational level is that properties satisfied by $\boldsymbol{\lambda}$ imply that:

$$\boldsymbol{\lambda} = P_\Lambda(\boldsymbol{\lambda} + r\sqrt{2}\tau_y\boldsymbol{D}(\boldsymbol{u})) \quad , \quad \forall r > 0 \tag{43}$$

where Λ is a close convex set of $L^2(\Omega)^{d\times d}$ and $P_\Lambda : L^2(\Omega)^{d\times d} \to \Lambda$ is the orthogonal projection operator, respectively defined by

$$\Lambda = \{\boldsymbol{\lambda} | \boldsymbol{\lambda} \in L^2(\Omega)^{d\times d}, |\boldsymbol{\lambda}| \le 1 \text{ in } \Omega\} \tag{44}$$

$$P_\Lambda(\boldsymbol{\lambda}) = \begin{cases} \boldsymbol{\lambda} & \text{if } |\boldsymbol{\lambda}| \le 1 \\ \boldsymbol{\lambda}/|\boldsymbol{\lambda}| & \text{instead} \end{cases} \tag{45}$$

The problem is also solved by an Uzawa-type algorithm where the projection step corresponds to the update of the Lagrange multiplier based on its own value and the value of the velocity field at the previous iteration. Glowinski (2003) showed that provided $r < \frac{\mu}{2\tau_y^2}$, the convergence of the iteration is guaranteed.

1.2 Regularization Methods

Essentially, the regularization technique involves writing the constitutive law in terms of apparent (or equivalent) viscosity and to approximate, i.e., to smooth, this viscosity such that the standard computation is feasible and the computed solution is close enough to the actual solution. By "computation is feasible", we mean that standard mathematical methods and solution algorithms for non-Newtonian purely viscous fluid flows can be applied right away. As a starting point, we replace (17) by:

$$\tau = 2\mu_e \boldsymbol{D}(\boldsymbol{u}) \tag{46}$$

$$\mu_e = \mu + \frac{\tau_y}{2\|\boldsymbol{D}(\boldsymbol{u})\|} \tag{47}$$

The above equation is obviously well suited to yielded regions, i.e., regions where the rate-of-strain tensor $\boldsymbol{D}(\boldsymbol{u})$ is non-zero. However, as $\|\boldsymbol{D}(\boldsymbol{u})\| \to 0$, i.e., $\|\tau\| \to \tau_y$, the equivalent viscosity μ_e diverges, preventing any standard computation to be attempted. The basic idea of regularization models is simple and natural: is there a way to keep the viscosity finite and to provide a computed solution which is close to the exact one? Introducing a regularization parameter ϵ, usually as small as possible, regularization models replace μ_e by $\mu_{e,\epsilon}$ such that $\mu_{e,\epsilon} \to \mu_e$ as $\epsilon \to 0$ and $\mu_{e,\epsilon} \in]0, \infty[$, $\forall \|\boldsymbol{D}(\boldsymbol{u})\| \in [0, \infty]$. The use or justification of regularized models may be presented with some slight variations depending on the reader's background:

- at the practical level, it might be viewed as a trick to make the computation feasible while not polluting too much the computed solution, provided ϵ is "small enough".
- at the physical level, some scientists state that regularized models might be even closer to the real behavior of the material since true "yield stress" fluids do not exist. They claim that in any case the material will strain, though very slightly. This would bring us back to the yield stress myth and the wild debate that many "rheologists" have been enjoying for more than thirty years, and this is precisely beyond the goal of this lecture (the interested readers may refer to Barnes and Walters (1985), Barnes (1999) and the lectures of G. Ovarlez over that course).

- at the mathematical level, regularization permits to smooth out the non-differentiable minimization problem presented in the previous section.

Among the assorted regularized functions employed in the literature, we have:

- the exponential model of Papanastasiou (1987):

$$\mu_{e,\epsilon} = \mu + \frac{\tau_y}{2\|\boldsymbol{D}(\boldsymbol{u})\|}(1 - e^{-2\|\boldsymbol{D}(\boldsymbol{u})\|/\epsilon}) \tag{48}$$

- the model of Bercovier and Engleman (1980):

$$\mu_{e,\epsilon} = \mu + \frac{\tau_y}{\sqrt{\epsilon^2 + 2\|\boldsymbol{D}(\boldsymbol{u})\|^2}} \tag{49}$$

- the simple algebraic model employed by Allouche et al. (2000) (and probably elsewhere previously):

$$\mu_{e,\epsilon} = \mu + \frac{\tau_y}{\epsilon + 2\|\boldsymbol{D}(\boldsymbol{u})\|} \tag{50}$$

A slightly different approach has been proposed by Beverly and Tanner (1992) known as the bi-viscosity model. Essentially, it consists in introducing a critical shear rate $\dot{\gamma}_c$, as small as possible, and a second viscosity $\bar{\mu}$, as high as possible, such that:

$$\boldsymbol{\tau} = \begin{cases} 2\bar{\mu}\boldsymbol{D} & \text{if } \|\boldsymbol{D}(\boldsymbol{u})\| \leq \dot{\gamma}_c \\ 2\mu\boldsymbol{D} + \dfrac{\boldsymbol{D}}{\|\boldsymbol{D}\|}\tau_y & \text{instead} \end{cases} \tag{51}$$

In other words, unyielded regions are replaced by very viscous zones delineated by the critical shear rate $\dot{\gamma}_c$. This model does not smooth the constitutive equation (it is still non-differentiable) but nevertheless improves the computation since the apparent viscosity can be computed for any strain rate magnitude.

Surprisingly, the idea of regularizing was first introduced by applied mathematicians, the ones who devised the Lagrange multiplier technique: Duvaut, Lions, Glowinski and Tremolieres, as a mean to prove the existence and uniqueness of the solution as well as producing the first proper error analysis. In the first three regularized models, $\dot{\gamma} \to 0$, $\mu_{e,\epsilon} \to \mu + \frac{\tau_y}{\epsilon}$, that means that they all have the same low shear rate limit. So, if in the Berverly and Tanner model, we set $\bar{\mu} = \mu + \frac{\tau_y}{\epsilon}$, all regularized models behave more or less in the same manner. If needed, regularized viscoplastic problems can also be written as minimization problems using a regularized functional. For instance, in the case of the Bercovier and Engelman model, the functional (34) reads:

$$\Gamma(\boldsymbol{v}, q) = \frac{1}{2}\int_\Omega \frac{\rho}{\Delta t}\boldsymbol{v}^2 dx + \int_\Omega \mu|\boldsymbol{D}(\boldsymbol{v})|^2 dx + \int_\Omega \sqrt{2}\tau_y\sqrt{|\boldsymbol{D}(\boldsymbol{v})|^2 + \epsilon^2} dx - \int_\Omega q\nabla\cdot\boldsymbol{v} dx - \int_\Omega \boldsymbol{f}\cdot\boldsymbol{v} dx \tag{52}$$

1.3 Ready-on-the-Shelf Algorithms

Although the Lagrange multiplier formulations are quite technical to derive, the corresponding solving algorithms can be implemented in a very straightforward and easy way. Regularized models render the whole flow problem equivalent to the problem for a generalized viscous fluid, with presumably higher viscosity gradients in the flow domain.

We give below the solution algorithms in the case of an isothermal, unsteady and incompressible flow of a Bingham fluid, i.e., problem (18)–(22). We present the decomposition-coordination variant of the Lagrange multiplier formulation that leads to the well known AL algorithm. It is the most popular variant among practitioners (for details on the orthogonal-projection variant, see, e.g., Glowinski and Wachs 2011). We also present the fixed-point variant of the regularized model combined to a L^2-projection operator-splitting (predictor-corrector) technique. For the purpose of illustration, we select the Papanastasiou model. The Newton-Raphson variant is quite straightforward to derive.

To increase the readibility, all algorithms are presented in a non-variational form. More "exotic" physical effects will be presented in Sect. 4.

ALG2: Augmented Lagrangian algorithm

- Initialisation of $\boldsymbol{u}^0$, p^0, $\boldsymbol{d}^0$, $\boldsymbol{\lambda}^0$
- Time loop $t^{n+1} = t^n + \Delta t, n \geq 0$
  - Initialisation: $\boldsymbol{u}^{n+1,0} = \boldsymbol{u}^n$, $p^{n+1,0} = p^n$, $\boldsymbol{d}^{n+1,0} = \boldsymbol{d}^n$ and $\boldsymbol{\lambda}^{n+1,0} = \boldsymbol{\lambda}^n$
  - Solve the velocity-pressure problem by an AL Uzawa algorithm: $k \geq 0$
    Solve the Stokes problem: find $\boldsymbol{u}^{n+1,k+1}$, $p^{n+1,k+1}$ such that

$$\rho\frac{\boldsymbol{u}^{n+1,k+1}}{\Delta t} - \nabla\cdot\left(\left(2\mu + r\right)\boldsymbol{D}(\boldsymbol{u}^{n+1,k+1})\right) + \nabla p^{n+1,k+1} = \nabla\cdot\left(\boldsymbol{\lambda}^{n+1,k} - r\boldsymbol{d}^{n+1,k}\right) + \rho\frac{\boldsymbol{u}^n}{\Delta t} \quad \text{in } \Omega \tag{53}$$

$$\nabla\cdot\boldsymbol{u}^{n+1,k+1} = 0 \quad \text{in } \Omega \tag{54}$$

$$\boldsymbol{u}^{n+1,k+1} = \boldsymbol{g} \quad \text{on } \partial\Omega \tag{55}$$

Evaluate the rate-of-strain tensor $\boldsymbol{d}^{n+1,k+1}$

$$\boldsymbol{d}^{n+1,k+1} = \begin{cases} 0 \quad \text{if } \|\boldsymbol{\lambda}^{n+1,k} + r\boldsymbol{D}(\boldsymbol{u}^{n+1,k+1})\| < \tau_y \\ \dfrac{1}{r}\left(1 - \dfrac{\tau_y}{\|\boldsymbol{\lambda}^{n+1,k} + r\boldsymbol{D}(\boldsymbol{u}^{n+1,k+1})\|}\right)\left(\boldsymbol{\lambda}^{n+1,k} + r\boldsymbol{D}(\boldsymbol{u}^{n+1,k+1})\right) \\ \text{if not} \end{cases} \tag{56}$$

Update the Lagrange multipliers $\boldsymbol{\lambda}^{n+1,k+1}$

$$\boldsymbol{\lambda}^{n+1,k+1} = \boldsymbol{\lambda}^{n+1,k} + r(\boldsymbol{D}(\boldsymbol{u}^{n+1,k+1}) - \boldsymbol{d}^{n+1,k+1}) \tag{57}$$

Convergence if

$$\Delta(\boldsymbol{u}, p)_{k+1} = \max_{\Omega} |\boldsymbol{u}^{n+1,k+1} - \boldsymbol{u}^{n+1,k}| + \max_{\Omega} |p^{n+1,k+1} - p^{n+1,k}| \le \epsilon_1 \tag{58}$$

$$\Delta \boldsymbol{D}_{k+} = \max_{\Omega} |\boldsymbol{D}(\boldsymbol{u}^{n+1,k+1}) - \boldsymbol{d}^{n+1,k+1}| \le \epsilon_2 \tag{59}$$

  - Set: $\boldsymbol{u}^{n+1} = \boldsymbol{u}^{n+1,k+1}$, $p^{n+1} = p^{n+1,k+1}$, $\boldsymbol{d}^{n+1} = \boldsymbol{d}^{n+1,k+1}$, $\boldsymbol{\lambda}^{n+1} = \boldsymbol{\lambda}^{n+1,k+1}$.

- END

FPAR: Fixed-point algorithm for the regularized model

- Initialisation of $\boldsymbol{u}^0$, p^0, $\boldsymbol{d}^0$, $\boldsymbol{\lambda}^0$
- Time loop $t^{n+1} = t^n + \Delta t$, $n \ge 0$
  - Initialisation : $\boldsymbol{u}^{n+1,0} = \boldsymbol{u}^n$, $p^{n+1,0} = p^n$
  - Solve the generalized viscous flow problem by a fixed-point operator-splitting algorithm: $k \ge 0$

    Compute the equivalent viscosity field using the velocity field from previous iteration k:

$$\mu^k_{e,\epsilon} = \mu + \frac{\tau_y}{2\|\boldsymbol{D}(\boldsymbol{u}^{n+1,k})\|}(1 - e^{-2\|\boldsymbol{D}(\boldsymbol{u}^{n+1,k})\|/\epsilon}) \tag{60}$$

Solve the advection-diffusion velocity predictor problem: find $\boldsymbol{u}^{n+1,*}$ such that

$$\rho\frac{\boldsymbol{u}^{n+1,*}}{\Delta t} - \nabla \cdot \mu^k_{e,\epsilon}\boldsymbol{D}(\boldsymbol{u}^{n+1,*}) = -\nabla p^n + \nabla \cdot \mu^k_{e,\epsilon}\boldsymbol{D}(\boldsymbol{u}^n) \quad \text{in } \Omega \tag{61}$$

$$\boldsymbol{u}^{n+1,*} = \boldsymbol{g} \quad \text{on } \partial\Omega \tag{62}$$

Enforce mass conservation by a L^2-projection (corrector step):
find ψ such that

$$\nabla^2 \psi = \frac{\rho}{\Delta t} \nabla \cdot \boldsymbol{u}^{n+1,*} \quad \text{in } \Omega, \quad \frac{\partial \psi}{\partial n} = 0 \quad \text{on } \partial\Omega \tag{63}$$

Update $\boldsymbol{u}^{n+1,k+1}$, $p^{n+1,k+1}$ as

$$\boldsymbol{u}^{n+1,k+1} = \boldsymbol{u}^{n+1,*} - \frac{\Delta t}{\rho} \nabla \psi \tag{64}$$

$$p^{n+1,k+1} = p^n + \psi \tag{65}$$

Convergence if

$$\Delta(\boldsymbol{u}, p)_{k+1} = \max_{\Omega} |\boldsymbol{u}^{n+1,k+1} - \boldsymbol{u}^{n+1,k}| + \max_{\Omega} |p^{n+1,k+1} - p^{n+1,k}| \leq \epsilon_1 \tag{66}$$

– Set: $\boldsymbol{u}^{n+1} = \boldsymbol{u}^{n+1,k+1}$, $p^{n+1} = p^{n+1,k+1}$ and
compute $\boldsymbol{\tau}^{n+1} = 2\mu_{e,\epsilon}^{k+1} \boldsymbol{D}(\boldsymbol{u}^{n+1})$

- END

1.4 Solution Methods: Advantages and Drawbacks

The least we can say is that the two solution methods are of equivalent complexity and actually not very complex. Both methods require to solve a generalized Stokes problem followed by explicit updates of different fields. Standard iterative or direct linear solvers can be used.

We can make the following comments about our 2 algorithms:

- both algorithms are of the fixed-point type, which means that their convergence rate is pretty low. This is the main drawback of these algorithms. The convergence of ALG2 has been mathematically proven for any value of the augmentation parameter r (see, e.g., Fortin and Glowinski 1983). There is no rigorous mathematical proof of the convergence of FPAR, so the only thing we can say is that it generally converges fine.
- In ALG2, we generally avoid to solve the Stokes sub-problem with an operator splitting-algorithm and instead solve it as another saddle-point problem (with $\nabla \cdot \boldsymbol{u} = 0$ as the constraint and the pressure p as the Lagrange multiplier). The motivation is to keep the problem fully implicit and fully coupled.
- In FPAR, the computed solution becomes closer to the solution of the actual Bingham model as the regularization parameter ϵ goes to 0, but then simultaneously the number of iterations for the fixed-point algorithm to converge increases as the dependence of the equivalent viscosity to the velocity gradient increases too.

- The most important comment is that FPAR does not solve the actual Bingham model. This prevents from determining the unyielded regions as the ones where $\|\boldsymbol{D}\| = 0$ as $\boldsymbol{D}$ is nowhere truly zero but close to zero only. Instead a criterion on the stress is used: $\|\boldsymbol{\tau}^{n+1}\| = 2\mu_{e,\epsilon}^{k+1}\|\boldsymbol{D}\| < \tau_y$.

In general, we favor ALG2 over FPAR for its rigorous mathematical derivation and its ability to solve the actual Bingham model. However, depending on the nature of the flow problem, it may be useful too to consider a regularized model and implement a variant of FPAR. The extension to a yield stress fluid model with variable plastic viscosity is quite straightforward. In FPAR, it represents an additional dependence of $\mu_{e,\epsilon}^{k}$ with $\dot{\gamma}$, i.e., $\boldsymbol{u}$ while in ALG2 it can be integrated in the iterative process in k as in FPAR (at each iteration k the viscous operator is recomputed with the velocity field from the previous iteration). We will come back to implementation and computing performances issues in Sect. 3.

2 Discretization Schemes for Visoplastic Fluid Flows: FDM, FVM, FEM, LBM

As pointed out in Sect. 1, the most popular solution algorithms require to solve a generalized Stokes problem followed by explicit updates of additional fields. There are hence no particular constraints or particular accuracy requirement in the selection of the discretization scheme for viscoplastic fluid flows. The knowledge available in the literature and the experience gained by practitioners have however led to a number of best practices. The choice of a discretization scheme depends on the nature of the problem and the objective/expectations of the researcher.

Both solution methods presented in Sect. 1 can be combined with any of the classical discretization scheme: Finite Difference (FDM), Finite Volume (FVM) and Finite Element (FEM). The lattice-Boltzmann method (LBM) has gained a remarkable popularity over the past 10 years. In the LBM, the conservation equations are reformulated using collision operators. This method was initially developed for Newtonian fluid flows. However, by modifying the collision operator, it can be extended to any purely viscous (and even viscoelastic) non-Newtonian material. Concerning viscoplastic materials, a Lagrange multiplier based formulation looks quite complicated (if not impossible) to establish. The only contributions in the literature of LBM for viscoplastic fluid flows used a regularized model (see, e.g., Derksen (2009); Prashant and Derksen 2011). Below we will shortly review the classical discretization schemes and comment on what is specific to viscoplastic fluid flows. The reader interested in LBM is referred to the vast literature available on the topic (Mohamad (2011) as a starting point, among many others).

2.1 Finite Difference (FD)

FD discretization is well suited to rectangular/box or block-rectangular/box geometries. They can be extended to non-rectangular/box geometries by using specific reconstructions close to boundaries, but this would take us far beyond the scope of this course. The basic principle is to approximate spatial differential operators by truncated Taylor expansions.

Let's illustrate this on a 1D case. Assuming a uniform discretization of the domain with a discretization step Δx, we can write the following Taylor expansions up to the 4th order:

$$v(x_i + \Delta x) = v_{i+1} = v_i + \Delta x \left(\frac{\partial v}{\partial x}\right)_i + \frac{\Delta x^2}{2}\left(\frac{\partial^2 v}{\partial x^2}\right)_i + \frac{\Delta x^3}{6}\left(\frac{\partial^3 v}{\partial x^3}\right)_i + \frac{\Delta x^4}{24}\left(\frac{\partial^4 v}{\partial x^4}\right)_i \tag{67}$$

$$v(x_i - \Delta x) = v_{i-1} = v_i - \Delta x \left(\frac{\partial v}{\partial x}\right)_i + \frac{\Delta x^2}{2}\left(\frac{\partial^2 v}{\partial x^2}\right)_i - \frac{\Delta x^3}{6}\left(\frac{\partial^3 v}{\partial x^3}\right)_i + \frac{\Delta x^4}{24}\left(\frac{\partial^4 v}{\partial x^4}\right)_i \tag{68}$$

Taking (67) – (68) and dividing by $2\Delta x$ yields a second order accurate approximation of $\left(\frac{\partial v}{\partial x}\right)_i$:

$$\left(\frac{\partial v}{\partial x}\right)_i = \frac{v_{i+1} - v_{i-1}}{2\Delta x} + \frac{\Delta x^2}{6}\left(\frac{\partial^3 v}{\partial x^3}\right)_i = \frac{v_{i+1} - v_{i-1}}{2\Delta x} + O(\Delta x^2) \tag{69}$$

Taking (67) + (68) and dividing by Δx^2 yields a second order accurate approximation of $\left(\frac{\partial^2 v}{\partial x^2}\right)_i$:

$$\left(\frac{\partial^2 v}{\partial x^2}\right)_i = \frac{v_{i+1} + v_{i-1} - 2v_i}{\Delta x^2} - \frac{\Delta x^2}{12}\left(\frac{\partial^4 v}{\partial x^4}\right)_i = \frac{v_{i+1} + v_{i-1} - 2v_i}{\Delta x^2} + O(\Delta x^2) \tag{70}$$

Using (67) alone and dividing by Δx yields a first order accurate approximation of $\left(\frac{\partial v}{\partial x}\right)_i$:

$$\left(\frac{\partial v}{\partial x}\right)_i = \frac{v_{i+1} - v_i}{\Delta x} + \frac{\Delta x}{2}\left(\frac{\partial^2 v}{\partial x^2}\right)_i + \cdots = \frac{v_{i+1} - v_i}{\Delta x} + O(\Delta x) \tag{71}$$

and so on. To derive the FD discrete form of the mass and momentum conservation equations, suffice to replace the different operators by their discrete analogs.

However, to avoid stability issues (often called checkerboard pressure), the location on a 2D or 3D grid of the velocity and pressure unknowns cannot be chosen freely. In general, either the pressure and velocity unknowns are half-staggered or correction terms are added to the mass conservation equations (Rhie and Chow interpolation). This is equivalent to satisfying the Babuska-Brezzi inf-sup condition in Finite Element schemes.

2.2 *Finite Volume (FV)*

The principle of FV discretization is to integrate conservation equations on a discrete control volume and to replace the volume integrals by surface integrals using the divergence theorem wherever possible. Unknowns are centered in their corresponding control volume and are assumed to be piecewise-constant (similar to a P0 or Q0 FE interpolation). Then the discretization procedure involves finding an approximation of the fluxes at the edges/faces of the control volume.

To illustrate the method, let's integrate the x momentum equation (without inertia and constant viscosity) over a control volume V_x:

$$\int_{V_x} \rho \frac{\partial u_x}{\partial t} dx - \int_{V_x} \mu \nabla^2 u_x dx + \int_{V_x} \frac{\partial p}{\partial x} dx = 0 \tag{72}$$

Let's discretize the time derivative term by a first order backward Euler scheme, treat all other terms implicitly and use the divergence theorem for the two last terms on the left-hand side, we get:

$$\int_{V_x} \rho \frac{u_x^{n+1}}{\Delta t} dx - \int_{S_x} \mu \nabla u_x^{n+1} \cdot \boldsymbol{n} dx + \int_{S_x} p^{n+1} \boldsymbol{n} \cdot \boldsymbol{e}_x dx = \int_{V_x} \rho \frac{u_x^n}{\Delta t} dx \tag{73}$$

where S_x is the boundary surface of V_x. Replacing volume and surface integrals by their discrete analogs, we get for node i:

$$\rho \frac{u_{x,i}^{n+1}}{\Delta t} |V_x| - \sum_j^N \mu \left(\nabla u_x^{n+1} \right)_j \cdot \boldsymbol{n} S_{x,j} + \sum_j^N p_j^{n+1} \boldsymbol{n} \cdot \boldsymbol{e}_x S_{x,j} = \rho \frac{u_{x,i}^n}{\Delta t} |V_x| \tag{74}$$

where N is the number of edges/faces of V_x and $|V_x|$ is the surface area/volume of V_x. The last step is to compute the edge/face values and gradients by a suitable interpolation scheme. As for FD and FE, the location of the pressure and velocity unknowns on the grid and the type of interpolation to reconstruct face values are required to satisfy certain stability properties.

2.3 Finite Element (FE)

The starting point of the FE method is to write a continuous variational formulation of the conservation equations and then to approximate the continuous functional spaces by their finite-dimensional discrete analogs. Functions belonging to the discrete spaces are polynomials of various degree defined in a polygonal/polyhedral sub-regions of the domain called a finite element. Their value is non-zero in the finite element and zero everywhere else in the domain.

Once again, for the sake of illustration, let's integrate the x momentum equation (without inertia and constant viscosity). To derive a variational formulation of this equation, we multiply by a test function v_x that vanishes on the boundary of the domain $\partial\Omega$ (while $u_x = g_x$ on $\partial\Omega$) and integrate over Ω, to get:

$$\int_\Omega \rho \frac{\partial u_x}{\partial t} v_x dx - \int_\Omega \mu \nabla^2 u_x v_x dx + \int_\Omega \frac{\partial p}{\partial x} v_x dx = 0 \tag{75}$$

where $u_x \in \mathcal{V}(\Omega) = \{w \in \mathcal{H}^1(\Omega) | w = g_x \; on \; \partial\Omega\}$, $v_x \in \mathcal{V}_0(\Omega) = \{w \in \mathcal{H}^1(\Omega) | w = 0 \; on \; \partial\Omega\}$ and $p \in \mathcal{L}_0^2(\Omega) = \{q \in \mathcal{L}^2(\Omega) | \int_\Omega q dx = 0\}$. Using the fact that v_x vanishes on $\partial\Omega$, discretizing the time derivative term by a first order backward Euler scheme, treating all other terms implicitly and using integration by part for the two last terms on the left-hand side, we get:

$$\int_\Omega \rho \frac{u_x^{n+1}}{\Delta t} v_x dx + \int_\Omega \mu \nabla u_x \cdot \nabla v_x dx - \int_\Omega p \frac{\partial v_x}{\partial x} dx = \int_\Omega \rho \frac{u_x^n}{\Delta t} v_x dx \tag{76}$$

Finally we introduce a discretisation (triangulation) $\mathcal{T}_h$ of Ω and approximate $\mathcal{V}(\Omega)$, $\mathcal{V}_0(\Omega)$ and $\mathcal{L}_0^2(\Omega)$ by their finite-dimensional analogs. For the sake of illustration, we consider triangular/tetrahedral elements and continuous quadratic and linear functions for velocity and pressure respectively. Hence, $\mathcal{V}(\Omega)$, $\mathcal{V}_0(\Omega)$ and $\mathcal{L}_0^2(\Omega)$ are replaced by:

$$\mathcal{W}_h(\Omega) = \{w_h \mid w_h \in \mathcal{C}^0(\Omega) \, , \; w_h|_T \in P^2 \, \forall \, T \in \mathcal{T}_h\} \tag{77}$$

$$\mathcal{W}_{0h}(\Omega) = \{w_h \in \mathcal{W}_h(\Omega) \mid w_h = 0 \; on \; \partial\mathcal{T}_h\} \tag{78}$$

$$\mathcal{P}_h(\Omega) = \{q_h \mid q_h \in \mathcal{C}^0(\Omega) \, , \; q_h|_T \in P^1 \, \forall \, T \in \mathcal{T}_h\} \tag{79}$$

where P^1 and P^2 are the spaces of polynomials of degree 1 and 2 respectively. Dropping the subscript h for readability, the discrete form of the x momentum equation at node i reads as follows:

$$\sum_{k \ni i} \sum_{j \in T_k} \left(\int_{T_k} \left(\frac{\rho}{\Delta t} w_i w_j + \mu \nabla w_i \cdot \nabla w_j \right) dx \right) u_{x,j}^{n+1}$$
$$- \sum_{k \ni i} \sum_{l \in T_k} \left(\int_{T_k} \frac{\partial w_i}{\partial x} q_l dx \right) p_l^{n+1} = \sum_{k \ni i} \sum_{j \in T_k} \left(\int_{T_k} \frac{\rho}{\Delta t} w_i w_j dx \right) u_{x,j}^n \tag{80}$$

In the above, we have chosen a compatible pair of finite elements for the velocity field and the pressure field called P2-P1, i.e., velocity basis/test functions are piecewise quadratic continuous functions and pressure basis/test functions are piecewise linear continuous functions. The compatibility condition is called in rigorous applied mathematics textbooks the Babuska-Brezzi inf-sup condition. In other words, the space of basis/test functions for the velocity field and the pressure field cannot be chosen independently. The list of compatible pairs can be found in many textbooks as, e.g., Brezzet al. (2008). Finite element functions can be continuous or discontinuous, conforming or non-conforming, low or high order. A choice of a suitable compatible pair of finite elements for the Stokes problem depend of the regularity of the solution, the expected accuracy and the geometric shape of the domain Ω.

2.4 Discretization Schemes for Viscoplastic Fluid Flows

Regardless of the rheological nature of the fluid and the flow regime (Stokes or inertial), the choice of a particular discretization schemes depends first on the geometric shape of the flow domain. In fact, FE and FV/Staggered Grid (SG) schemes are suitable for rectangular/box or block-rectangular/box geometries only. Conversely, unstructured cell-centered and vertex-centered FV can be used for any kind of geometry. Finally, FE are by nature unstructured in the sense that the discretization of the governing equations does not depend a priori on the shape of the domain.

Let's turn our attention to the specific case of viscoplastic material. The first comment to make is the poor regularity of the solution, emphasized in the early papers of the 70 s by Glowinski (1974), Bristeau and Glowinski (1974). In case a regularized model is used, there is nothing special to underline as the problem is recast into a generalized viscous problem. Thus, any discretization scheme for a viscous fluid that satisfies the appropriate stability condition can be chosen.

In case the actual Bingham model is solved with a variant of ALG2, the poor regularity of the solution makes any attempt to use high-order schemes to improve the accuracy of the solution not worth. Actually, the experience has shown that low-order schemes for the velocity field, i.e., linear interpolation, perform better than high-order schemes as, e.g., quadratic interpolation, in the context of FE discretization. The important question that arises from the classical Lagrange multiplier formulation of the fluid momentum equation (53), recalled here for the sake of clarity:

$$\rho\frac{\boldsymbol{u}^{n+1,k+1}}{\Delta t} - \nabla\cdot\left(\left(2\mu + r\right)\boldsymbol{D}\left(\boldsymbol{u}^{n+1,k+1}\right)\right) + \nabla p^{n+1,k+1} = \nabla\cdot\left(\boldsymbol{\lambda}^{n+1,k} - r\boldsymbol{d}^{n+1,k}\right) + \rho\frac{\boldsymbol{u}^{n}}{\Delta t}$$

is the discretization of the additional fields $\boldsymbol{d}$ and $\boldsymbol{\lambda}$. In FE jargon, $\boldsymbol{D}(\boldsymbol{u})$ and $\boldsymbol{d}$ should belong to the same space for the following reason: this equation should be valid for a Newtonian fluid ($\tau_y = 0$) for any value of r. When $\tau_y = 0$, the Lagrange multiplier field is uniformly zero in the domain, hence to recover the exactly same Newtonian solution as from a standard Newtonian algorithm, the terms $\nabla \cdot r\boldsymbol{D}(\boldsymbol{u})$ on the right-hand side and $\nabla \cdot r\boldsymbol{d}$ on the left-hand side should cancel out. This happens only if $\boldsymbol{D}(\boldsymbol{u})$ and $\boldsymbol{d}$ belong to the same space. Then, having $\boldsymbol{D}(\boldsymbol{u})$, $\boldsymbol{d}$ and $\boldsymbol{\lambda}$ in the same space or discretized the same way avoid any interpolation in steps 2 and 3 of ALG2 and $\boldsymbol{d}$ and $\boldsymbol{\lambda}$ can hence be updated point-wise.

Finally, although other choices of discretization scheme are possible, we recommend the following:

- rectangular/box or block-rectangular/box geometries without mesh refinement: the FV/SG scheme with the d_{ij} and λ_{ij} components located on the grid where the D_{ij} components are, as shown in Fig. 1 in 2D.
- for more complex geometries or rectangular/box or block-rectangular/box geometries with mesh refinement: the P1-P1isoP2 FE for the pressure-velocity fields and the P0isoP2 FE for the $\boldsymbol{d}$ and $\boldsymbol{\lambda}$ fields. In 2D, P1isoP2 is a triangular element constructed by splitting it into 4 sub-triangles using the center of the edges. Linear basis functions are defined over each sub-triangle (the equivalent in 3D involves splitting a tetrahedron into 8 sub-tetrahedra). P0isoP2 follows the same construction rules except that basis functions over subtriangles are constant functions. This combination P1isoP2-P1-P0isoP2 satisfies the Babuska-Brezzi inf-sup stability condition. Finally, since the derivative of a linear function is a constant function, we do have $\boldsymbol{D}(\boldsymbol{u})$ and $\boldsymbol{d}$ consistently discretized.

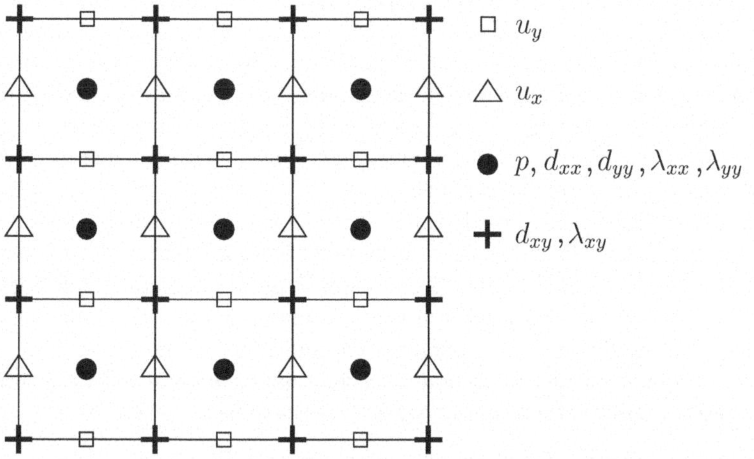

Fig. 1 Finite Volume/Staggered Grid discretization for the 2D discretization of viscoplastic fluid flows

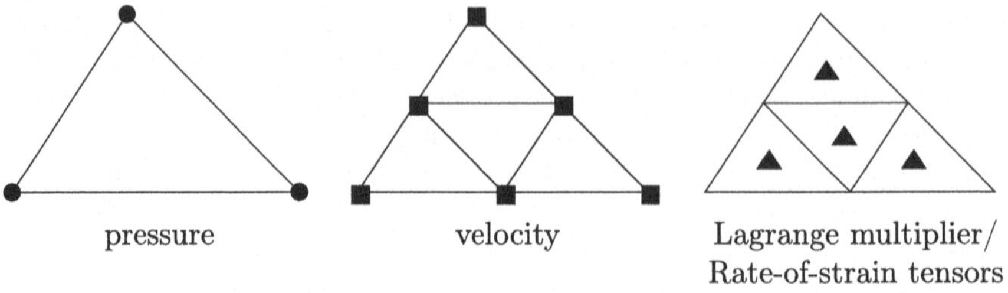

Fig. 2 P1-P1isoP2-P0isoP2 finite elements for the 2D discretization of viscoplastic fluid flows

3 Better and Faster: Capturing the Yield Surfaces and High Performance Computing

One of the main challenges in computing viscoplastic fluid flows is the accurate determination of the unyielded regions or equivalently the curves/surfaces delineating yielded and unyielded regions, called the yield surfaces. In theory, unyielded regions are characterized by the two following equivalent criteria:

1. $\|\boldsymbol{\tau}\| < \tau_y$, or
2. $\boldsymbol{D}(\boldsymbol{u}) = 0$

When the model is regularized, criterion 2 cannot be used anymore. In fact, in unyielded regions, the equivalent viscosity is extremely high but not infinite. As a consequence, $\boldsymbol{D}(\boldsymbol{u})$ is never truly zero anywhere in these regions and only the stress criterion 1 makes sense. However, using the regularized stress/strain-rate relation, we can determine a critical strain-rate $\dot{\gamma}_c = 2\|\boldsymbol{D}\|_c$ that satisfies $\mu_e(\dot{\gamma}_c)\dot{\gamma}_c = \tau_y$ and determine unyielded regions with a stain-rate criterion $2\|\boldsymbol{D}\| < \dot{\gamma}_c$.

When the actual Bingham model is solved with the Lagrange multiplier algorithm ALG2, criterion 2 is replaced by $\boldsymbol{d} = 0$ as $\boldsymbol{d}$ represents the actual rate-of-strain tensor in the flow and is explicitly constructed to zero by (56) in unyielded regions. This facilitates the determination of unyielded regions as the criterion does not rely on any a priori additional parameter as the regularization parameter of regularized models. This is one of the advantages of Lagrange multiplier algorithms, and in particular ALG2, over regularization algorithms.

Regardless of the model, the location of unyielded regions can be determined a posteriori only, once the stress field or the rate-of-strain field has been computed. In other words, unyielded regions are unknowns in the problem. While for some classes of problem, an initial uniform mesh or a mesh constructed with an a priori knowledge of the flow can be sufficient to accurately compute the solution, other classes of problem require to accurately capture the location of the yield surfaces. This could be achieved by Adaptive Mesh Refinement for:

1. unstructured meshes for FE or cell/vertex-centered FV,
2. structured mesh with quadtree/octree data structures.

The former has been used successfully with FE by many authors and was pioneered by Saramito and Roquet (2001), Roquet and Saramito (2003, 2008). It has been mainly implemented for 2D steady-state flows, see Rheolef (2015). The flow solver is coupled to BAMG (Bidimensional Anisotropic Mesh Generator) from the FreeFem++ package (2018), Hecht (2012). BAMG adapts a mesh based on a certain metric field calculated from the computed solution on the initial mesh. The cycle of mesh adaption is as follows:

- Create an initial mesh M^0
- Compute the flow problem solution s^0 on that initial mesh M^0
- Mesh adaption cycle: $k \geq 0$

1. Compute the metric field c^k based on s^k
2. Adapt the mesh: create a new mesh M^{k+1} using c^k
3. Compute the flow problem solution s^{k+1} on the new mesh M^{k+1}
4. Stop the cycle if the solution quality criteria are met, otherwise $k = k + 1$
5. END

The question of what is the right metric field to adapt the mesh is central to the quality of the mesh adaption process. It strongly depends of the nature of the problem. A common trend to many engineering problems is to refine the mesh in regions of high gradients of a certain relevant field. In Fluid Mechanics, regions of high velocity gradients are good candidates as, e.g., in boundary layers. However, in viscoplastic fluid flows, we are interested in accurately capturing the yield surfaces, which do not correspond to the regions of high velocity gradients. To refine the mesh around

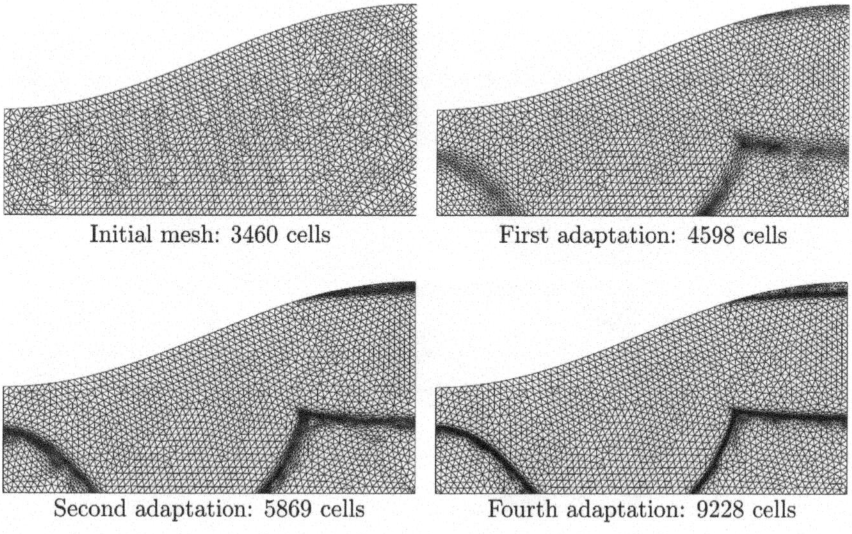

Fig. 3 Mesh adaption cycle with BAMG for the 2D steady flow in a wavy channel (courtesy of Ali Roustaei)

the yield surfaces, the simplest metric field is an indicator that takes a value of 0 in yielded regions and 1 in unyielded regions. As a consequence, the gradient of that metric field is theoretically infinite close to the yield surfaces.

In Fig. 3, the authors used this simple indicator 0/1 (yielded/unyielded) and specified a maximum grid size everywhere else in the flow domain. In Fig. 4, they constructured a composite criterion using both the indicator 0/1 to locate yield surfaces and another metric field to locate regions of high velocity gradients. The difficulty is the determination of the appropriate relative weights of these two criteria in the composite criterion used by BAMG. But overall, once the right values are determined (probably simply by trying, as these values strongly depend on the problem studied), it is visible that the mesh quality is higher in Fig. 4 than in Fig. 3.

To the best of our knowledge, quadtree/octree structured adaptive mesh refinement (AMR) has never been used for viscoplastic fluid flows, although it should perform quite nicely. The principle of quadtree structured AMR is illustrated in Fig. 5. One of the major assets of quadtree/octree structured AMR (SAMR) in engineering problems is its ability to embed a complex geometry/domain into a simpler structured one as, e.g., a box, and to impose boundary condition using local reconstruction as,

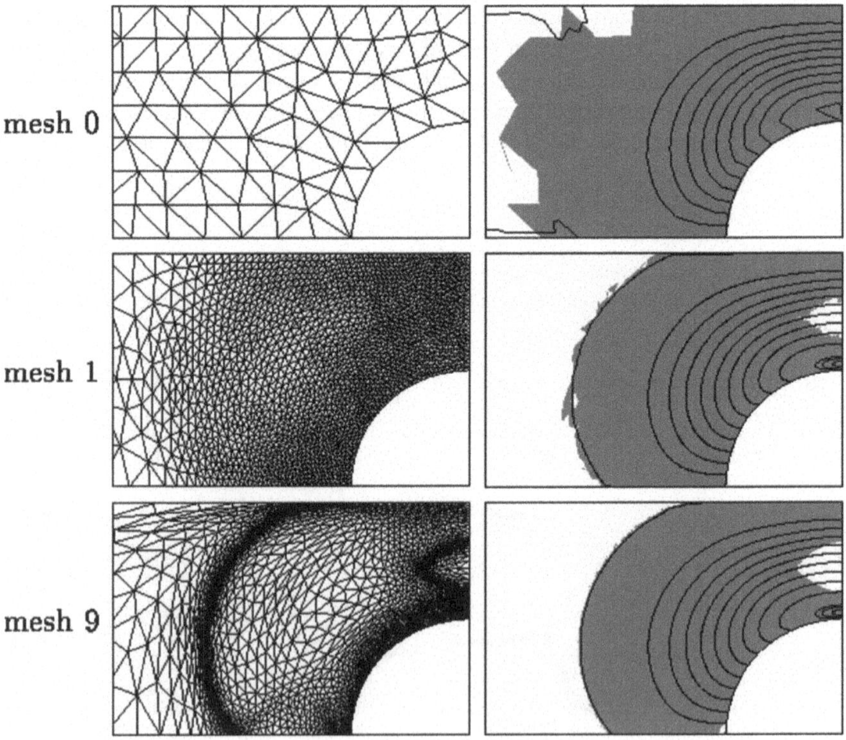

Fig. 4 Mesh adaption cycle with BAMG for the 2D steady flow of a settling cylinder in a 2:1 channel (courtesy of Pierre Saramito)

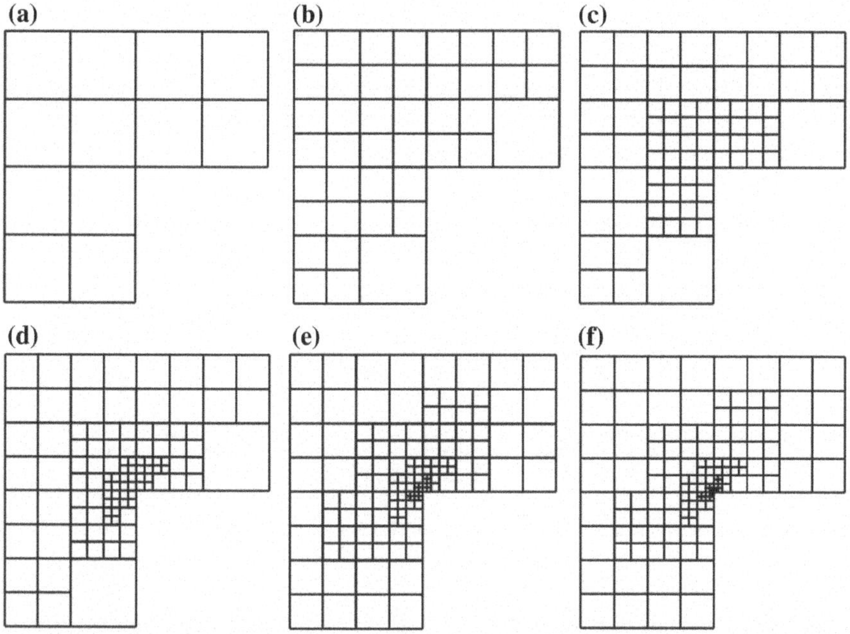

Fig. 5 Quad-tree Adaptive Mesh Refinement cycle around a singularity in a L-shaped domain

e.g., Immersed Boundary method (IBM) in Fluid Mechanics. The mesh is then refined close to the boundary using quadtree/octree refinement to guarantee that the real complex geometry is properly accounted for. Quadtree/octree SAMR could be used right away for viscoplastic fluid flows in simple geometry as, e.g., in 2D or 3D lid-driven cavity, or could be combined with IBM for more complex geometries or problems with moving boundaries.

Finally, once again to the best of our knowledge, there has been no report so far in the literature on the use AMR (regardless of unstructured or structured) for the solution of time-dependent viscoplastic flows. This is also true for 3D problems, both for the steady-state problem and for the time-dependent problem. It is essentially a high performance computing and parallel software programming issue that we will shortly address in the rest of this Lecture.

The two algorithms presented in Sect. 1 or their variants are fixed-point algorithms. Their convergence is pretty slow in general. For the regularized model, the most efficient way to speed up the convergence is to solve the non-linear problem by a Newton/Raphson algorithm, and not a fixed-point algorithm. Actually, if the regularization parameter ε is too small, the number of iterations for the fixed-point algorithm to converge becomes so large that the computation is almost impracticable and switching to a Newton/Raphson algorithm is mandatory. For the actual viscoplastic model, the available Lagrange multiplier algorithms are by default fixed-point algorithms. Recent developments by groups in France and New-Zealand have

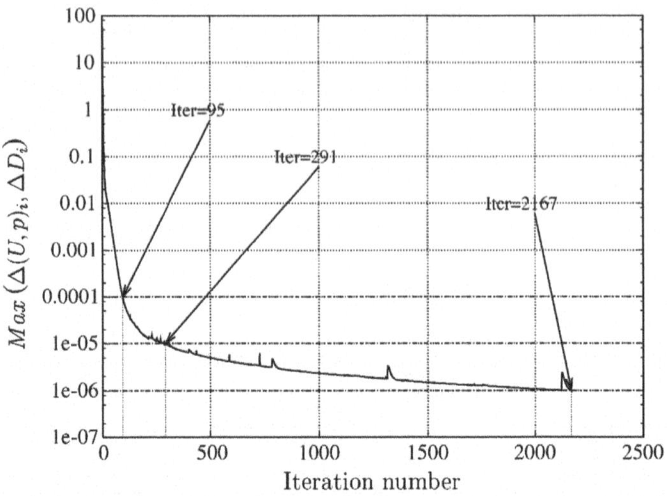

Fig. 6 Typical convergence of ALG2 (courtesy of Guillaume Vinay)

contributed to significantly accelerate the Lagrange multiplier algorithms by slightly reformulating the problem and/or combining them with a Newton/Raphson solution procedure (Saramito 2016; Treskatis 2016).

The main advantage of a Newton/Raphson algorithm is its quadratic convergence. However, it requires a good initial guess and in case this initial guess is not good enough, the convergence might fail. At each iteration, the Jacobian matrix needs to be recomputed, which also adds up to the whole computing time. Fixed-point algorithms are more robust but slower. In case of ALG2, the convergence has been rigorously proven for any value of the augmentation parameter r, but once again ALG2 is not very fast (Fig. 6).

Let's consider the solution of a Bingham fluid flow problem.

Solution with a regularized model:

Positive	Negative
⊕ Easy to implement	⊖ Slow to converge
⊕ Fixed-point iteration for moderately small ε	⊖ Switch to Newton/Rahpson algorithm for very small ε
⊕ Can be used within any generalized viscous solver (as, e.g., SIMPLE in Fluent)	⊖ Velocity matrix cannot be factorized as it changes over the iterations
⊕ Apart from bigger matrices, 3D is not tougher than 2D	
⊕ Hershel-Bulkley model is not tougher than Bingham model	

Solution with the actual model and ALG2:

Positive	Negative
⊕ Guaranteed convergence for any r	⊖ Slow to converge
⊕ Not really tougher to implement	⊖ Convergence rate depends on r
⊕ Velocity matrix is constant	⊖ Optimal value of r is unknown
⊕ In 2D, the velocity matrix can be Cholesky factorized once and for all	⊖ Cholesky factorization is not possible in 3D as the matrix is too big
	⊖ Cholesky factorization is not possible either if Hershel-Bulkley model is considered as the plastic viscosity depends on $\dot{\gamma}_c = 2\|\boldsymbol{D}\|_c$

Although we favor ALG2 as it solves the actual viscoplastic model, the full algorithm exhibits 3 levels of nested loops: (i) the outer loop to enforce the constraint $\boldsymbol{D} - \boldsymbol{d} = \boldsymbol{0}$, (ii) the inner loop to solve the Stokes problem and hence to enforce the constraint $\nabla \cdot \boldsymbol{u} = 0$ and (iii) the iterative solution of the velocity matrix system. If the problem is 2D and the Bingham model is considered (constant plastic viscosity), the velocity matrix can be Cholesky factorized at the beginning of the computation (the matrix changes neither in time (if the time step Δt is kept constant) nor over the iteration of ALG2). But in the general case, an iterative solution method of the preconditioned conjugate gradient (PCG) algorithm is required. The smaller the time step magnitude is, the more diagonally dominant the velocity matrix is and the faster the convergence of the PCG is for steady-state problems, the velocity matrix is however not very diagonally dominant. So directly computing the 3D solution of a steady-state viscoplastic flow problem is the hardest problem in terms of convergence rate. It is very often advisable to compute the steady-state solution as the long-time solution of a time-dependent problem.

The regularized algorithm has one less level of nested loop. In fact, the outer loop does not exist conceptually as the model is regularized. Whether the total number of iterations as well as the total wall clock time (actual computing time) in the regularized algorithm is smaller or larger than in ALG2 is not necessarily easy to predict. There has not been any systematic comparison in the literature of the computing time associated to both methods but in general the computing time associated to ALG2 is much larger.

From a purely High Performance Computing standpoint, it is worthwhile to note that at the end of the day, the major part of the computing time is spent at solving linear matrix systems, which in turn involves performing vector-vector and matrix-vector products (the point-wise update of $\boldsymbol{d}$ and $\boldsymbol{\lambda}$ as well as the update of the regularized viscosity field should normally be negligible with respect to solving linear matrix systems). The convergence rate being what it is, another angle of attack to lower the computing time is to use MPI-parallelized numerical codes and run jobs on supercomputers. In other words, the computing time essentially depends on how fast these vector-vector and matrix-vector products can be performed and how well the implementation scales on large supercomputers. The good news in that area is that extremely scalable and efficient linear algebra libraries can be freely downloaded from the web and simply plugged to one's favorite flow solver. We strongly recommend the reader to look at the following libraries, among others:

- PETSc (2018), a complete, well maintained and powerful linear algebra library,
- HYPRE (2018), for its parallel multigrid iterative linear solvers,
- MUMPS (2018), for its parallel direct linear solvers,

Highly scalable implementations combined to the next generation of solution algorithms Saramito (2016), Treskatis (2016) should normally lead to a real change of paradigm in computations of viscoplastic fluid flows.

4 Additional Effects: Compressibility, Heat Transfer, Elasticity, Thixotropy, Two-Phase Flows

Many real flows involving viscoplastic materials also exhibit aditional properties related to (i) the weak compressibility of the materials, (ii) the thermo-dependence of the rheological behavior of the materials in relation to the presence of strong heat transfer in the flow, (iii) the time-dependent rheological properties of the materials (reversible elasticity or reversible/irreversible microstructure build-up/breakdown) and (iv) the presence of more than a single fluid in the flow. The goal here is not to review these additional features of the flow from a physical viewpoint as this will be done by other lecturers but to address and illustrate the computational issues related to them.

To get started, we will introduce the fundamental notion of weak/loose coupling in the solution method for a set of coupled governing equations. Let's consider two scalar fields u and v solving the following set of PDEs:

$$\frac{\partial u}{\partial t} = F(u, \nabla^2 u, v, uv) \tag{81}$$

$$\frac{\partial v}{\partial t} = G(v, \nabla^2 v, u, uv) \tag{82}$$

with appropriate initial and boundary conditions, where F and G are arbitrary (potentially non-linear) functions of u, $\nabla^2 u$ and v and u, $\nabla^2 v$ and v respectively. Assuming that we look for an "as implicit as possible" solution method, the type of solution method depends strongly on the form of F and G. In practice, non-linearities in u and v separately can be treated by linearization or a Newton/Raphson algorithm while terms involving a product of both can be treated by explicit weak/loose coupling in the following way:

$$\frac{u^{n+1} - u^n}{\Delta t} = F(u^{n+1}, \nabla^2 u^{n+1}, v^n, u^{n+1} v^n) \tag{83}$$

$$\frac{v^{n+1} - v^n}{\Delta t} = F(u^n, \nabla^2 v^{n+1}, v^{n+1}, u^n v^{n+1}) \tag{84}$$

In case the equations are physically weakly coupled or the time step Δt is selected small enough such that u and v do not change "too much" over a time step, weak coupling solution methods are cheap, easy-to-implement and efficient.

Let's illustrate this on the incompressible flow of a Newtonian fluid with heat transfer. In addition to $\nabla \cdot \boldsymbol{u} = 0$, the fluid velocity $\boldsymbol{u}$ and temperature Θ solve the following conservation equations:

$$\rho_0 \left(\frac{\partial \boldsymbol{u}}{\partial t} + \boldsymbol{u} \cdot \nabla \boldsymbol{u} \right) - \mu \nabla^2 \boldsymbol{u} + \nabla p = -\rho_0 \beta \boldsymbol{g} (\Theta - \Theta_0) \tag{85}$$

$$\rho_0 C_p \left(\frac{\partial \Theta}{\partial t} + \boldsymbol{u} \cdot \nabla \Theta \right) - \kappa \nabla^2 \Theta = 0 \tag{86}$$

where $\rho = \rho_0(1 - \beta(\Theta - \Theta_0))$ is assumed to obey the classical Boussinesq equation. If we assume $\beta = 0$ and linearize $\boldsymbol{u} \cdot \nabla \boldsymbol{u}$, then the two equations are one-way coupled only and we can solve them in the following sequence:

1. Solve the Navier-Stokes problem as follows:

$$\rho_0 \left(\frac{\boldsymbol{u}^{n+1}}{\Delta t} + \boldsymbol{u}^n \cdot \nabla \boldsymbol{u}^{n+1} \right) - \mu \nabla^2 \boldsymbol{u}^{n+1} + \nabla p^{n+1} = \rho_0 \frac{\boldsymbol{u}^n}{\Delta t} \tag{87}$$

$$\nabla \cdot \boldsymbol{u}^{n+1} = 0 \tag{88}$$

2. Solve the temperature equation as follows:

$$\rho_0 C_p \left(\frac{\Theta^{n+1}}{\Delta t} + \boldsymbol{u}^* \cdot \nabla \Theta^{n+1} \right) - \kappa \nabla^2 \Theta^{n+1} = \rho_0 C_p \frac{\Theta^n}{\Delta t} \tag{89}$$

where $\boldsymbol{u}^*$ is a (generally linear) combination of $\boldsymbol{u}^n$ and $\boldsymbol{u}^{n+1}$. In practice, the usual choices are $\boldsymbol{u}^n$, $\boldsymbol{u}^{n+1}$ or $0.5(\boldsymbol{u}^n + \boldsymbol{u}^{n+1})$. Note that for $\beta = 0$, problem (87)–(88) is independent of problem (89) while the reverse is not true. Hence problem (87)–(88) is solved first, and then problem (89) can be solved too. In other words, once the velocity field is known, the temperature field can be computed.

Now if we assume $\beta \neq 0$ and/or $\mu = \mu(\Theta)$, and linearize $\boldsymbol{u} \cdot \nabla \boldsymbol{u}$, the fluid velocity $\boldsymbol{u}$ and temperature Θ conservation equations are two-way coupled and we can solve them in a weak coupling way as follows:

1. Solve the Navier-Stokes problem as follows:

$$\rho_0\left(\frac{\boldsymbol{u}^{n+1}}{\Delta t}+\boldsymbol{u}^n\cdot\nabla\boldsymbol{u}^{n+1}\right)-\nabla\cdot 2\mu(\Theta^n)\boldsymbol{D}(\boldsymbol{u}^{n+1})+\nabla p^{n+1}$$
$$=-\rho_0\beta\boldsymbol{g}(\Theta^n-\Theta_0)+\rho_0\frac{\boldsymbol{u}^n}{\Delta t} \tag{90}$$

$$\nabla\cdot\boldsymbol{u}^{n+1}=0 \tag{91}$$

2. Solve the temperature equation as follows:

$$\rho_0 C_p\left(\frac{\Theta^{n+1}}{\Delta t}+\boldsymbol{u}^*\cdot\nabla\Theta^{n+1}\right)-\kappa\nabla^2\Theta^{n+1}=\rho_0 C_p\frac{\Theta^n}{\Delta t} \tag{92}$$

In general, heat transfer, thixotropy and two-fluid flow can be reasonably well computed using such a strategy, provided the physical coupling is "not too strong" and/or the time step magnitude is small enough. Compressibility requires to slightly modify the solution of the coupled mass and momentum conservation equations as the mass conservation equation does not simplify to the divergence free of the velocity field anymore. We give below the solution algorithms for the case of a viscoplastic fluid flow coupled to these four additional effects (heat transfer, thixotropy, two-fluid and compressibility).

4.1 Heat Transfer

The weak-coupling based solution algorithm, where for the sake of variety we treat the advective/convective terms fully explicity, simply reads:

- Initialisation of $\boldsymbol{u}^0$, p^0, $\boldsymbol{d}^0$, $\boldsymbol{\lambda}^0$, Θ^0
- Time loop $t^{n+1}=t^n+\Delta t, n\geq 0$
  - Solve the mass and momentum conservation equations
    Initialisation : $\boldsymbol{u}^{n+1,0}=\boldsymbol{u}^n$, $p^{n+1,0}=p^n$, $\boldsymbol{d}^{n+1,0}=\boldsymbol{d}^n$ and $\boldsymbol{\lambda}^{n+1,0}=\boldsymbol{\lambda}^n$
    Solve the velocity-pressure problem by an AL Uzawa algorithm: $k\geq 0$
    · Solve the Stokes problem: find $\boldsymbol{u}^{n+1,k+1}$, $p^{n+1,k+1}$ such that

$$\rho_0\frac{\boldsymbol{u}^{n+1,k+1}}{\Delta t}-\nabla\cdot\left(\left(2\mu(\Theta^n)+r\right)\boldsymbol{D}(\boldsymbol{u}^{n+1,k+1})\right)+\nabla p^{n+1,k+1}$$
$$=\nabla\cdot\left(\boldsymbol{\lambda}^{n+1,k}-r\boldsymbol{d}^{n+1,k}\right)-\rho_0\boldsymbol{u}^n\cdot\nabla\boldsymbol{u}^n \tag{93}$$
$$-\rho_0\beta\boldsymbol{g}(\Theta^n-\Theta_0)+\rho\frac{\boldsymbol{u}^n}{\Delta t}\quad\text{in }\Omega$$

$$\nabla \cdot \boldsymbol{u}^{n+1,k+1} = 0 \quad \text{in } \Omega \tag{94}$$

$$\boldsymbol{u}^{n+1,k+1} = \boldsymbol{g} \quad \text{on } \partial\Omega \tag{95}$$

· Evaluate the rate-of-strain tensor $\boldsymbol{d}^{n+1,k+1}$

$$\boldsymbol{d}^{n+1,k+1} = \begin{cases} 0 \text{ if } \|\boldsymbol{\lambda}^{n+1,k} + r\boldsymbol{D}(\boldsymbol{u}^{n+1,k+1})\| < \tau_y \\ \dfrac{1}{r}\left(1 - \dfrac{\tau_y}{\|\boldsymbol{\lambda}^{n+1,k} + r\boldsymbol{D}(\boldsymbol{u}^{n+1,k+1})\|}\right)\left(\boldsymbol{\lambda}^{n+1,k} + r\boldsymbol{D}(\boldsymbol{u}^{n+1,k+1})\right) \\ \text{if not} \end{cases} \tag{96}$$

· Update the Lagrange multipliers $\boldsymbol{\lambda}^{n+1,k+1}$

$$\boldsymbol{\lambda}^{n+1,k+1} = \boldsymbol{\lambda}^{n+1,k} + r(\boldsymbol{D}(\boldsymbol{u}^{n+1,k+1}) - \boldsymbol{d}^{n+1,k+1}) \tag{97}$$

· *Convergence if*

$$\Delta(\boldsymbol{u}, p)_{k+1} = \max_{\Omega} |\boldsymbol{u}^{n+1,k+1} - \boldsymbol{u}^{n+1,k}| + \max_{\Omega} |p^{n+1,k+1} - p^{n+1,k}| \leq \epsilon_1 \tag{98}$$

$$\Delta \boldsymbol{D}_{k+} = \max_{\Omega} |\boldsymbol{D}(\boldsymbol{u}^{n+1,k+1}) - \boldsymbol{d}^{n+1,k+1}| \leq \epsilon_2 \tag{99}$$

Set: $\boldsymbol{u}^{n+1} = \boldsymbol{u}^{n+1,k+1}$, $p^{n+1} = p^{n+1,k+1}$, $\boldsymbol{d}^{n+1} = \boldsymbol{d}^{n+1,k+1}$, $\boldsymbol{\lambda}^{n+1} = \boldsymbol{\lambda}^{n+1,k+1}$.

– Solve the temperature equation

$$\rho_0 C_p \frac{\Theta^{n+1}}{\Delta t} - \kappa \nabla^2 \Theta^{n+1} = -\rho_0 C_p \boldsymbol{u}^{n+1} \cdot \nabla \Theta^n + \rho_0 C_p \frac{\Theta^n}{\Delta t} \tag{100}$$

- END

4.2 Thixotropy

Most thixotropic models (see the Lecture of G. Ovarlez) describe the changes in the rheological behavior of the material by introducing an additional parameter related to its microstructure, with viscosity and yield-stress being a function of that parameter. Typically, the so-called scalar structure parameter, denoted λ_s for instance, time evolution is governed by the following constitutive equation:

$$\frac{\partial \lambda_s}{\partial t} + \boldsymbol{u} \cdot \nabla \lambda_s = f(\lambda_s, \dot{\gamma}) \tag{101}$$

where $f(\lambda_s, \dot{\gamma})$ is a function of the local generalized shear rate and accounts for the breakdown and buildup of the material. The form of the function f leads to different models in the literature. Among many others, the model of Houska reads as follows:

$$\frac{\partial \lambda_s}{\partial t} + \boldsymbol{u} \cdot \nabla \lambda_s = a(1 - \lambda_s) - b\lambda_s \dot{\gamma}^m \tag{102}$$

where a is the structure build-up parameter, b the structure breakdown parameter and m a fitting parameter without any particular physical significance. In addition, the actual viscosity and yield stress depend linearly on the structure parameter λ_s as:

$$\mu = \mu_0 + \lambda_s \mu_1 \tag{103}$$

$$\tau_y = \tau_{y0} + \lambda_s \tau_{y1} \tag{104}$$

Very straightforwardly, we can construct a weak coupling-based solution algorithm as follows:

- Initialisation of $\boldsymbol{u}^0$, p^0, $\boldsymbol{d}^0$, $\boldsymbol{\lambda}^0$, λ_s^0
- Time loop $t^{n+1} = t^n + \Delta t, n \geq 0$
  - Solve the mass and momentum conservation equation
    Initialisation : $\boldsymbol{u}^{n+1,0} = \boldsymbol{u}^n$, $p^{n+1,0} = p^n$, $\boldsymbol{d}^{n+1,0} = \boldsymbol{d}^n$
    and $\boldsymbol{\lambda}^{n+1,0} = \boldsymbol{\lambda}^n$
    Solve the velocity-pressure problem by an AL Uzawa algorithm: $k \geq 0$
    · Solve the Stokes problem: find $\boldsymbol{u}^{n+1,k+1}$, $p^{n+1,k+1}$ such that

$$\rho \frac{\boldsymbol{u}^{n+1,k+1}}{\Delta t} - \nabla \cdot \left(\left(2\mu_0 + 2\lambda_s^n \mu_1 + r \right) \boldsymbol{D}(\boldsymbol{u}^{n+1,k+1}) \right) + \nabla p^{n+1,k+1}$$
$$= \nabla \cdot \left(\boldsymbol{\lambda}^{n+1,k} - r\boldsymbol{d}^{n+1,k} \right) - \rho \boldsymbol{u}^n \cdot \nabla \boldsymbol{u}^n + \rho \frac{\boldsymbol{u}^n}{\Delta t} \quad \text{in } \Omega \tag{105}$$

$$\nabla \cdot \boldsymbol{u}^{n+1,k+1} = 0 \quad \text{in } \Omega \tag{106}$$

$$\boldsymbol{u}^{n+1,k+1} = \boldsymbol{g} \quad \text{on } \partial\Omega \tag{107}$$

· Evaluate the rate-of-strain tensor $\boldsymbol{d}^{n+1,k+1}$

$$\boldsymbol{d}^{n+1,k+1} = \begin{cases} 0 \ \text{ if } \|\boldsymbol{\lambda}^{n+1,k} + r\boldsymbol{D}(\boldsymbol{u}^{n+1,k+1})\| < \tau_{y0} + \lambda_s^n \tau_{y1} \\ \dfrac{1}{r}\left(1 - \dfrac{\tau_y}{\|\boldsymbol{\lambda}^{n+1,k} + r\boldsymbol{D}(\boldsymbol{u}^{n+1,k+1})\|}\right)\left(\boldsymbol{\lambda}^{n+1,k} + r\boldsymbol{D}(\boldsymbol{u}^{n+1,k+1})\right) \\ \text{if not} \end{cases} \tag{108}$$

· Update the Lagrange multipliers $\boldsymbol{\lambda}^{n+1,k+1}$

$$\boldsymbol{\lambda}^{n+1,k+1} = \boldsymbol{\lambda}^{n+1,k} + r(\boldsymbol{D}(\boldsymbol{u}^{n+1,k+1}) - \boldsymbol{d}^{n+1,k+1}) \tag{109}$$

· *Convergence if*

$$\Delta(\boldsymbol{u}, p)_{k+1} = \max_{\Omega} |\boldsymbol{u}^{n+1,k+1} - \boldsymbol{u}^{n+1,k}| + \max_{\Omega} |p^{n+1,k+1} - p^{n+1,k}| \leq \epsilon_1 \tag{110}$$

$$\Delta \boldsymbol{D}_{k+} = \max_{\Omega} |\boldsymbol{D}(\boldsymbol{u}^{n+1,k+1}) - \boldsymbol{d}^{n+1,k+1}| \leq \epsilon_2 \tag{111}$$

Set: $\boldsymbol{u}^{n+1} = \boldsymbol{u}^{n+1,k+1}$, $p^{n+1} = p^{n+1,k+1}$, $\boldsymbol{d}^{n+1} = \boldsymbol{d}^{n+1,k+1}$, $\boldsymbol{\lambda}^{n+1} = \boldsymbol{\lambda}^{n+1,k+1}$.

– Solve the structure parameter equation

$$\frac{\lambda_s^{n+1}}{\Delta t} - a(1 - \lambda_s^{n+1}) + b\lambda_s^{n+1}\dot{\gamma}^m = -\boldsymbol{u}^{n+1} \cdot \nabla\lambda_s^n + \frac{\lambda_s^n}{\Delta t} \tag{112}$$

- END

4.3 *Two-Fluid Flow*

The easiest way to compute a flow with two fluids is to track the interface between the two fluid phases with a simple indicator c, often called in the literature colour or concentration function. This method is referred to as the Volume-Of-Fluid (VOF) method (other approaches include Level-Set, Front Tracking, etc). In the VOF approach, c is equal to 1 and phase 1 and 0 in phase 2, and satisfies the following conservation equation:

$$\frac{\partial c}{\partial t} + \boldsymbol{u} \cdot \nabla c = D_c \nabla^2 c \tag{113}$$

where D_c in the interphase diffusion coefficient. D_c is set to 0 in case of immiscible fluids. However, even in that case, in the vicinity of the interface, the colour function c takes values between 0 and 1 as the sharp transition from fluid 1 to fluid 2 is smoothed out at the discrete level. In general, the interface position is determined as the contour $c = 0.5$.

The actual viscosity and yield stress in the domain depend on the value of the concentration function c. Very often, a simple arithmetic average is used as follows:

$$\mu = c\mu_1 + (1 - c)\mu_2 \tag{114}$$

$$\tau_y = c\tau_{y1} + (1 - c)\tau_{y2} \tag{115}$$

Once again, constructing a weak coupling solution algorithm is straightforward. For the sake of variety, we treat the advective term $\boldsymbol{u} \cdot \nabla \boldsymbol{u}$ implicitly and include it in the AL saddle point iterative procedure. The algorithm reads as follows:

- Initialisation of $\boldsymbol{u}^0$, p^0, $\boldsymbol{d}^0$, $\boldsymbol{\lambda}^0$, c^0
- Time loop $t^{n+1} = t^n + \Delta t, n \geq 0$
  - Solve the mass and momentum conservation equations
    Initialisation: $\boldsymbol{u}^{n+1,0} = \boldsymbol{u}^n$, $p^{n+1,0} = p^n$, $\boldsymbol{d}^{n+1,0} = \boldsymbol{d}^n$ and $\boldsymbol{\lambda}^{n+1,0} = \boldsymbol{\lambda}^n$
    Solve the velocity-pressure problem by an AL Uzawa algorithm: $k \geq 0$
    · Solve the Stokes problem: find $\boldsymbol{u}^{n+1,k+1}$, $p^{n+1,k+1}$ such that

$$\rho\frac{\boldsymbol{u}^{n+1,k+1}}{\Delta t} - \nabla\cdot\left(\left(2(c^n\mu_1 + (1-c^n)\mu_2) + r\right)\boldsymbol{D}(\boldsymbol{u}^{n+1,k+1})\right) + \nabla p^{n+1,k+1} = \nabla\cdot\left(\boldsymbol{\lambda}^{n+1,k} - r\boldsymbol{d}^{n+1,k}\right) - \rho\boldsymbol{u}^{n+1,k}\cdot\nabla\boldsymbol{u}^{n+1,k} + \rho\frac{\boldsymbol{u}^n}{\Delta t} \quad \text{in } \Omega \tag{116}$$

$$\nabla\cdot\boldsymbol{u}^{n+1,k+1} = 0 \quad \text{in } \Omega \tag{117}$$

$$\boldsymbol{u}^{n+1,k+1} = \boldsymbol{g} \quad \text{on } \partial\Omega \tag{118}$$

    · Evaluate the rate-of-strain tensor $\boldsymbol{d}^{n+1,k+1}$

$$\boldsymbol{d}^{n+1,k+1} = \begin{cases} 0 \ \text{ if } \|\boldsymbol{\lambda}^{n+1,k} + r\boldsymbol{D}(\boldsymbol{u}^{n+1,k+1})\| < c^n\tau_{y1} + (1-c^n)\tau_{y2} \\ \dfrac{1}{r}\left(1 - \dfrac{\tau_y}{\|\boldsymbol{\lambda}^{n+1,k} + r\boldsymbol{D}(\boldsymbol{u}^{n+1,k+1})\|}\right)\left(\boldsymbol{\lambda}^{n+1,k} + r\boldsymbol{D}(\boldsymbol{u}^{n+1,k+1})\right) \\ \text{if not} \end{cases} \tag{119}$$

    · Update the Lagrange multipliers $\boldsymbol{\lambda}^{n+1,k+1}$

$$\boldsymbol{\lambda}^{n+1,k+1} = \boldsymbol{\lambda}^{n+1,k} + r(\boldsymbol{D}(\boldsymbol{u}^{n+1,k+1}) - \boldsymbol{d}^{n+1,k+1}) \tag{120}$$

    · *Convergence if*

$$\Delta(\boldsymbol{u}, p)_{k+1} = \max_{\Omega}|\boldsymbol{u}^{n+1,k+1} - \boldsymbol{u}^{n+1,k}| + \max_{\Omega}|p^{n+1,k+1} - p^{n+1,k}| \leq \epsilon_1 \tag{121}$$

$$\Delta\boldsymbol{D}_{k+} = \max_{\Omega}|\boldsymbol{D}(\boldsymbol{u}^{n+1,k+1}) - \boldsymbol{d}^{n+1,k+1}| \leq \epsilon_2 \tag{122}$$

    Set: $\boldsymbol{u}^{n+1} = \boldsymbol{u}^{n+1,k+1}$, $p^{n+1} = p^{n+1,k+1}$, $\boldsymbol{d}^{n+1} = \boldsymbol{d}^{n+1,k+1}$, $\boldsymbol{\lambda}^{n+1} = \boldsymbol{\lambda}^{n+1,k+1}$.
  Solve the colour function equation

$$\frac{c^{n+1}}{\Delta t} - D_c\nabla^2 c^{n+1} = -\boldsymbol{u}^{n+1}\cdot\nabla c^n + \frac{c^n}{\Delta t} \tag{123}$$

- END

4.4 Compressibility

When the flow is slightly compressible, the full mass conservation equation (2) has to be solved. For the sake of clarity, we recall it here:

$$\frac{\partial \rho}{\partial t} + \nabla \cdot \rho \boldsymbol{u} = 0$$

If we assume that the density ρ is a function of pressure p only, we can rewrite the mass conservation equation as follows:

$$\chi_\Theta \left(\frac{\partial p}{\partial t} + \boldsymbol{u} \cdot \nabla p \right) + \nabla \cdot \boldsymbol{u} = 0 \tag{124}$$

where χ_Θ is the isothermal compressibility coefficient defined as:

$$\chi_\Theta = \frac{1}{\rho} \left(\frac{\partial \rho}{\partial p} \right)_\Theta \tag{125}$$

Also, the Bingham constitutive equation is slightly altered to reflect the effect of compressibility on the plastic viscosity. In fact, the compressible Bingham model reads as follows:

$$\begin{cases} \boldsymbol{\tau} = 2\mu(\boldsymbol{D} - \frac{1}{3}(\nabla \cdot \boldsymbol{u})\boldsymbol{I}) + \frac{\boldsymbol{D}}{\|\boldsymbol{D}\|}\tau_y & \text{if } \|\boldsymbol{\tau}\| > \tau_y \\ \boldsymbol{D} = 0 & \text{if } \|\boldsymbol{\tau}\| \leq \tau_y \end{cases} \tag{126}$$

In this case, a simple weak-coupling solution algorithm cannot be constructed as the full Navier-Stokes problem (mass and momentum conservation) is altered. Hence we need to do a bit more than simple weak-coupling. Actually, we can construct a weakly compressible variant of ALG2 by:

1. writing the discrete in time form of the problem using a first-order backward Euler scheme for the time derivative and treating advective terms explicitly (although include them in the AL saddle point iterative procedure is also conceivable as shown in Sect. 4.3),
2. introducing a slightly different functional than (37)

In fact, the appropriate functional reads as follows:

$$\begin{aligned} \tilde{\mathcal{L}}_r(\boldsymbol{v}, q, \boldsymbol{q}, \boldsymbol{\alpha}) = & \frac{1}{2} \int_\Omega \frac{\rho}{\Delta t} \boldsymbol{v}^2 dx + \int_\Omega \mu |\boldsymbol{D}(\boldsymbol{v})|^2 dx + \int_\Omega \sqrt{2}\tau_y |\boldsymbol{q}| dx \\ & - \int_\Omega q \nabla \cdot \boldsymbol{v} dx - \int_\Omega \boldsymbol{f} . \boldsymbol{v} dx - \frac{1}{3} \int_\Omega |\nabla \cdot \boldsymbol{v}|^2 dx + \frac{1}{2} \int_\Omega \frac{\chi_\Theta}{\Delta t} q^2 dx \\ & \int_\Omega h q dx + \int_\Omega \boldsymbol{\alpha} . (\boldsymbol{D}(\boldsymbol{v}) - \boldsymbol{q}) dx + \frac{r}{2} \int_\Omega |\boldsymbol{D}(\boldsymbol{u}) - \boldsymbol{q}|^2 dx \end{aligned} \tag{127}$$

where $\boldsymbol{f} = \frac{\rho}{\Delta t}\boldsymbol{u}^n - \rho\boldsymbol{u}^n \cdot \nabla\boldsymbol{u}^n$ and $h = \frac{\chi_\Theta}{\Delta t}p^n - \chi_\Theta\boldsymbol{u}^n \cdot \nabla p^n$. Now writing the optimality conditions in terms of each variable, we recover the conservation and constitutive equations. In a similar fashion to the derivation of ALG2, we can construct a compressible variant of ALG2 as follows:

- Initialisation of $\boldsymbol{u}^0$, p^0, $\boldsymbol{d}^0$, $\boldsymbol{\lambda}^0$
- Time loop $t^{n+1} = t^n + \Delta t, n \geq 0$
  - Initialisation : $\boldsymbol{u}^{n+1,0} = \boldsymbol{u}^n$, $p^{n+1,0} = p^n$, $\boldsymbol{d}^{n+1,0} = \boldsymbol{d}^n$ and $\boldsymbol{\lambda}^{n+1,0} = \boldsymbol{\lambda}^n$
  - Solve the velocity-pressure problem by an AL Uzawa algorithm: $k \geq 0$
    Solve the Stokes problem: find $\boldsymbol{u}^{n+1,k+1}$, $p^{n+1,k+1}$ such that

$$\begin{aligned}\rho\frac{\boldsymbol{u}^{n+1,k+1}}{\Delta t} - \nabla\cdot\left(\left(2\mu + r\right)\boldsymbol{D}(\boldsymbol{u}^{n+1,k+1}) - \frac{2}{3}(\nabla\cdot\boldsymbol{u}^{n+1,k+1})\boldsymbol{I}\right)&\\ + \nabla p^{n+1,k+1} = \nabla\cdot\left(\boldsymbol{\lambda}^{n+1,k} - r\boldsymbol{d}^{n+1,k}\right)&\\ - \rho\boldsymbol{u}^n\cdot\nabla\boldsymbol{u}^n + \rho\frac{\boldsymbol{u}^n}{\Delta t} \quad \text{in } \Omega&\end{aligned} \tag{128}$$

$$\frac{\chi_\Theta}{\Delta t}p^{n+1,k+1} + \nabla\cdot\boldsymbol{u}^{n+1,k+1} = -\chi_\Theta\boldsymbol{u}^n\cdot\nabla p^n + \frac{\chi_\Theta}{\Delta t}p^n \quad \text{in } \Omega \tag{129}$$

$$\boldsymbol{u}^{n+1,k+1} = \boldsymbol{g} \quad \text{on } \partial\Omega \tag{130}$$

Evaluate the rate-of-strain tensor $\boldsymbol{d}^{n+1,k+1}$

$$\boldsymbol{d}^{n+1,k+1} = \begin{cases} 0 \;\; \text{if } \|\boldsymbol{\lambda}^{n+1,k} + r\boldsymbol{D}(\boldsymbol{u}^{n+1,k+1})\| < \tau_y \\ \frac{1}{r}\left(1 - \frac{\tau_y}{\|\boldsymbol{\lambda}^{n+1,k} + r\boldsymbol{D}(\boldsymbol{u}^{n+1,k+1})\|}\right)\left(\boldsymbol{\lambda}^{n+1,k} + r\boldsymbol{D}(\boldsymbol{u}^{n+1,k+1})\right) \\ \text{if not} \end{cases} \tag{131}$$

Update the Lagrange multipliers $\boldsymbol{\lambda}^{n+1,k+1}$

$$\boldsymbol{\lambda}^{n+1,k+1} = \boldsymbol{\lambda}^{n+1,k} + r(\boldsymbol{D}(\boldsymbol{u}^{n+1,k+1}) - \boldsymbol{d}^{n+1,k+1}) \tag{132}$$

Convergence if

$$\Delta(\boldsymbol{u}, p)_{k+1} = \max_{\Omega}|\boldsymbol{u}^{n+1,k+1} - \boldsymbol{u}^{n+1,k}| + \max_{\Omega}|p^{n+1,k+1} - p^{n+1,k}| \leq \epsilon_1 \tag{133}$$

$$\Delta\boldsymbol{D}_{k+} = \max_{\Omega}|\boldsymbol{D}(\boldsymbol{u}^{n+1,k+1}) - \boldsymbol{d}^{n+1,k+1}| \leq \epsilon_2 \tag{134}$$

  - Set: $\boldsymbol{u}^{n+1} = \boldsymbol{u}^{n+1,k+1}$, $p^{n+1} = p^{n+1,k+1}$, $\boldsymbol{d}^{n+1} = \boldsymbol{d}^{n+1,k+1}$, $\boldsymbol{\lambda}^{n+1} = \boldsymbol{\lambda}^{n+1,k+1}$.
- END

Similarly to the incompressible case, the underlying theory may appear a little complicated by the resulting algorithm is pretty simple to implement. Only the Stokes problem (step 1 of ALG2) is slightly different but is not really tougher to solve. Actually, if we write the Stokes problem at the matrix level, it reads as follows:

$$\begin{pmatrix} \boldsymbol{A} & \boldsymbol{B} \\ \boldsymbol{B}^t & \boldsymbol{C} \end{pmatrix} \begin{pmatrix} \boldsymbol{u} \\ p \end{pmatrix} = \begin{pmatrix} \tilde{\boldsymbol{f}} \\ \tilde{h} \end{pmatrix} \tag{135}$$

where $\boldsymbol{A}$ is a $N \times N$ square symmetric positive definite matrix, $\boldsymbol{B}$ a $N \times M$ rectangular matrix, $\boldsymbol{C}$ a $M \times M$ diagonal matrix and $(\boldsymbol{u}, \tilde{\boldsymbol{f}}) \in \mathbb{R}^N$ and $(p, \tilde{h}) \in \mathbb{R}^M$.

If χ_Θ is set to 0, the matrix $\boldsymbol{C}$ is 0 and we recover the standard incompressible case that can be solved as a saddle point problem (with $\nabla \cdot \boldsymbol{u} = 0$ as the constraint and p as the Lagrange multiplier) using a PCG/Uzawa algorithm. This PCG/Uzawa algorithm can be easily generalized to a non-zero $\boldsymbol{C}$. For instance, the general PCG/Uzawa algorithm suggested in Vinay et al. (2007) is valid both when $\boldsymbol{C}$ is zero (incompressible case) and when $\boldsymbol{C}$ is non-zero. This is very practical as the same solution algorithm can be used for both compressible and incompressible cases. It allows as well to address fluid/fluid flows in which a phase is compressible and the other phase is incompressible, by combining the above solution algorithm to a VOF method and a weak coupling approach.

The various algorithms presented above apply to a large variety of flows. The reader is now equipped with the necessary material to compute many different viscoplastic flows with additional physical effects. Although we presented AL type algorithms, note that the idea of weak coupling can be applied to regularized models too.

5 How to Capture the No-Flow Limit and Finite Time Decay in Single and Two-Phase Flows

Among the main singular properties of viscoplastic fluid flows, the limit of no-flow when the applied stresses do not exceed the yield stress (static) or the return to rest in finite time in a flow where the applied stresses vary from above to below the yield stress (dynamic) is an important property that any efficient computational method is expected to properly capture.

Regularized models transform the actual viscoplastic problem into a purely viscous problem with a high viscosity in unyielded regions. Because the problem is now purely viscous, fully quiescent situations or finite time decay can not be approximated properly. This necessitates to introduce a certain criterion to decide that a low velocity or low rate-of-strain tensor (in magnitude) well represents a quiescent flow, and this criterion enables one to determine a time that approximates finite-time decay. In general, regularized models are not good candidates for no-flow and finite time decay, even in a simple situation as the flow in a pipeline in which the pressure is suddenly suppressed. It is even more questionable in more challenging situations. Conversely, AL type solution methods solve the actual Bingham model and possess all the nice properties to compute reliably and accurately no-flow and finite time decay.

In the following we illustrate these nice properties in the case of (i) natural convection in a viscoplastic flow and (ii) settling of a solid particle in a viscoplastic material. In both cases, the solution method is required to be implicit (or as much implicit as possible). The backward Euler scheme has proven to be particular suited to that task. In case (ii), we show that standard weak coupling strategies are not well adapted as they do not solve the right force balance on the particle and that a particular solution method is required.

5.1 Natural Convection in a Viscoplastic Flow

We illustrate here the ability to compute finite time decay resulting in a fully quiescent flow using an ALG2-weak coupling strategy to solve the coupled mass, momentum and temperature equations in a square cavity filled with a viscoplastic fluid. Results have been published in Karimfazli et al. (2015) and we acknowledge the courtesy of the authors to use these results in this course.

The problem can be written in the usual dimensionless form involving the following dimensionless numbers:

- Rayleigh number: $Ra = \dfrac{\rho_0^2 C_p g \beta \Delta T L^3}{\mu \kappa} = Gr.Pr$
- Prandtl number: $Pr = \dfrac{\mu C_p}{\kappa}$
- Bingham number: $B = \dfrac{\tau_y}{\rho_0 g \beta \Delta T L}$

The problem is illustrated in Fig. 7. In short, the main feature of that problem is that there exists a critical Bingham number B_{cr} above which the buoyant stresses cannot overcome the yield stress and the flow is quiescent (stationary). B_{cr} can be determined both analytically and computationally to be 0.03125. Interestingly, B_{cr} is independent of Ra and Pr. For $Pr \in [1:100]$, varying Ra and $B < B_{cr}$ leads to different flow patterns with variable flow intensity and heat transfer at the wall (measured dimensionlessly by a Nusselt number Nu), as plotted in Fig. 8.

Now the main asset of our solution algorithm is the ability to compute properly finite time decay. The model situation is as follows: we initialize the flow with a flow field for $B < B_{cr}$, and then at $t = 0^+$ we set $B > B_{cr}$. In this situation, we expect that the flow returns to rest (quiescent/stationary flow) in finite time. Figure 9 shows the predictions of our solution method for various $B > B_{cr}$. The flow indeed remarkably returns to rest in a finite time and once the flow is suppressed the temperature profile relaxes to the conductive limit.

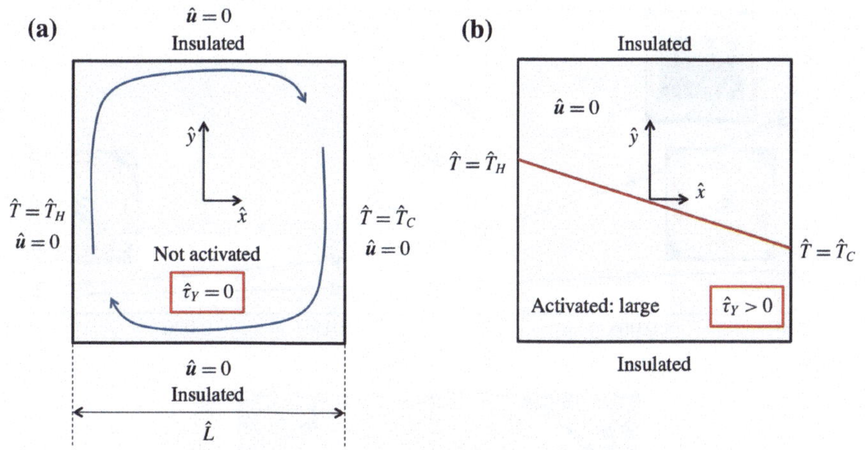

Fig. 7 Natural convection in a viscoplastic fluid: domain geometry and boundary conditions

5.2 *Settling of a Solid Particle in a Viscoplastic Fluid*

A second illustration pertains to the settling of a heavier solid particle in a yield stress fluid. In a Newtonian fluid, regardless of the density ratio and the viscosity of the fluid, if the particle is heavier than the fluid, the particle settles. Actually, in Stokes flow (i.e., without inertia), the settling velocity v_s of a single particle in an unbounded domain filled with a Newtonian fluid can be analytically predicted and is given by the following equation:

$$v_s = \frac{1}{18}\frac{(\rho_s - \rho_f)gd^2}{\mu} \tag{136}$$

where ρ_s is the solid density, ρ_f the fluid density and d the particle diameter. In a viscoplastic fluid, it is quite intuitive that the yield stress will resist the particle settling and hence there exists a critical ratio between the net weight and the yield stress resistance under which the particle does not settle anymore and the fluid is quiescent. In other words, if the particle net weight does not overcome the yield stress resistance, the fluid and the particle are motionless.

To get no-flow or return to rest in finite time, we need to solve the particle force balance accurately. This involves getting the balance between net weight and yield stress resistance implicitly. To achieve that, a simple weak coupling or operator-splitting approach of the following form:

1. solve the fluid problem,
2. using the velocity and pressure fields, compute the hydrodynamic force and solve the particle problem.

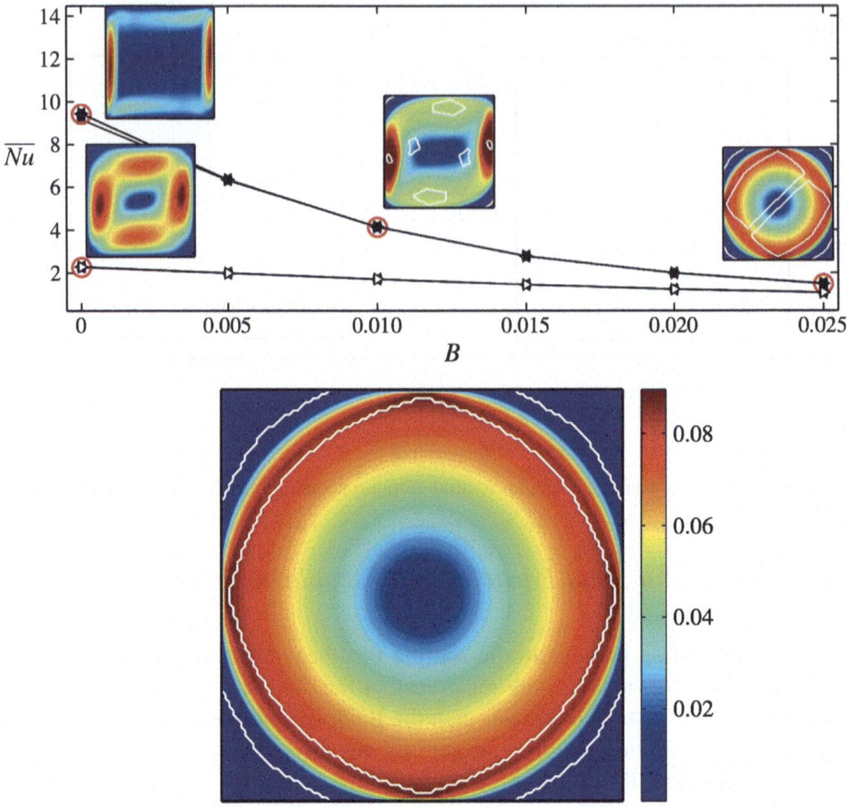

Fig. 8 Natural convection in a viscoplastic fluid: flow pattern and heat transfer at the wall as a function of Pr, Ra and $B < B_{cr}$. For $B \to B_{cr}$, the flow is close to a rigid body motion with a thin boundary layer of sheared viscoplastic fluid

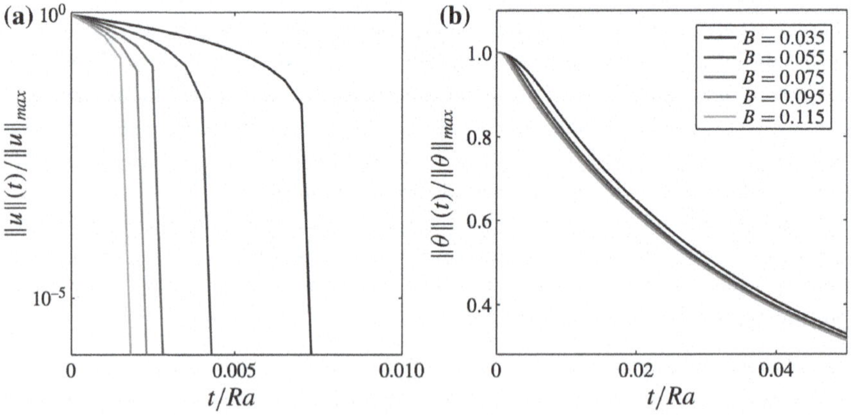

Fig. 9 Natural convection in a viscoplastic fluid: finite time decay

is not appropriate and we need to do something a little more advanced, i.e., to solve the problem in a more coupled way.

At the spatial level, we use a fixed-grid Distributed Lagrange multiplier/Fictitious domain (DLM/FD) approach to impose the rigid body motion $\boldsymbol{u} = \boldsymbol{U} + \boldsymbol{\omega} \times \boldsymbol{r}$ inside the region occupied by the particle $P(t)$. This constraint is relaxed by introducing a Lagrange multiplier field λ_m. To solve the viscoplastic flow, we use our favorite ALG2 algorithm but we embed the DLM/FD problem into the saddle point iterative process. We only decouple the pressure contribution to the force exerted on the particle as solving the fully coupled problem is very complicated. The resulting solution method is a combination of a DLM/FD-ALG2 algorithm to a L2-projection method to enforce mass conservation.

Using the particle diameter d as the characteristic length scale and $U_c = \frac{(\rho_s-\rho_f)gd^2}{\mu}$ as the characteristic velocity, d/U_c as the characteristic time scale and $\mu U_c/d$ as the characteristic pressure and stress scale, the problem can be written in the usual dimensionless form involving the following dimensionless numbers:

- Reynolds number: $Re = \dfrac{\rho_f U_c d}{\mu}$
- density ratio: $\rho_r = \dfrac{\rho_s}{\rho_f} > 1$
- Bingham number: $B = \dfrac{\tau_y}{(\rho_s - \rho_f)gd}$

Introducing $|P|$ the particle scaled volume, $\boldsymbol{J}$ the particle scaled inertia tensor and $\boldsymbol{e}_g$ the unit vector in the direction of gravity, the solution algorithm reads as follows:

- Initialisation of $\boldsymbol{u}^0$, p^0, $\boldsymbol{d}^0$, $\boldsymbol{U}^0$, $\boldsymbol{\omega}^0$, $\boldsymbol{\lambda}^0$, $\boldsymbol{\lambda_m}^0$
- Time loop $t^{n+1} = t^n + \Delta t, n \geq 0$
  - Initialisation of: $\boldsymbol{u}^{n+1/2,0} = \boldsymbol{u}^n$, $\boldsymbol{d}^{n+1,0} = \boldsymbol{d}^n$, $\boldsymbol{U}^{n+1,0} = \boldsymbol{U}^n$, $\boldsymbol{\omega}^{n+1,0} = \boldsymbol{\omega}^n$, $\boldsymbol{\lambda}^{n+1,0} = \boldsymbol{\lambda}^n$, $\boldsymbol{\lambda_m}^{n+1,0} = \boldsymbol{\lambda_m}^n$
  - Solve the viscoplastic-DLM/FD coupled problem with the following AL algorithm: $k \geq 0$

    Solve the following DLM/FD problem:
    find $\boldsymbol{u}^{n+1/2,k+1}$, $\boldsymbol{U}^{n+1,k+1}$, $\boldsymbol{\omega}^{n+1,k+1}$, $\boldsymbol{\lambda}_m^{n+1,k+1}$ such that:

$$
\begin{aligned}
Re\frac{\boldsymbol{u}^{n+1/2,k+1} - \boldsymbol{u}^n}{\Delta t} + \nabla p^n - (2+r)\nabla \cdot \boldsymbol{D}(\boldsymbol{u}^{n+1/2,k+1}) + \boldsymbol{\lambda}_m^{n+1,k+1} = \\
- Re(\boldsymbol{u}^n . \nabla \boldsymbol{u}^n) + \nabla \cdot (\boldsymbol{\lambda}^{n+1,k} - r\boldsymbol{d}^{n+1,k})
\end{aligned} \tag{137}
$$

$$
\left((\rho_r - 1) Re|P| \frac{\boldsymbol{U}^{n+1,k+1} - \boldsymbol{U}^n}{\Delta t} - |P|\boldsymbol{e}_g \right) - \int_P \boldsymbol{\lambda}_m^{n+1,k+1} d\boldsymbol{x} = \boldsymbol{0} \tag{138}
$$

$$
(\rho_r - 1) Re\boldsymbol{J}\frac{\boldsymbol{\omega}^{n+1,k+1} - \boldsymbol{\omega}^n}{\Delta t} - \int_P \boldsymbol{r} \times \boldsymbol{\lambda}_m^{n+1,k+1} . d\boldsymbol{x} = -(\rho_r - 1) Re\boldsymbol{\omega}^n \times \boldsymbol{J}\boldsymbol{\omega}^n \tag{139}
$$

$$
\boldsymbol{u}^{n+1/2,k+1} - (\boldsymbol{U}^{n+1,k+1} + \boldsymbol{\omega}^{n+1,k+1} \times \boldsymbol{r}) = \boldsymbol{0} \; in \; P(t) \tag{140}
$$

Update $\boldsymbol{d}^{n+1,k+1}$ by:

$$\boldsymbol{d}^{n+1,k+1} = \begin{cases} 0 \quad \text{if } |\boldsymbol{\lambda}^{n+1,k} + r\boldsymbol{D}(\boldsymbol{u}^{n+1/2,k+1})| < B \\ \dfrac{1}{r}\left(1 - \dfrac{B}{|\boldsymbol{\lambda}^{n+1,k} + r\boldsymbol{D}(\boldsymbol{u}^{n+1/2,k+1}))|}\right)\left(\boldsymbol{\lambda}^{n+1,k} + r\boldsymbol{D}(\boldsymbol{u}^{n+1/2,k+1}))\right) \\ \text{if not} \end{cases} \tag{141}$$

Update the Lagrange multipliers $\boldsymbol{\lambda}_b^{n+1,k+1}$

$$\boldsymbol{\lambda}^{n+1,k+1} = \boldsymbol{\lambda}^{n+1,k} + r(\boldsymbol{D}(\boldsymbol{u}^{n+1/2,k+1}) - \boldsymbol{d}^{n+1,k+1}) \tag{142}$$

Convergence if

$$\Delta(\boldsymbol{u}, p)^{k+1} = \max_{\Omega} |\boldsymbol{u}^{n+1/2,k+1} - \boldsymbol{u}^{n+1/2,k}| \leq \epsilon_1 \tag{143}$$

$$\Delta \boldsymbol{D}^{k+1} = \max_{\Omega} |\boldsymbol{D}(\boldsymbol{u}^{n+1/2,k+1}) - \boldsymbol{d}^{n+1,k+1}| \leq \epsilon_2 \tag{144}$$

- Set: $\boldsymbol{u}^{n+1/2} = \boldsymbol{u}^{n+1/2,k+1}$, $\boldsymbol{U}^{n+1} = \boldsymbol{U}^{n+1,k+1}$, $\boldsymbol{\omega}^{n+1} = \boldsymbol{\omega}^{n+1,k+1}$, $\boldsymbol{d}^{n+1} = \boldsymbol{d}^{n+1,k+1}$, $\boldsymbol{\lambda}^{n+1} = \boldsymbol{\lambda}^{n+1,k+1}$, $\boldsymbol{\lambda_m}^{n+1} = \boldsymbol{\lambda_m}^{n+1,k+1}$
- Solve the degenerated Stokes problem to enforce fluid velocity
- divergence free: find $\boldsymbol{u}^{n+1}$ and p^{n+1} such that

$$Re\frac{\boldsymbol{u}^{n+1} - \boldsymbol{u}^{n+1/2}}{\Delta t} + \nabla(p^{n+1} - p^n) = \boldsymbol{0} \tag{145}$$

$$\nabla \cdot \boldsymbol{u}^{n+1} = 0 \tag{146}$$

- END

As an illustration of how the algorithm performs, we consider the following simple 2D test case: a single circular particle settling in a 4:1 channel. In an unbounded domain, we can derive an analytical estimate of the critical Bingham number $B_{cr} \simeq 0.0658$ above which motion is suppressed. We initialize the flow field at $t = 0$ to the Newtonian flow field, i.e., for $B = 0$, and at $t = 0^+$ we set $B > B_{cr}$. Figure 10 shows the flow pattern of the return to rest of the flow. Although there are some fascinating features in this flow, our goal is here mainly to comment on the performance of the solution algorithm. In Fig. 11, we plot the time evolution of the settling velocity. It is remarkable that the classical weak coupling solution method never predicts a fully zero particle settling velocity, regardless of the magnitude of the time step. Obviously, reducing the time step magnitude reduces the magnitude of the error but qualitatively the intrinsic error results from the decoupling. In constrast, the semi-implicit solution algorithm captures the return to rest in finite time with a residual settling velocity of the order of the convergence criteria. In addition, the rate-of-strain is fully zero everywhere in the flow domain, which enables one to undoubtedly determine that the flow is back to a quiescent situation.

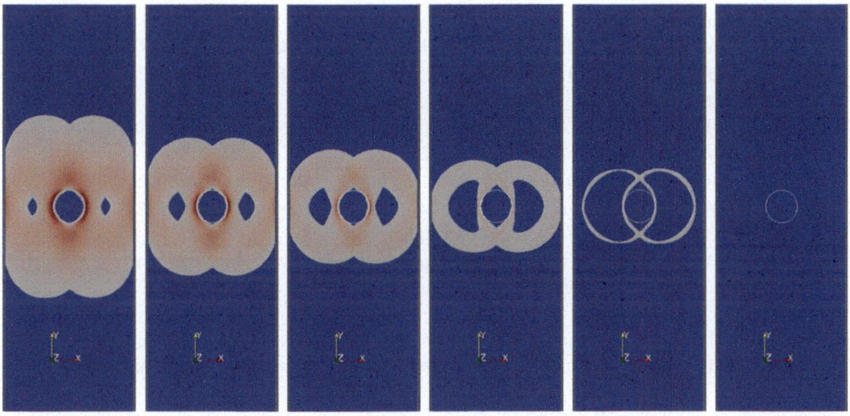

$t/\mathrm{Re} = 0.012$ $t/\mathrm{Re} = 0.050$ $t/\mathrm{Re} = 0.100$ $t/\mathrm{Re} = 0.160$ $t/\mathrm{Re} = 0.190$ $t/\mathrm{Re} = 0.200$

Fig. 10 Circular cylinder settling in a viscoplastic fluid: flow pattern of the return to rest in finite time

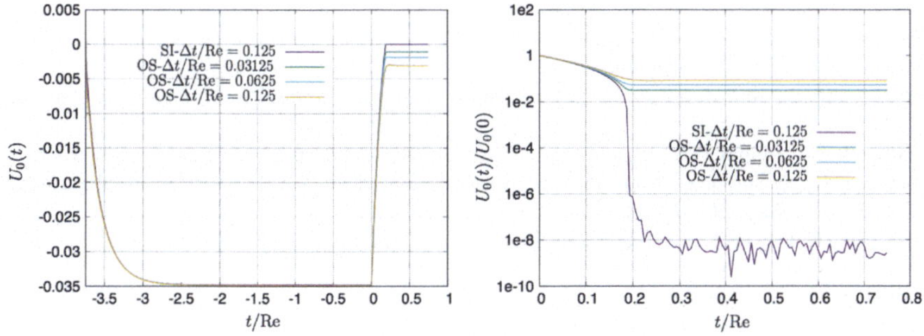

Fig. 11 Circular cylinder settling in a viscoplastic fluid: performance of our semi-implicit solution method versus a classical weak coupling solution method

References

Allouche, M., Frigaard, I. A., & Sona, G. (2000). Static wall layers in the displacement of two visco-plastic fluids in a plane channel. *Journal of Fluid Mechanics*, *424*, 243–277.

Barnes, H. A. (1999). The yield stress - a review or π-everything flows? *Journal of Non-Newtonian Fluid Mechanics*, *81*, 133–178.

Barnes, H. A., & Walters, K. (1985). The yield stress myth? *Rheologica Acta*, *24*, 324–326.

Bercovier, M., & Engleman, M. (1980). A finite-element method for incompressible non-Newtonian flows. *Journal of Computational Physics*, *36*, 313–326.

Beverly, C. R., & Tanner, R. I. (1992). Numerical analysis of three dimensional bingham plastic flow. *Journal of Non-Newtonian Fluid Mechanics*, *42*, 82–115.

Brezzi, F., Boffi, D., Demkowicz, L., Durán, R. G., Falk, R. S., & Fortin, M. (2008). *Mixed finite elements, compatibility conditions, and applications*. Springer.

Bristeau, M. O., & Glowinski, R. (1974). Finite element analysis of the unsteady flow of a visco-plastic fluid in a cylindrical pipe. In J. T. Oden, O. C. Zienkiewicz, R. H. Gallagher, & C. Taylor (Eds.), *Finite Element Methods in Flow Problems* (pp. 471–488). Hunstville: University of Alabama Press.

Derksen J. J., & Prashant. (2009). Simulations of complex flow of thixotropic liquids. *Journal of Non-Newtonian Fluid Mechanics*, *160*(2), 65–75.

Duvaut, G., & Lions, J. L. (1976). *Inequalities in mechanics and physics*. Berlin: Springer.

Fortin, M., & Glowinski, R. (1982). *Méthodes de Lagrangien augmenté*. Dunod-Bordas: Application à la résolution numérique de problèmes aux limites.

Fortin, M., & Glowinski, R. (1983). *Augmented Lagrangian methods: Application to the numerical solution of boundary value problems*. Amsterdam: North-Holland.

FreeFem++, http://www.freefem.org/ff++.

Glowinski, R. (2003). *Finite element methods for incompressible viscous flow*. In P. G. Ciarlet, & J. L. Lions (Eds.), *Handbook of Numerical Analysis* (Vol. IX, pp. 3–1176). Amsterdam: North-Holland.

Glowinski, R., & Le Tallec, P. (1989). *Augmented Lagrangian and operator-splitting methods in nonlinear mechanics*. SIAM Studies in Applied Mathematics.

Glowinski, R. (1974). Sur l'écoulement d'un fluide de Bingham dans une conduite cylindrique. *Journal de Mécanique*, *13*(4), 601–621.

Glowinski, R., & Wachs, A. (2011). On the numerical simulation of viscoplastic fluid flow. *Handbook of Numerical Analysis*, *16*(1), 483–718.

Glowinski, R., Lions, J. L., & Tremolieres, R. (1976). *Analyse Numérique des inéquations variationnelles* (Vol. 2). Paris: Dunod.

Hecht, F. (2012). New development in FreeFem++. *Journal of Numerical Mathematics*, *20*(3–4), 251–265.

HYPRE, http://computation.llnl.gov/projects/hypre-scalable-linear-solvers-multigrid-methods.

Ionescu, I. R., & Sofonea, M. (1986). The blocking property in the study of the Bingham fluid. *International Journal of Engineering Science*, *24*, 289–297.

Karimfazli, I., Frigaard, I. A., & Wachs, A. (2015). A novel heat transfer switch using the yield stress. *Journal of Fluid Mechanics*, *783*, 526–566.

Mohamad, A. A. (2011). *Lattice Boltzmann method: Fundamentals and engineering applications with computer codes*. Springer Science & Business Media.

MUMPS, http://mumps.enseeiht.fr/.

Papanastasiou, T. C. (1987). Flow of materials with yield. *Journal of Rheology*, *31*, 385–404.

PETSc, http://www.mcs.anl.gov/petsc/.

Prashant, & Derksen, J. J. (2011). Direct simulations of spherical particle motion in bingham liquids. *Computers & Chemical Engineering*, *35*(7), 1200–1214.

Roquet, N., & Saramito, P. (2003). An adaptive finite element method for Bingham fluid flows around a cylinder. *Computational Methods Applied Mechanics and Engineering*, *192*, 3317–3341.

Roquet, N., & Saramito, P. (2008). An adaptive finite element method for viscoplastic flows in a square pipe with stick-slip at the wall. *Journal of Non-Newtonian Fluid Mechanics*, *155*(3), 101–115.

Saramito, P. (2015). *Efficient C++ finite element computing with Rheolef*. CNRS-CCSD ed., http://cel.archives-ouvertes.fr/cel-00573970.

Saramito, P. (2016) A damped Newton algorithm for computing viscoplastic fluid flows. *Journal of Non-Newtonian Fluid Mechanics, 238*, 6–15.

Saramito, P., & Roquet, N. (2001). An adaptive finite element method for viscoplastic fluid flows in pipes. *Computational Methods Applied Mechanics and Engineering*, *190*, 5391–5412.

Szabo, P., & Hassager, O. (1992). Flow of viscoplastic fluids in eccentric annular geometries. *Journal of Non Newtonian Fluid Mechanics*, *45*, 149–169.

Treskatis, T., Moyers-Gonzalez, M. A., & Price, C. J. (2016). An accelerated dual gradient method for applications in viscoplasticity. *Journal of Non-Newtonian Fluid Mechanics, 238*, 115–130.

Vinay, G., Wachs, A., & Agassant, J. F. (2007). Numerical simulation of weakly compressible bingham flows: the restart of pipeline flows of waxy crude oils. *Journal of Non Newtonian Fluid Mechanics*, *145*, 78–91.

Rheometry of Visco-Plastic Fluids

Guillaume Ovarlez

1 Introduction

This chapter deals with experimental methods for the measurement of the rheological properties of materials, with a focus on yield stress fluids. It is based on Ovarlez (2011). Section 2 provides the basis of rheometry. Standard equations for viscosimetric flows are given together with the main hypotheses used in their derivation. Their application to visco-plastic fluids is discussed. The third section deals with the measurements of the macroscopic properties of simple yield stress fluids. Several methods are given to measure their solid viscoelastic properties, their yield stress, and their visco-plastic properties. The fourth section is devoted to the measurements of the properties of thixotropic yield stress fluids. Appropriate procedures are proposed to characterize the dependence of their behaviour on shear history and on their aging at rest. The concept of critical shear rate is introduced. The phenomena of shear localisation and shear banding are finally discussed in the fifth section, together with possible material inhomogeneities. Their impact on rheological measurements is also commented.

G. Ovarlez (✉)
Université de Bordeaux, CNRS, Solvay, LOF - UMR 5258, 33608 Pessac, France
e-mail: guillaume.ovarlez@u-bordeaux.fr

G. Ovarlez and S. Hormozi (eds.), *Lectures on Visco-Plastic Fluid Mechanics*,
CISM International Centre for Mechanical Sciences 583,
https://doi.org/10.1007/978-3-319-89438-6_4

2 Rheometry

Simple shear. In order to predict the behaviour of materials in any complex flow, which a priori involves strains in all directions of space, it is necessary to know the relationship between the deviatoric stress tensor τ_{ij} and the strain rate tensor d_{ij} (which is, for example, $\tau_{ij} = 2\eta d_{ij}$ for a Newtonian fluid of viscosity η)[1]. Such a tensional relationship is hard to derive from experimental measurements. The objective of rheometry is to simplify the problem as much as possible by studying the material flow properties in a case where there is only one non-zero component of the strain rate tensor, namely simple shear. Another simple case is extensional flow, which provides a complementary means of characterisation of materials but is not addressed here (see Macosko 1994). Note that other methods have been recently developed to investigate the material behaviour in controlled three-dimensional flows (Jop et al. 2006; Ovarlez et al. 2010).

Thanks to its simplicity, simple shear allows one to characterise and understand the flow behaviour of complex fluids. Moreover, such characterisation is sufficient to predict the flow behaviour in many practical situations (pipe flows, flows down a slope, etc.). Nevertheless, in more complex cases, the derived behaviour still has to be generalised in a tensorial form.

In simple shear (Fig. 1), shear occurs through the relative motion of parallel layers of the material. Noting x the direction of flow and y the direction of the spatial variations of the velocity $V_x(y)$, the only non-zero component of the strain rate tensor in simple shear is $d_{xy} = d_{yx}$. It is then convenient to define the shear rate as $\dot{\gamma} = 2d_{xy} = dV_x/dy$ which is thus the velocity gradient. If $\dot{\gamma}$ is constant then simple shear is said to be homogeneous.

Three components of the stress tensor can be non-zero in simple shear: the shear stress $\tau = \tau_{xy}$ and the first and second normal stress differences $N_1 = \tau_{xx} - \tau_{yy}$, $N_2 = \tau_{yy} - \tau_{zz}$. The purpose of rheometry is to characterise the dependence of the shear stress τ and, when relevant, of the normal stress differences, on the shear strain γ and/or the shear rate $\dot{\gamma}$ in simple shear.

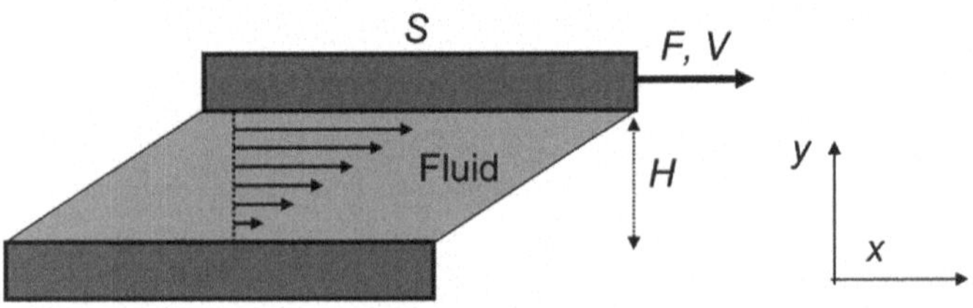

Fig. 1 Schematic representation of a simple shear experiment

[1] For the sake of simplicity, the discussion in this section is based on the case of simple constitutive laws of the form $\tau_{ij} = f(d_{ij})$. This can be easily generalised to cases where the behaviour also depends on the strain tensor and on the strain history.

Homogeneous simple shear is theoretically achieved by putting a sample of the material between two infinite parallel plates separated by a distance H (named 'gap') and imposing a difference of velocity V between the two plates. The material behaviour is then characterised by measuring the relationship between two macroscopic quantities: the force $F(V)$ that has to be applied to enforce a flow characterised by the velocity V. In the ideal conditions of the homogeneous simple shear of Fig. 1, the shear rate and the shear stress are obtained through $\dot{\gamma} = V/H$ and $\tau = F/S$ (S being the plates' surface, supposed to tend to infinity). The macroscopic measurements $F(V)$ then provide directly the constitutive law $\tau(\dot{\gamma})$. In some materials, a non-zero normal force F_N has to be applied to maintain a constant gap during shear; in the ideal conditions of Fig. 1, F_N then provides a measurement of the total normal stress $\sigma_{zz} = F_N/S$ as a function of $\dot{\gamma}$. As two normal stress differences may be non-zero, it is seen here that a single experiment in simple shear is not sufficient to obtain these two quantities. It is shown below that the use of two different geometries allows the two quantities to be measured.

In order to impose simple shear with a finite geometry, most experiments use rotational flows: this avoids having boundary effects in the flow direction. The main problems of rheometry are then to:

1 find conditions in which homogeneous simple shear can be achieved;
2 relate in a general case (homogeneous and inhomogeneous shear), the measured macroscopic quantities (rotational velocity, torque, normal force) to local quantities (shear rate and stresses) and thus to the material constitutive behaviour $\tau(\dot{\gamma})$.

It can be seen below that, in many cases, only approximate equations exist. It is thus vital to know their domain of validity and to be able to choose the equations appropriate for a given material in a given situation.

The three main geometries that achieve simple shear flow are cone and plate, parallel plate and Couette (Fig. 2). Simple shear is also achieved in capillary flow and inclined plane geometry. These are not studied here but details can be found in Macosko (1994) and Coussot (2005).

Below we present the basic equations and their origin, and discuss the application to complex fluids; detailed calculations and in-depth analyses can be found in Macosko (1994).

Cone and plate. In a cone and plate geometry, the material is put between a cone and a disk of the same symmetry axis and the same radius R. The cone tip is truncated to avoid friction between the cone and the bottom plate. The cone virtual tip is supposed to touch the plate. The bottom plate is fixed, and shear is imposed by rotating the cone around the symmetry axis at a velocity Ω, resulting in a torque T.

In these conditions, in spherical coordinates, the only non-zero component of the strain rate tensor is $d_{\theta\varphi}$, which a priori depends on the polar angle θ only. From the stress equilibrium equation, it can then be shown that the shear stress $\tau = \tau_{\theta\varphi}$ is roughly homogeneous for small cone angles: the shear stress variation from the plate to the cone is of order of 0.5% for a cone angle $\alpha_0 = 4°$, which explains why cone of angles lower than 4° are always chosen. Because the shear stress τ is homogeneous and uniform in the whole gap, for materials which are characterised

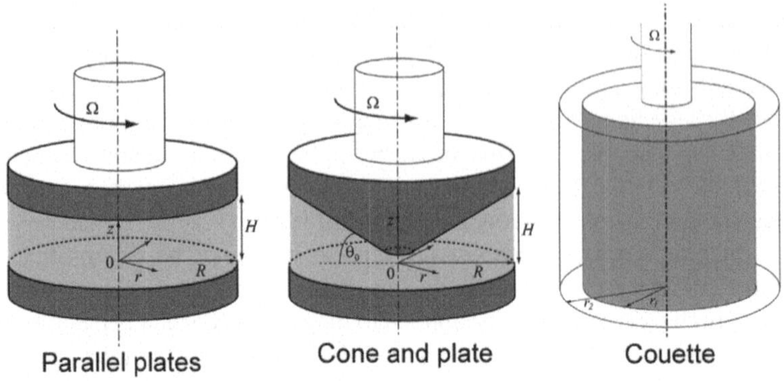

Fig. 2 Standard geometries: parallel plates, cone and plate, Couette. Reprinted with permission from Coussot (2005). Copyright (2005)

by a univalent relationship between stresses and strain rates (which excludes shear-banding materials), the shear rate $\dot{\gamma}$ is constant in the whole gap. With a no-slip boundary condition at the cone and plate surfaces, the shear rate can then be computed simply. The velocity at a radial position r on the cone surface is Ωr; the velocity is zero on the plate surface, while the gap size at this radial position is $r\ \tan\alpha_0$ leading to $\dot{\gamma} \approx \Omega r \big/ (r\ \tan\alpha_0) \approx \Omega \big/ \alpha_0$. This equation simply relates the shear rate to the applied macroscopic rotational velocity Ω. The torque T resulting from a constant shear stress τ exerted on the cone surface is $T = \int_0^R 2\pi r^2 (\cos\alpha_0)^2 \tau dr \approx 2\pi R^3 \tau \big/ 3$. Note that the measurement of the normal force F_z exerted on the cone also provides:

$$N_1 = \tau_{\varphi\varphi} - \tau_{\theta\theta} = 2F_z/(\pi R^2) \tag{1}$$

To conclude, the main advantage of the cone and plate geometry is that it achieves almost perfect homogeneous simple shear, provided the angle is small and that slip is avoided on the surfaces. Then, from the torque/rotational velocity relationship $T(\Omega)$, one can simply extract the constitutive behaviour $\tau(\dot{\gamma})$ thanks to reliable equations that are true whatever the material constitutive law is:

$$\tau = \frac{3}{2}\frac{T}{\pi R^3} \tag{2}$$

$$\dot{\gamma} = \frac{\Omega}{\alpha_0} \tag{3}$$

This means that, together with thin gap Couette geometry (see below), this geometry is in general the best choice for material characterisation.

However, it should be noted that the size of the truncature of the cone tip is usually a few tens of microns and that the gap at the edge for a 2°/40 mm diameter cone is 0.6 mm. This implies that this geometry cannot be used for many suspensions; it can be used for dispersions whose elements' size is of a few microns at most.

Moreover, special care should be taken with shear-banding materials, for which two material states may coexist for a given shear stress. In this case, the approximate homogeneity of stress for small angles is still valid (it comes from the stress equilibrium equation, independently of the material behaviour), but the shear rate needs not be homogeneous anymore (a basic analysis of this case is found in Sects. 4 and 5; details are found in Coussot 2005; Ovarlez et al. 2009). Other problems may occur with thixotropic systems: at yield, the material may experience strong rejuvenation near the cone (where the shear stress is slightly higher) and stay solid near the plate; subsequent analysis of the properties from macroscopic measurements in such conditions are then likely to be incorrect.

Parallel plates. In a parallel plate geometry, the material is put between two disks of same symmetry axis and of same radius R, separated by a distance H (named the 'gap'). The bottom plate is fixed and shear is imposed by rotating the upper plate around the symmetry axis at velocity Ω, resulting in a torque T.

In these conditions, in cylindrical coordinates, the only non-zero component of the strain rate tensor is $d_{z\theta}$ which a priori depends on both r and z. From the stress equilibrium equation it can then be shown that the shear stress $\tau(r) = \tau_{z\theta}(r)$ does not depend on z. This is the only information one has about the shear stress: in contrast with a cone and plate geometry, $\tau(r)$ cannot be computed directly from the torque $T = \int_0^R 2\pi r^2 \tau(r) dr$ only. For materials characterised by a univalent relationship between stresses and strain rates, this also means that the shear rate $\dot{\gamma}(r)$ depends only on the radial position r. With a no slip boundary condition at both plate surfaces, the velocity at a radial position r is Ωr on the upper plate and zero on the lower plate, leading to:

$$\dot{\gamma}(r) = \frac{\Omega r}{H} \tag{4}$$

In other words, simple shear is achieved at any radial position r in a parallel plate geometry, but shear is inhomogeneous along the radial direction. Equation 4 also means that, from the knowledge of the applied macroscopic rotational velocity Ω, the local shear rate in the gap is known whatever the material constitutive law is.

To compute the shear stress from the torque measurements, there are two possibilities. First, it can be shown that the shear stress $\tau(R)$ at the edge of the geometry $(r = R)$ can be computed exactly as:

$$\tau(R) = \frac{T}{2\pi R^2}\left[3 + \frac{\Omega}{T}\frac{\partial T}{\partial \Omega}\right] \tag{5}$$

When associated with the shear rate $\dot{\gamma}(R) = \Omega R / H$ computed at the same radial position R, Eq. 5 provides $\tau(\dot{\gamma})$ data from $T(\Omega)$ data. However, as this expression involves the derivative of the torque with respect to the rotational velocity, many accurate $T(\Omega)$ data have to be recorded to allow for the computation of $\tau(\dot{\gamma})$ data.

It is thus not very convenient to use. Another possibility consists of using expressions that are strictly valid only for a given material and to use them as an approximation for any material. For example, for a viscous fluid $\tau(r) = \eta\dot{\gamma}(r)$, one gets $\tau(r) = 2rT / \pi R^4$. Combining this equation with $\dot{\gamma}(r) = \Omega r / H$ makes it possible to extract a $\tau(\dot{\gamma})$ data from a single $T(\Omega)$ measurement. These couple of data can a priori be computed at any radial position r. This position can be chosen so as to minimise the error made in the material characterisation when using these equations for non-Newtonian materials. It has been shown that the error is minimised at $r \approx 3R/4$ (more precisely at $r \approx 0.76R$) for materials of constitutive law $\tau \prec \dot{\gamma}^n$ with $n < 1.4$; combining $\tau(3R/4)$ and $\dot{\gamma}(3R/4)$ then leads to a correct estimate of their behaviour within 2% error (Macosko 1994). This explains why the choice of computing the shear stress and shear rate at $3R/4$ is made by default in most rheometer softwares.

Note that the measurement of the normal force exerted on the upper plate also provides $\tau_{\theta\theta} + \tau_{rr} - 2\tau_{zz} = N_1 - N_2$ (Macosko 1994). If normal stress differences are important, full determination of the stress tensor in simple shear then implies a need to use both parallel plate and cone and plate geometries.

To conclude, parallel plate geometries are characterised by an inhomogeneous but controlled shear rate $\dot{\gamma}(r) = \Omega r / H$ in most situations provided slip is avoided on the surfaces. Standard approximate equations can be used in many situations to directly relate a torque/rotational velocity measurement $T(\Omega)$ to a data point of the constitutive behaviour through:

$$\dot{\gamma} = \frac{3}{4}\frac{\Omega R}{H} \tag{6}$$

$$\tau = \frac{3}{2}\frac{T}{\pi R^3} \tag{7}$$

It should be noted that the shear rate computed in that way is the exact value at $3R/4$ whatever the material constitutive law is, whereas the shear stress is exact only for Newtonian fluids.

Finally, note that in contrast to a cone and plate geometry, the gap value is not fixed by the theory but can be chosen by the user. It can be as high as a few millimetres (the main limitation is that gravity effects have to remain negligible to allow for the above equations to be derived; Coussot 2005). Parallel plates can then be used for suspensions whose elements size is of 200 μm at most (as mortars with fine particles). Shear being inhomogeneous, special care should be taken with yield stress fluids at low shear rate: shear localisation is then likely to appear (Coussot 2005) and is more difficult to take into account than in a Couette geometry.

Couette geometry. In a Couette geometry, the material is put between two coaxial cylinders of radii R_i (inner cylinder) and R_o (outer cylinder). In most cases, the outer cylinder is fixed and shear is imposed by rotating the inner cylinder of height H around the symmetry axis at a velocity Ω, resulting in a torque T. Note that the distance of the inner cylinder to the bottom of the cup is usually chosen to be much larger than the gap size to ensure that the contribution to the torque from shear at the bottom is as negligible as possible; these end effects can be corrected with appropriate procedures if necessary (Macosko 1994).

In these conditions, in cylindrical coordinates, the only non-zero component of the strain rate tensor is $d_{r\theta}$ which a priori depends on r only. From the stress equilibrium equation, it can be shown that the shear stress $\tau(r) = \tau_{r\theta}(r)$ is known everywhere in the gap and is given by:

$$\tau(r) = \frac{T}{2\pi H r^2} \tag{8}$$

i.e. the shear stress is inhomogeneous and controlled. On the other hand the shear rate $\dot{\gamma}(r)$ is a priori unknown. This is different from the case of a parallel plate geometry, where the shear rate is inhomogeneous and controlled while the shear stress is unknown.

Two cases have to be distinguished: thin gap Couette geometry and wide gap Couette geometry. A Couette geometry is arbitrarily said to have a thin gap when the shear stress can be considered as homogeneous within 1%. From $\tau(r) = T/2\pi H r^2$, this means that $(R_o^2 - R_i^2)/R_i^2 < 1\%$ i.e. that the gap size $R_o - R_i$ is smaller than 0.5% of the inner cylinder radius R_i. It then implies that, for materials characterised by a univalent relationship between stresses and strain rates (which excludes shear-banding materials), the shear rate is almost constant in the whole gap. With a no slip boundary condition at the surface of both cylinders, the shear rate can then be computed simply as:

$$\dot{\gamma} = \Omega \frac{R_i}{R_o - R_i} \tag{9}$$

while the shear stress is:

$$\tau = T/2\pi H R_i^2 \tag{10}$$

Other cases are referred to as wide gap Couette geometries. In this case, as for a parallel plate geometry, an exact equation exists (Coussot 2005):

$$2T\frac{\partial \Omega}{\partial T} = \left[\dot{\gamma}(\tau(R_i)) - \dot{\gamma}(\tau(R_o))\right] \tag{11}$$

However, Eq. 11 is not very convenient to use as it does not provide directly the value of the shear rate at a given position, but the difference between the shear rates at the inner and outer cylinders. Nevertheless, it is used in the simple case of yield stress fluids close to yielding, where shear localisation occurs (see Sect. 5). It then reduces to

$$2T\partial\Omega/\partial T = \dot{\gamma}(\tau(R_i)) \tag{12}$$

which automatically takes shear localisation into account (Jarny et al. 2008; Ovarlez et al. 2008; Coussot et al. 2009).

In general cases, it is more usual to use approximate equations. As for a parallel plate geometry, one can compute the shear rate distribution for Newtonian materials and use the $\dot{\gamma}(r)$ equation obtained in that way as an approximation for any material. To extract $\tau(\dot{\gamma})$ data from a single $T(\Omega)$ measurement, a conventional choice is to compute the shear stress $\tau(r)$ and the shear rate $\dot{\gamma}(r)$ in the middle of the gap. Standard equations used to relate a torque/rotational velocity data $T(\Omega)$ to a constitutive behaviour data point $\tau(\dot{\gamma})$ are finally, when computed in the middle of the gap $\bar{R} = (R_o + R_i)/2$:

$$\dot{\gamma} = \Omega \frac{R_i^2 R_o^2}{\bar{R}^3 (R_o - R_i)} \tag{13}$$

$$\tau = T \big/ 2\pi H \bar{R}^2 \tag{14}$$

Other approximate equations can be used, depending on the radial position where shear stress and shear rate are computed; at the inner radius, one gets $\dot{\gamma} = \Omega R_o^2 \big/ \bar{R}(R_o - R_i)$ and $\tau = T \big/ 2\pi H R_i^2$. Note that flows in a Couette geometry do not provide straightforwardly normal stress measurements: to get such information, specific normal force probes have to be put on the outer cylinder surface (Deboeuf et al. 2009) or the top free surface deformation has to be characterized (Boyer et al. 2011).

The main advantage of Couette geometries for complex fluids is that the gap can be large. It is thus the preferred geometry for suspensions made of large particles such as fresh concrete. However, unless the inner cylinder radius is very large, a wide gap implies a strongly inhomogeneous shear stress distribution which poses the problems of shear localisation and of shear-induced migration—see below.

Vane tool. Concentrated suspensions are often studied within a vane-in-cup geometry; see Barnes and Nguyen (2001) for a review. In this geometry, the inner cylinder of the Couette geometry is replaced by a vane tool, which consists of N thin blades centred on a thin shaft (usually, $N = 4$ or 6). Compared with the coaxial cylinders Couette geometry, it a priori offers two main advantages. First, it allows the study of the properties of structured materials such as colloidal gels with minimal disturbance of the material structure during the insertion of the tool. Second, it is supposed to avoid wall slip (Saak et al. 2001) because the material sheared in the gap of the geometry is sheared by the (same) material that is trapped between the blades. Consequently, it is widely used to study the behaviour of pasty materials containing large particles as fresh concrete (Koehler et al. 2006; Estellé et al. 2008; Wallevik 2008).

The streamlines in a vane-in-cup geometry are generally not circles. This implies that the above equations for a Couette geometry cannot a priori be applied in this case. In practice, for a Newtonian material, the torque exerted on a vane tool is lower than that exerted on a cylinder of same inner radius driven at a same rotational velocity; it is typically 10–30% lower for standard geometries (Sherwood and Meeten 1991; Ovarlez et al. 2011). Nevertheless, a practical method, known as the Couette analogy

(Bousmina et al. 1999; Estellé et al. 2008), can still be used to infer constitutive laws from macroscopic measurements with a vane tool. It basically consists of calibrating the geometry with linear (Hookean or Newtonian) materials. The equivalent inner radius $R_{i,eq}$ of a vane-in-cup geometry is then defined as the radius of the inner cylinder of a Couette geometry that would have the same torque/shear stress and velocity/shear rate conversion factors for a linear material. For any material, all macroscopic data are then analysed as if the material was sheared in a Couette geometry of inner cylinder radius $R_{i,eq}$. The nonlinearity (which affects the flow field) is sometimes accounted for as it is in a standard Couette geometry (Estellé et al. 2008). This approach may finally provide constitutive law measurements within a good approximation (Baravian et al. 2002).

Note that for suspensions of coarse particles, particle depletion may be induced very rapidly by shear near the blades. It results in a kind of wall slip near the blades (Ovarlez et al. 2011), whereas the vane tool is mainly used to avoid this phenomenon. This suggests that a coaxial cylinders geometry with properly roughened surfaces is preferable, when possible, when studying flows of pasty materials with large particles. A detailed analysis is provided in Sect. 5.

Geometry factors for yield stress measurements. Some of the equations derived in parallel plate and Couette geometries are only approximate: they are chosen to provide the best possible characterisation of a wide range of materials with very different nonlinear rheological properties. When studying a given kind of material, a question that arises is the accuracy of these equations for specific quantities. Below we discuss the case of yield stress measurements in yield stress fluids when a torque T_y is measured at yield.

In a parallel plate geometry, at yield $\tau(r) = \tau_Y$ everywhere on the upper plate. Together with $T = \int_0^R 2\pi r^2 \tau(r) dr$ this yields:

$$\tau_Y = \frac{3}{2} \frac{T_y}{\pi R^3} \tag{15}$$

It is remarkable that Eq. 15, which is exact here for yield stress measurements, is the same as Eq. 7, which is optimal for flow properties. Equation 7 can then be used reliably together with Eq. 6 to evaluate all material properties.

In a Couette geometry, yield first occurs on the inner cylinder, where stress is maximal. This implies that at yield $\tau(R_i) = \tau_Y$, and thus:

$$\tau_Y = T_y / 2\pi H R_i^2 \tag{16}$$

This means that, in a wide gap Couette geometry, the best choice of geometry factors for yield stress measurements may not be the same as those for flow properties (see for example Eqs. 13 and 14).

In a vane-in-cup geometry, cylindrical symmetry seems to be recovered in yield stress fluids near yielding (Keentok et al. 1985; Yan and James 1997; Savarmand et al. 2007); this point is discussed in more detail in Sect. 5. This would mean that $R_{i,eq} = R_i$ can be used for yield stress measurements; the cylindrical yield surface thus contributes to the yield torque T_y by a value $\tau_Y 2\pi H R_i^2$. However, bottom effects are never negligible at yield with a vane tool: the contribution to T_y from the bottom yield surface can be computed as in a parallel plate geometry as $\tau_Y 2\pi R_i^3/3$. This finally yields:

$$\tau_Y = \frac{T_y}{2\pi H R_i^2 (1 + R_i/3H)} \tag{17}$$

If the vane tool is fully embedded in the material, another $R_i/3H$ factor has to be added in Eq. 17 to account for the contribution of the top surface to the torque.

Experimental problems. Most of the experimental problems found in rheometric experiments come from the main hypotheses made when computing shear stress/shear rate relationships from torque/rotational velocity measurements. First, one has to ensure that the flow and the material are homogeneous; these specific issues are dealt with in Sect. 5. Then, a no-slip boundary condition is assumed at the interfaces of both tools. Rough surfaces are usually used to ensure there is no wall slip in the experiments. Moreover, the absence of wall slip in parallel plate and Couette geometries can be checked by performing experiments with two different gaps (Coussot 2005). If different results with two different gaps are obtained, this likely means that the no-slip hypothesis made in the analysis is not correct, i.e. there is wall slip.

Since most rheometers control the applied torque, another important problem in rheometry arises from the fact that the applied torque is the sum of the torque applied to the material and of the inertial torque applied to accelerate/decelerate the rotating tool. One thus has to be particularly careful when applying ramps and oscillations (see Sect. 3). Although the inertial torque can theoretically be accounted for, this may pose accuracy problems when it is dominant. This makes high frequency oscillatory measurements particularly difficult. For more details on experimental problems, see Coussot (2005).

3 Characterisation of Simple Yield Stress Fluids

Yield stress are materials that flow steadily only when the applied shear stress τ in simple shear is higher than a yield stress τ_Y. In addition, the behaviour of 'simple' yield stress fluids is basically independent of the flow history. Different methods are presented below devoted to characterizing the flow properties of yield stress fluids when $\tau \geq \tau_Y$ and their solid properties when $\tau \leq \tau_Y$, and to determining the value of τ_Y accurately.

Steady flow properties. Steady flow properties in simple shear are characterised by the stationary relationship $\tau = f(\dot{\gamma})$ between the shear stress τ and the shear rate $\dot{\gamma}$. These data can be obtained in both stress controlled and shear rate controlled experiments.

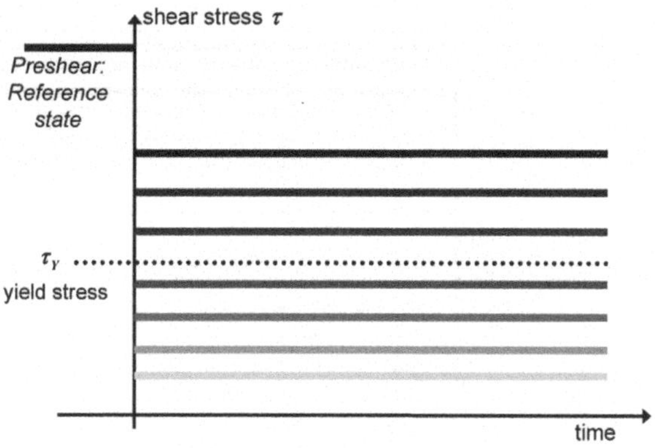

Fig. 3 Schematic representation of creep experiments

Creep tests. Creep tests consist of applying a series of constant shear stresses τ_i over a long period of time, usually after a preshear (Fig. 3). For, $\tau_i \geq \tau_Y$ a yield stress fluid reaches steady flow at a shear rate $\dot{\gamma}_i$ after a time t_{eq} which depends on τ_i and the material (Fig. 4a). The data $\tau_i = f(\dot{\gamma}_i)$ then define the flow curve of the material (Fig. 4b). For $\tau_i < \tau_Y$, one should observe unsteady flow with a vanishing shear rate $\dot{\gamma}_i(t) \overset{t\to\infty}{\to} 0$ (Fig. 4b), which defines the yield stress τ_Y.

Of course, a series of constant shear rates can also be applied until a constant shear stress is reached. These last experiments are commented on in more detail in the case of thixotropic materials (Sect. 4).

But although performing creep tests is a very accurate method for characterising materials, it is quite time consuming and does not provide a lot of data.

Ramps and sweeps. The independence of the behaviour of simple yield stress fluids on flow history usually implies that, when the applied shear stress (or shear rate) is changed abruptly, they ideally reach their new steady-state instantaneously (typically within one second); this is the case in Fig. 4a, except when the applied stress is very close to the yield stress. The practical consequences of this are that:

1. a preshear is not necessary in principle, although a first preshear is still preferable just after loading to ensure a reproducible initial state;
2. a series of applied shear stresses τ_i can be imposed in any order;
3. the value of τ_i can be changed (almost) as quickly as desired to obtain a large set of $\tau_i = f(\dot{\gamma}_i)$ data.

This results in much faster experiments than creep experiments and in more data. The main limitations to the rate of change of τ_i are the rheometer inertia and the material inertia.

Based on these principles, two experiments are usually performed: ramps and sweeps (Fig. 5). Stress ramps consist of continuously changing (upward or downward) the value of $\tau(t)$ in time and in recording $\dot{\gamma}(t)$ to give $\tau = f(\dot{\gamma})$; equivalently, shear rate ramps $\dot{\gamma}(t)$ can be applied. Two kinds of ramps are usually distinguished: linear ramps, in which the applied shear stress or shear rate increases (or decreases) linearly with time, and logarithmic ramps in which the value of the shear stress or

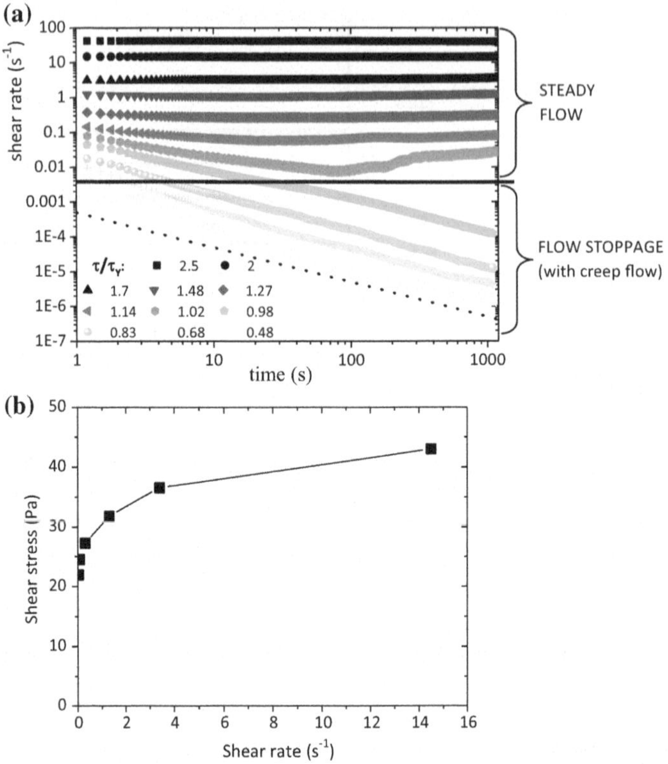

Fig.. 4 **a** Shear rate versus time for creep tests performed on a model yield stress fluid (Carbopol gel); the dimensionless applied shear stresses are given in the legend. **b** Steady-state shear stress versus shear rate data extracted from Fig. 4a. Adapted from Balmforth et al. (2014)

Fig. 5 Schematic representation of linear stress sweep and stress ramp experiments

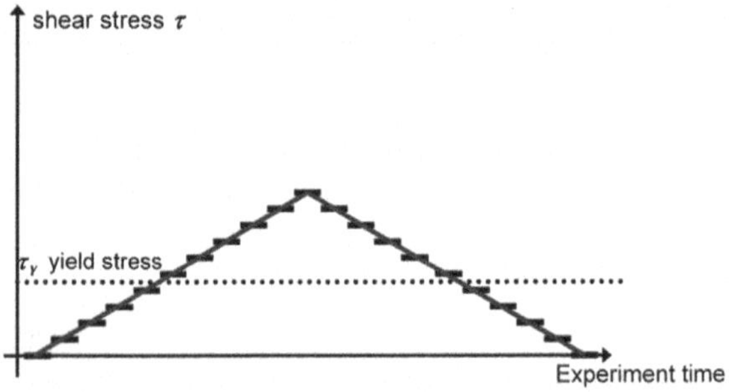

shear rate is multiplied by a constant value at any second (i.e. during a logarithmic shear rate ramp, the same time is spent in each decade of shear rates, and the shear rate evolution in time is $\dot{\gamma} = \beta * 10^{\alpha t}$ with α and β free parameters of the ramp).

Usually, an up-ramp is followed by a down-ramp. This allows to show that the 'steady' behaviour of a 'simple' yield stress fluid is actually obtained by checking the

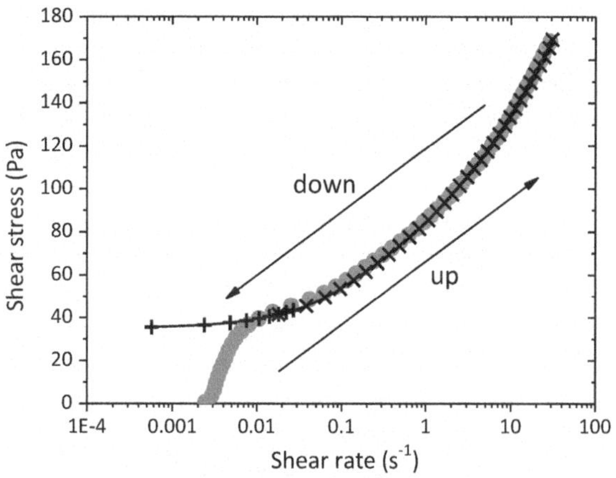

Fig. 6 Shear stress versus shear rate for linear—up (squares) and down (crosses)—stress ramps for a model yield stress fluid (a concentrated emulsion)

reproducibility of the results obtained during the up- and down-ramps. If a hysteretic response is found, this usually means that the material is thixotropic (see Sect. 4). Note that ramps performed too quickly may also generate discrepancies between the results obtained during the up- and down-ramps because of the geometry and material inertia; a simple way to avoid (and test) this effect is to decrease the rate of change of shear stress/shear rate, which minimises the impact of inertia.

An alternative to ramps is sweeps, in which a series of increasing/decreasing constant shear stresses τ_i (or shear rates $\dot{\gamma}_i$) are applied during very short time periods (typically ten seconds). The advantage of sweeps is that they allow checking at each step that the behaviour is steady.

The results of typical linear (up and down) stress ramps are shown in Fig. 6 for a model yield stress fluid (a concentrated emulsion). It is observed that the same behaviour is obtained during the up- and down-ramps (except for low shear rate). This shows that the steady behaviour $\tau(\dot{\gamma})$ is obtained and thus justifies the use of ramps.

The behaviour at low rate deserves a comment. During the first (up) ramp, the material is initially at rest and thus in its solid regime where its behaviour is basically elastic $\tau = G\gamma$. This means that, when applying a linear ramp $\tau = \alpha t$, as long as $\tau < \tau_Y$, the (elastic) material is strained at a constant rate $\dot{\gamma}(\alpha, G) = \alpha/G$.

This explains the first part of the curve during the up-ramp in which a roughly constant shear rate is measured at any shear stress $\tau < \tau_Y$. Afterwards, the material flows and the $\tau(\dot{\gamma})$ data correspond to the flow behaviour. During the down-ramp, the same $\tau(\dot{\gamma})$ curve is described until $\tau = \tau_Y$; when $\tau < \tau_Y$, the material progressively restores its elastic energy which results in a negative rate of strain (absent from the curve in Fig. 6 where the shear rate is in logarithmic coordinates). A simple way to check that data correspond to the flow behaviour is to perform the same experiment at two different ramp rates: only that part of the curve corresponding to a solid regime depends on this rate. Of course, the same features are observed when applying up/down shear rate ramps. This implies that, if only a short time is available to perform experiments, then a down-ramp should be applied, starting from a high shear stress (or shear rate), rather than an up-ramp.

Finally let us comment on the use of linear/logarithmic shear rate ramps. As yield stress fluids usually have a nonlinear behaviour of the form $\tau = \tau_Y + K\dot{\gamma}^n$, if an accurate characterisation of the behaviour (in particular of the yield stress) is required, then it is better to have data for several decades of shear rates. In this case, one has to apply a logarithmic shear rate ramp and to plot shear stress versus shear rate in a logarithmic scale (Fig. 6). On the other hand, if only the behaviour in a limited range of $\dot{\gamma}$ is of interest, a linear ramp can be applied: it will provide enough data for the characterisation and will be faster to perform than a logarithmic ramp.

Elastic modulus measurements. In their solid regime, yield stress fluids usually behave as viscoelastic materials. At sufficiently low strain, their behaviour is linear and their elastic modulus is defined as $G = \tau/\gamma$ where τ is the steady shear stress that has to be applied to maintain a given steady shear strain γ. However, when strained, these materials also exhibit viscous behaviour and their elastic response is not instantaneous. In order to evaluate viscous and elastic effects separately, the classical experiment consists of applying strain oscillation:

$$\gamma(t) = \gamma_0 \sin \omega t \tag{18}$$

and measuring the stress response $\tau(t)$. When the material is probed in its linear regime[2], $\tau(t)$ is of the form:

$$\tau(t) = G'(\gamma_0 \sin \omega t) + G''(\gamma_0 \cos \omega t) \tag{19}$$

As elastic effects are in phase with the strain whereas viscous effects are in phase with the strain rate (and thus in quadrature with the strain), this defines an elastic modulus G' and a viscous modulus G''. Alternatively, stress oscillations can be applied and strain oscillations measured, which does not change the definition of G' and G''.

To understand the interpretation of these coefficients, it is useful to consider the simplest model of viscoelastic solid, which is that of Kelvin–Voigt (Fig. 7), symbolised as a spring and a damper in parallel.

In this model, the relationship between the shear stress and the shear strain is $\tau = G\gamma + \eta\dot{\gamma}$. For the above applied strain oscillations, this yields $\tau = G\gamma_0 \sin \omega t + \eta\gamma_0\omega \cos \omega t$. For the Kelvin–Voigt model, and for that model only, G' and G'' have a simple interpretation as $G'(\gamma_0, \omega) = G$, $G''(\gamma_0, \omega) = \eta\omega$. However, in general, G' and G'' have a complex dependence on both γ_0 and ω (see for example Derec et al. 2003). It is thus usual to perform amplitude sweeps (where oscillations of increasing γ_0 are applied) to check in which range of strains the material has a linear

[2]When the behaviour is nonlinear, the response is not sinusoidal, though one can still define G' and G'' through Fourier analysis: they are the coefficients of the first harmonic in the Fourier series decomposition of the response $\tau(t)$ (for details see Wilhelm et al. 1998; Ewoldt et al. 2008). In this case, while G'' still encodes all the dissipative properties of the material, G' has in general no simple quantitative meaning.

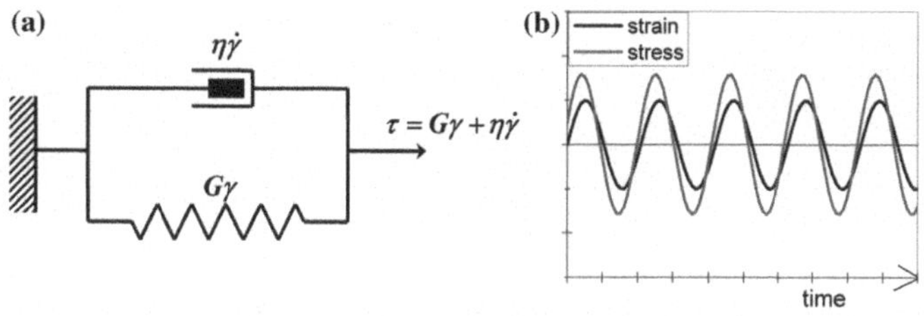

Fig. 7 **a** Kelvin–Voigt model; **b** typical oscillatory experiment

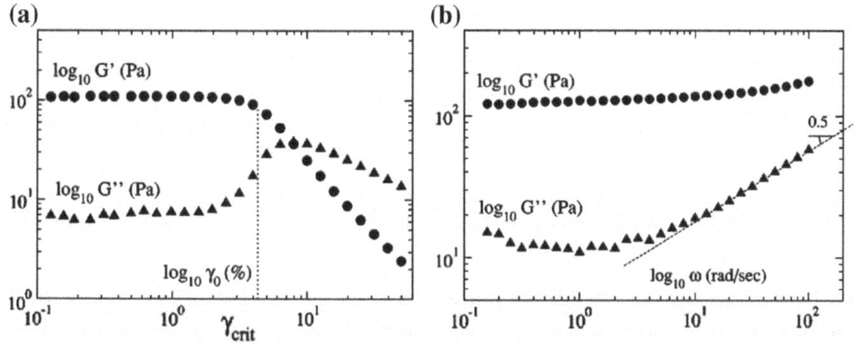

Fig. 8 **a** *G'* and *G"* versus strain measured in a dense colloidal suspension with a yield stress during a strain sweep oscillatory experiment; **b** *G'* and *G"* versus frequency in the linear regime, measured in the same suspension during a frequency sweep oscillatory experiment. Reprinted with permission from Derec et al. (2003). Copyright (2003) by the American Physical Society

behaviour: this defines a critical strain below which G' is constant (Fig. 8a). In this regime, frequency sweeps (where oscillations of increasing ω are applied) may then be performed to better understand the dissipation properties of the material.

Whereas G' and G'' may depend on shear history for complex fluids (in particular on the time elapsed since the beginning of the measurement), these are constant material properties in simple yield stress fluids and no specific preparation is a priori required to perform these measurements. It is worth noting that, in yield stress fluids, G' is usually found to depend only slightly on ω (Fig. 8b); this implies that an experiment at low strain and at a single frequency is basically sufficient to characterise the elastic properties of yield stress fluids in their solid regime.

Note finally that the linearity of the response can be easily checked by looking at the raw response to oscillations. If a sinusoidal response is obtained when a sinusoidal oscillation is imposed, then the *G'* measurement can be trusted as the elastic modulus of the material without performing strain sweep.

Yield stress measurements. Simple yield stress fluids are characterised by a single yield stress τ_Y. Starting from a liquid state, flow stops when the shear stress τ is decreased below τ_Y; starting from a solid state, flow starts when τ is increased above τ_Y. Several methods exist for the determination of τ_Y.

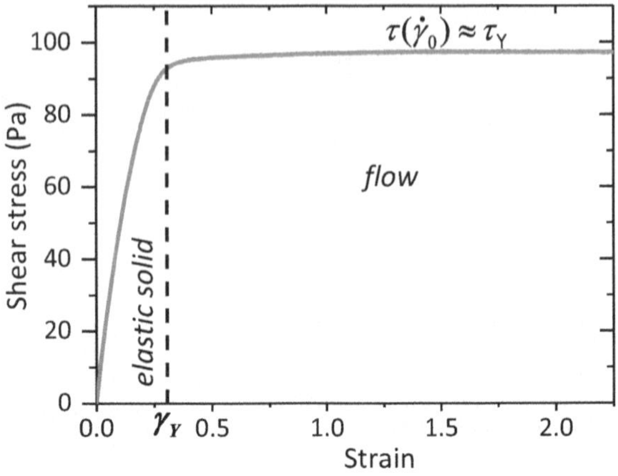

Fig. 9 Shear stress versus shear strain during a yield stress measurement at low constant shear rate for a model yield stress fluid (a concentrated emulsion)

A first method simply consists of analysing the flow curves $\tau(\dot{\gamma})$ obtained thanks to the experiments described earlier (Figs. 4b and 6): the yield stress is the value τ_Y of the shear stress at vanishing shear rate $\tau(\dot{\gamma}) \overset{\dot{\gamma}\to 0}{\to} \tau_Y$. This means that data at a sufficiently low shear rate have to be measured; as pointed out before, the best method of measuring $\tau(\dot{\gamma})$ in these circumstances consists of applying a decreasing logarithmic shear rate ramp. A decreasing stress ramp can also be applied: in this case, it has to be slow enough because small stress changes lead to large shear rate changes at the approach of τ_Y.

When only the yield stress is of interest, it can be measured directly by applying a constant low shear rate $\dot{\gamma}_0$ to the material, starting from rest. A typical measurement is displayed in Fig. 9. Two regimes are observed: a roughly linear increase of the shear stress at low strain, followed by a stress plateau. Initially, the material is a solid: the first regime then corresponds to the elastic straining of the material $\tau \simeq G\gamma \simeq G\dot{\gamma}_0 t$. When the shear stress reaches the yield stress, i.e. when the strain is equal to the yield strain $\gamma_Y \simeq \tau_Y/G$, the material turns to a fluid and quickly reaches its stationary state, characterised by a resistance to flow $\tau(\dot{\gamma}_0)$. The fact that a plateau is observed is typical of simple yield stress fluids. For a sufficiently low value of $\dot{\gamma}_0$, this plateau is identified with the yield stress as $\tau(\dot{\gamma}_0) \overset{\dot{\gamma}_0\to 0}{\to} \tau_Y$. 'Low' shear rate means that viscous effects are negligible, but depends on the material. For a material behaviour of the form $\tau = \tau_Y + K\dot{\gamma}^n$, this means that $\dot{\gamma}_0 << (\tau_Y/K)^{1/n}$. Typical values for standard yield stress fluids are $\dot{\gamma}_0 = 10^{-3}$–$10^{-2}\,\mathrm{s}^{-1}$.

Finally, stress (or strain) amplitude sweeps in oscillatory experiments may also provide an evaluation of the yield stress. G' and G'' measured during such an experiment are shown in Fig. 8a. At low stress, in their linear regime, yield stress materials display dominant–constant–elastic properties with $G' >> G''$. On increasing stress, the material leaves its linear regime, G' decreases and G'' increases. At a given shear stress, which is often identified with the yield stress τ_Y, G' and G'' cross; this crossover approximately corresponds to a peak in G'', which subsequently decreases on increasing stress. In this last regime, $G'' > G'$ consistently with the interpretation that $\tau > \tau_Y$ and that the material is in its liquid regime. The decrease of G'' with increasing stress for $\tau > \tau_Y$ is a signature of the shear-thinning properties of the material. Note that such determination of the yield stress is rather qualitative.

4 Characterisation of Thixotropic Yield Stress Fluids

Thixotropic materials have a shear-history-dependent behaviour. After a flow at high shear rate, these materials progressively thicken in low shear flow; if they subsequently experience a high shear flow, they progressively get more fluid and go back to their initial state: their evolutions are reversible. A specificity of thixotropic yield stress fluids is that they may exhibit a solid behaviour only after a resting period; their elastoplastic properties then show aging. It is thus convenient to describe separately their flow properties and their solid properties.

Flow properties. This section begins by describing the most basic characterisation of thixotropy. It then explains how the most important characteristics of thixotropic materials are determined: dynamic yield stress, critical shear rate, steady-state behaviour and characteristic time to reach steady state in simple situations.

Thixotropic loop. The simplest experiment which aims at evidencing thixotropy is the so-called 'thixotropic loop' test (see for example Toorman 1997; Coussot et al. 2002). A thixotropic loop is the hysteretic response obtained in a shear stress versus shear rate plot (Fig. 10) when a material is submitted to a cyclic shear history—typically, a series of up and down shear rate ramps, cf. Fig. 5. The observation of a hysteretic response shows that the material behaviour depends on shear history—in contrast with simple yield stress fluids (see Fig. 6). For a given procedure, the amplitude of the hysteresis gives an indication of the importance of the material dependence on shear history. However, it is very difficult to extract any quantitative information from such test. Indeed, throughout the whole shear history in this test, the material structure evolves in a way which depends complexly on the parameters used (ramp rate, linear ramp versus logarithmic ramp, etc.,). It is worth noting in particular that the response to a series of up/down ramps evolves in time: it stabilises after several loops (Toorman 1997), and the 'steady loop' is only characteristic of the loop parameters, it does not provide any intrinsic characteristics of the material. To better characterise thixotropic yield stress fluids, it is better to perform creep tests and shear rate step tests. These are described below.

Creep tests. Creep tests (see Sect. 3) are the first important tests to perform as they allow the determination of the dynamic yield stress and provide evidence of the viscosity bifurcation phenomenon specific to thixotropic materials (Coussot et al. 2002, 2006).

Before applying a given constant shear stress τ_i, it is important to preshear the material at a high shear rate for long enough to ensure that a steady state is reached. In these conditions, all tests always start from the same destructured state of the material. Creep tests provide the dynamic yield stress of the material, namely the

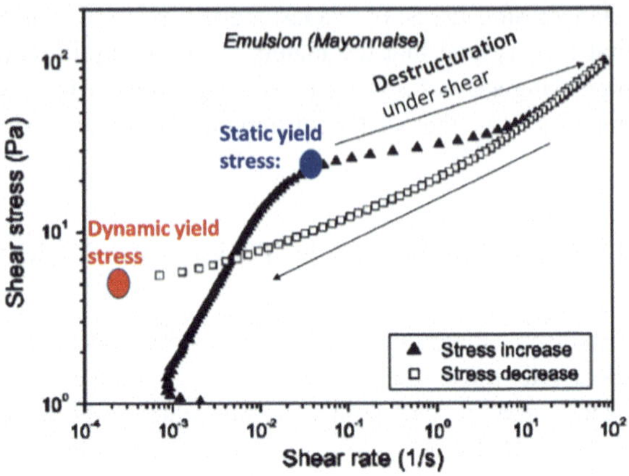

Fig. 10 Thixotropic loop: response of a thixotropic material (mayonnaise) to a series of up/down ramps. Adapted from Coussot et al. (2002)

value $\tau_{Y,d}$ below which the material progressively stops flowing and above which a steady flow can be reached. Compared with simple yield stress fluids, the specificities of the response to creep tests of thixotropic materials are that

1. the time $t_{eq}(\tau_i)$ for which the material reaches steady-state may be long;
2. the steady shear rate achieved at shear stresses τ_i close to $\tau_{Y,d}$ is usually higher than a critical value $\dot{\gamma}_c$ i.e. $\dot{\gamma}_i(\tau_i) \overset{\tau_i \to \tau_{Y,d}}{\to} \dot{\gamma}_c$.

This contrasts with the behaviour of simple yield stress fluids where $\dot{\gamma}_i(\tau_i) \overset{\tau_i \to \tau_{Y,d}}{\to} 0$. In other words, the material steady viscosity tends to a finite value when the applied stress tends to $\tau_{Y,d}$, which led this phenomenon being called 'viscosity bifurcation'.

An example of the response of a thixotropic material to creep tests is shown in Fig. 11a and steady state flow curves obtained in creep tests experiments for various thixotropic yield stress fluids are depicted in Fig. 11b. It is clearly seen that flow starts only above a critical shear rate $\dot{\gamma}_c$, which is a characteristics of the material.

The fact that it is impossible to impose a stable flow at a shear rate lower than $\dot{\gamma}_c$ in thixotropic materials is of major importance for controlled shear rate experiments: ignorance of this feature may indeed lead to experimental errors. In a controlled shear rate mode, a macroscopic shear rate is achieved by imposing the rotational velocity of the tool, which is computed using standard equations of rheometry (see Sect. 2). These standard equations are based on the assumption of a homogeneous flow in the gap of the geometry. However, from the above observations, this macroscopic shear rate $\dot{\Gamma}$ corresponds to the true (local) shear rate of a homogeneous flow only if it is higher than $\dot{\gamma}_c$. In cases where $\dot{\Gamma}$ is lower than $\dot{\gamma}_c$, the material behaviour is not compatible with the prescribed shear rate and a phenomenon of 'shear banding' is observed: the material splits into two regions in the gap, a region flowing at $\dot{\gamma}_c$, and a non flowing region (Coussot 2005; Ovarlez et al. 2009) (Fig. 12). The relative width of the two regions is such that the average shear rate equals $\dot{\Gamma}$. Characterisations obtained for macroscopic shear rates lower than $\dot{\gamma}_c$ are thus meaningless as they correspond to the response of an inhomogeneous material. Moreover, if such a macroscopic shear rate

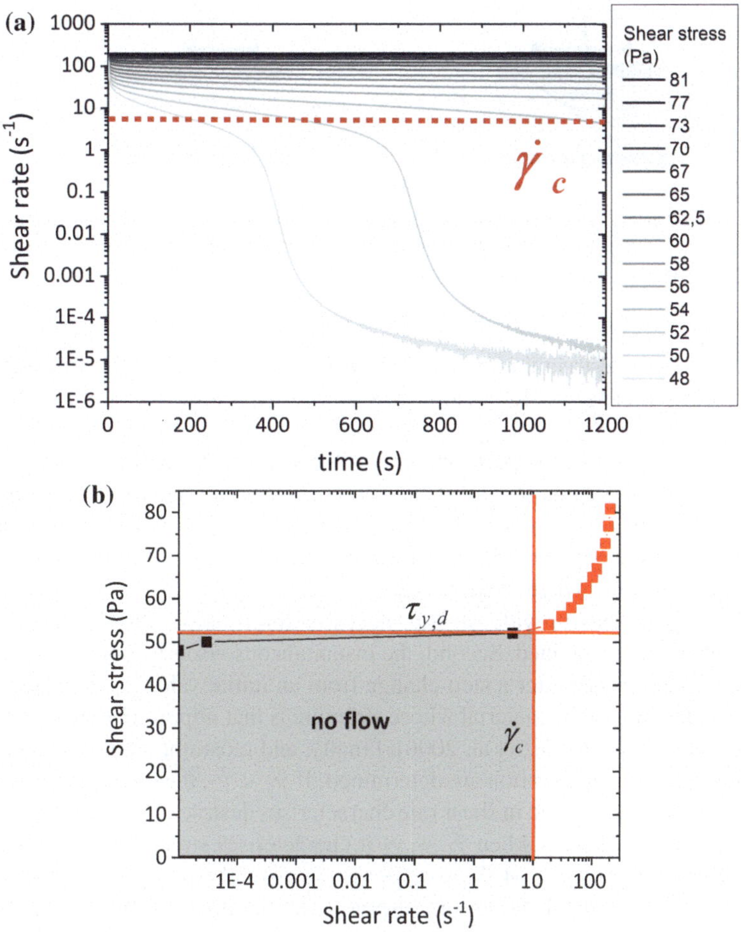

Fig. 11 **a** Shear rate versus time in creep experiments for a bentonite suspension. **b** Steady state shear stress versus shear rate from Fig. 11a. No steady flow is observed below a well-defined critical shear rate

is imposed once, it may affect all subsequent measurements as the material history is spatially inhomogeneous.

To conclude, let us stress that only prior determination of the critical shear rate in a controlled stress mode allows the domain of validity of controlled shear rate experiments to be determined.

Shear rate step tests. Although a first insight into the kinetics of material structuration under shear can be obtained from creep tests described above through the time necessary to reach steady state, the response to shear rate step tests is often thought to be more relevant. Indeed, most structural models are written as function of the shear rate $\dot{\gamma}$, as the characteristic time $1/\dot{\gamma}$ seems to fix the timescales of

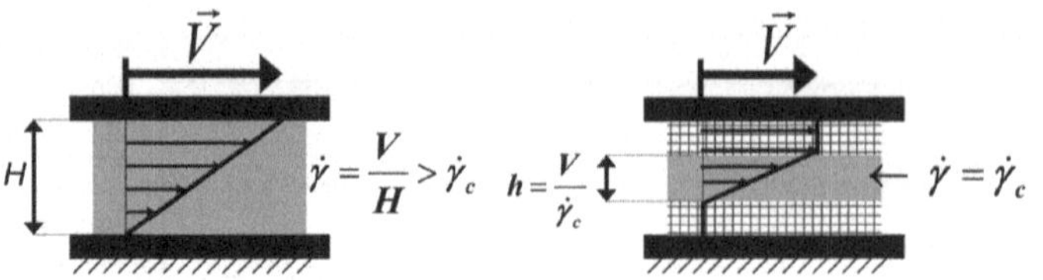

Fig. 12 Sketch of the velocity profiles in a simple shear experiment. Left: when the macroscopic shear rate $\dot{\gamma} = V/H$ is higher than the critical shear rate; right: when the macroscopic shear rate $\dot{\gamma} = V/H$ is lower than the critical shear rate

structuration and destructuration under shear (Mujumdar et al. 2002; Roussel et al. 2004; Dullaert and Mewis 2005), although a few exceptions exist (de Souza Mendes 2009).

A shear rate step test is performed as follows (Fig. 13): starting from a steady state at a given shear rate $\dot{\gamma}_1$, the shear rate is changed abruptly to a new value $\dot{\gamma}_2$ until a new steady state is reached. The transient response is studied by plotting the shear stress (or the apparent viscosity) versus time, the origin of time being defined as the instant of the change in shear rate.

These experiments provide several pieces of information. First, of course, the steady behaviour is obtained. Second, the instantaneous viscosities measured at various shear rates $\dot{\gamma}_2$ just after a step-change from an initial value $\dot{\gamma}_1$ characterise the intrinsic behaviour of the material whose structure is that imposed by steady shear at $\dot{\gamma}_1$ (Toorman 1997; Ovarlez et al. 2006). Finally, and most importantly, the kinetics of structuration/destructuration are determined. If $\dot{\gamma}_1 < \dot{\gamma}_2$, the transient shear stress response to the step-change in shear rate characterises destructuration (or breakdown of the structure), whereas when $\dot{\gamma}_1 > \dot{\gamma}_2$ it characterises structuration (or build-up of the structure). Analysis of these responses enables the parameters of structural models to be determined. When the response can be modelled by a simple exponential, a characteristic time of structuration/destructuration is extracted (Mujumdar et al. 2002; Coussot 2005). It seems that these characteristic times mostly depend on the target shear rate $\dot{\gamma}_2$. An example of the response of a thixotropic material to step-changes in shear rate is shown in Fig. 14.

Solid properties. At rest, thixotropic materials become structured in time; they are also said to age. At a macroscopic level, this implies that their elastic modulus and static yield stress increase with the time they spend at rest. This section introduces methods devoted to characterising these evolutions.

Procedure to obtain a well-defined aging solid state. Before characterising the structuration at rest of thixotropic materials, it is crucial to achieve a well-defined initial state. It is a priori possible to turn the material to a solid by starting from any liquid state and by applying any shear stress history below the dynamic yield stress. However, it has been shown that the values of the macroscopic properties of the material in its solid state depend strongly on the shear stress applied during flow stoppage (Ovarlez and Chateau 2008) and may also depend on the shear rate history applied in the liquid state (Rueb and Zukoski 1997).

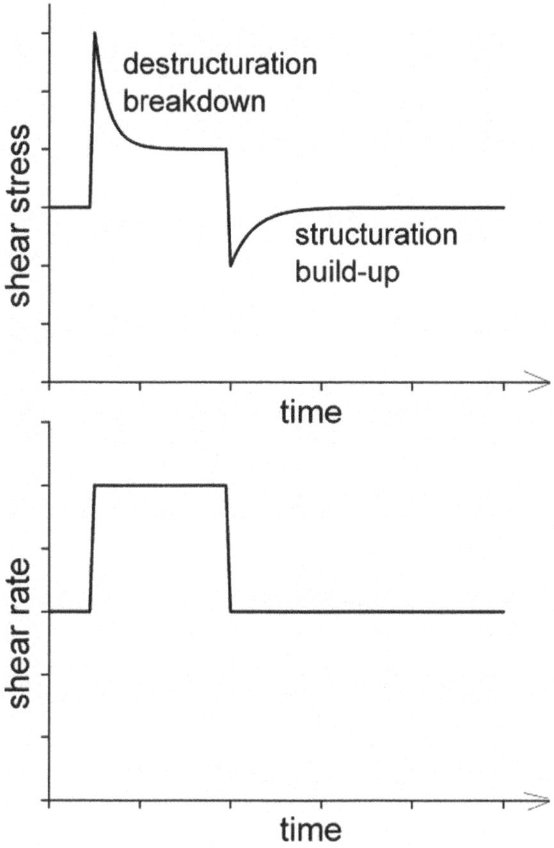

Fig. 13 Representation of the stress response to step-changes in shear rate

In order to study the material structuration at rest independently of these effects (which can be studied separately), a procedure consists of:

1. preshearing 'strongly' the material, always at a same high shear rate until a steady state is reached, in order to always start the experiments from a reproducible destructured state;
2. applying a zero stress to the material (this defines a period during which the material turns from a fluid to a solid and subsequently ages).

Rest is thus here defined as a period during which a zero stress is applied (Ovarlez and Chateau 2008). Note that it is very different to apply a zero shear rate after a preshear: if the material has the possibility to store elastic energy, its solid state is ill-defined as the applied shear stress at rest may then take any value between 0 and the dynamic yield stress.

Elastic modulus. Once the material is at rest, its structuration can be studied thanks to oscillatory experiments in the linear regime (see Sect. 3). From the definition of

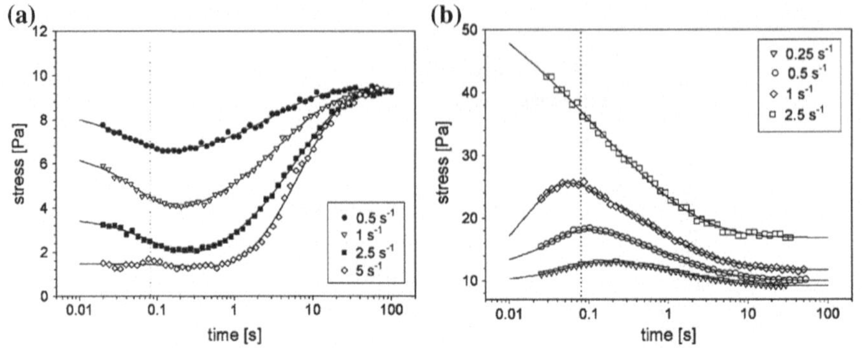

Fig. 14 Stress response to step-changes in shear rate in a thixotropic suspension: **a** when the shear rate is decreased from various values (see legend) towards 0.25 s^{-1}; **b** when the shear rate is increased from various values (see legend) towards 5 s^{-1}. Reprinted with permission from Dullaert and Mewis (2005). Copyright (2005) by the Society of Rheology

Fig. 15 Elastic modulus versus time spent at rest for a thixotropic bentonite suspension (adapted from Ovarlez and Chateau 2008)

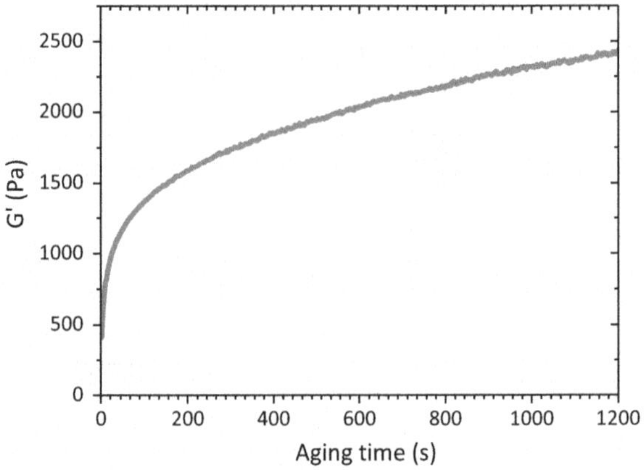

rest given above, it means that stress (and not strain) oscillations should be applied. In this case, the stress is indeed zero on average. One then has to plot the elastic modulus value versus time. A signature of structuration is found in the continuous increase of the elastic modulus in time (Derec et al. 2003; Coussot et al. 2006; Ovarlez and Chateau 2008) (Fig. 15).

It should be checked that it is a non-perturbative experiment, i.e., that applying these oscillations does not change the material state in time. This can be done by comparing the elastic modulus obtained after a long resting time in two situations: when oscillations are applied during all the resting period and when oscillations are applied only at the end of the resting period. It seems that as long as these measurements are performed in the linear regime of the materials, they are indeed nonperturbative (Ovarlez and Chateau 2008); on the other hand, oscillations applied in a non-linear regime have been observed to modify the structuration kinetics (Viasnoff and Lequeux 2002).

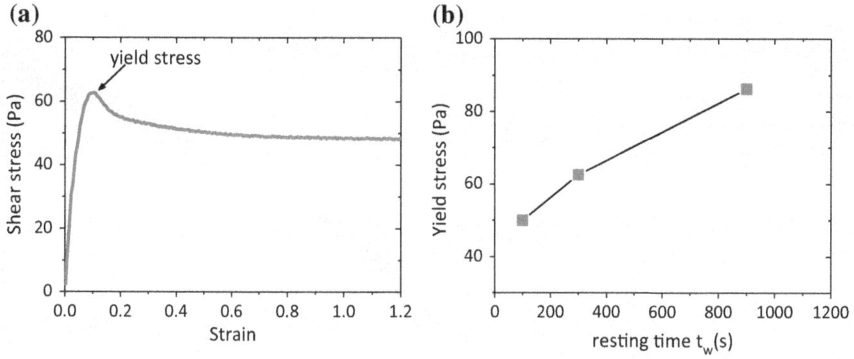

Fig. 16 **a** Shear stress versus strain for a yield stress measurement at low shear rate in a bentonite suspension. **b** Static yield stress versus resting time. Adapted from Ovarlez and Chateau (2008)

Static yield stress. Another consequence of structuration at rest is the increase of the static yield stress with the resting time: it is harder and harder in time to restart the flow of a thixotropic yield stress fluid. The static yield stress evolution can be studied by applying the yield stress measurement procedure described in Sect. 3 after a preshear and various resting periods (of different duration) at zero stress (as stated above). In such an experiment (Fig. 16a), the shear stress first increases roughly linearly with strain, corresponding to the elastic straining of the material in its solid regime; in agreement with Fig. 15, the initial slope of the shear stress increase versus strain (i.e. the elastic modulus) is higher for higher resting time. At some yield strain (usually of the order of 0.1 and poorly dependent of the resting time), a peak is observed and is followed by a slow decrease of the shear stress. The peak defines the static yield stress, the decrease being a signature of destructuration of the material under slow shear in the liquid regime. In this last regime, for measurements performed at a same given low shear rate, the shear stress should always tend to the same plateau value (close to the dynamic yield stress) whatever the resting period duration before the measurement. Checking that the same plateau value is reached in a series of static yield stress measurements is a good test of the reproducibility of the procedure.

In order to study the time dependence of the static yield stress, this procedure (preshear/rest/measurement) has to be repeated many times for various resting times, with each experiment providing only a single point of the static yield stress versus time relationship. In contrast with the elastic modulus measurement method, this method is indeed perturbative as the material is destructured by shear. Static yield stress measurements in a thixotropic suspension are displayed versus resting time in Fig. 16b and show a clear increase in time. Note that in many systems it is observed that the relative increase of the elastic modulus in time and that of the static yield stress are quantitatively similar (see Fig. 17). In this case, the elastic modulus experiment should be preferred as a systematic method of characterisation of structuration as it is much more rapid, being nonperturbative.

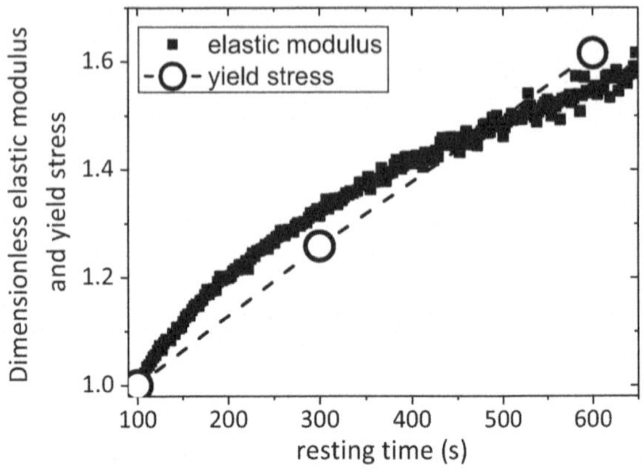

Fig. 17 Dimensionless elastic modulus and dimensionless yield stress versus resting time for a bentonite suspension. Adapted from Ovarlez and Chateau (2008)

However, counterexamples to this behaviour exist. Recent experiments (Roussel et al. 2012) suggest that the linear elastic modulus and the yield stress of cement pastes may find different origins and may then have different time evolutions. Two structuration mechanisms may indeed be at work, one linked to early age hydration (the signature of which is found on the elastic modulus) and another linked to rearrangements of the colloidal network (which impacts the yield stress).

5 Techniques to Study Local Flow Properties

Whatever care is taken, problems may occur during rheological measurements. This is indicated by the strong discrepancy between the values of the rheological parameters measured in the same materials using various rheometers (Ferraris and Brower 2001, 2004). These discrepancies must find their origin in the violation of one or several of the main hypotheses at the basis of the conversion of the macroscopic data (torque, rotational velocity) into local data (shear stress, shear rate), namely:

1. the no slip boundary condition on the geometry walls;
2. the homogeneity of the flow, and
3. the homogeneity of the material.

Indeed, wall slip leads to overestimation of the shear rate in the material and thus to underestimation of its apparent viscosity. Flow inhomogeneities (localisation) lead to underestimation of the true (local) shear rate and thus to overestimation of the material apparent viscosity. Material inhomogeneities (migration and sedimentation) generally lead to overestimation of the material volume fraction and thus to underestimation of its apparent viscosity.

Wall slip can be mostly avoided thanks to the use of roughened surfaces. In contrast, one has a priori no control of the flow and material homogeneity. These characteristics of the flow cannot be checked without a means of visualisation. Below we discuss the principle of local information measurements and show how they can help to obtain access to the local and intrinsic constitutive behaviour. We then illustrate their application to the analysis of flow inhomogeneities and of material inhomogeneities. We finally present specific issues with a vane-in-cup geometry.

Experimental methods. The basic idea is to couple a well-controlled rheometric experiment, which provides macroscopic information about the flow (the velocity and the torque at the boundary), and a probe of the local material velocity and of the local material composition. Such probe has, of course, to be non-perturbative, which makes the experiments challenging.

Several methods exist:

- In optical techniques (see for example Barentin et al. 2004), the motion of tracers is followed. In suspensions, the tracers are usually the constitutive elements of the material. A velocity field can be obtained either by tracking individual particles ('particle tracking' technique) or by tracking zones containing several particles thanks to image correlation ('particle image velocimetry' technique). The particle local concentration can also be measured directly. The main drawbacks of these techniques are their limitation to transparent materials and the need to seed the material with particles (which may be slightly perturbative) when the constitutive elements cannot be chosen as tracers. Note that other optical techniques exist such as those based on dynamic light scattering (Salmon et al. 2003).
- In acoustic techniques (see for example Manneville et al. 2004), a local velocity field can be obtained by sending ultrasonic pulses into the flowing material and correlating successive scattered signals (this provides a signature of the scatterer displacements). The scatterers are sometimes the material constitutive elements, but are more usually seeded particles, which may make the technique slightly perturbative. The main drawbacks of this technique are that the scattering properties of the material have to be compatible with it, and that the spatial attenuation of the signal does not allow measurements in wide gap geometries.
- In nuclear magnetic resonance (NMR) techniques (see for example Raynaud et al. 2002; Rodts et al. 2004), magnetic fields are used to interact with a quantum mechanical property of the hydrogen nuclei in the material, namely the spin. It is a nonintrusive technique, as no seeding is required and as the spins can be handled independently of any other material property. Depending on the sequences of magnetic fields applied to the material in time, one can get information about the local hydrogen density (which provides the particle volume fraction in a suspension) or the local material velocity. The main drawback of such a technique is of course its cost and its difficulty of implementation. Its main advantage is that it can a priori be applied to any material, without any seeding, and that it is provides information on the bulk of the material in wide gap geometries.

Local constitutive law determination. From these techniques, the local velocity in the material $V(\vec{r})$ is known everywhere, allowing computation of the local strain rate. In order to obtain the local constitutive behaviour of the material, one then needs to know the local stress value, which implies in particular a torque measurement in rheometric experiments. As discussed in Sect. 2, in simple shear experiments, the

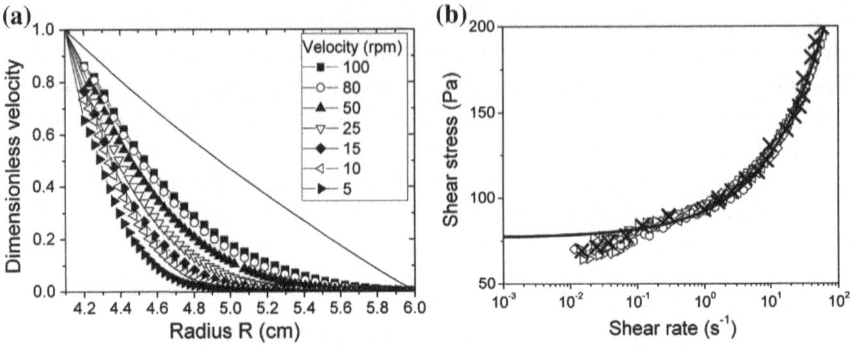

Fig. 18 **a** Dimensionless velocity as a function of the radial position within the gap of a wide gap Couette geometry (R_i=4.15 cm, R_o=6 cm) for a concentrated emulsion (NMR measurements) for various rotation velocity of the inner cylinder (see legend). **b** Local constitutive law inferred from the velocity profiles of Fig. 18a (empty symbols) and data obtained using Eq. 11 (crosses). Figures from Ovarlez et al. (2008)

stress distribution is known independently of the material behaviour in a Couette geometry and in a cone and plate geometry, but not in a parallel plate geometry (it is also known in capillary flows and inclined plane flows).

Below we illustrate the coupling of rheometric experiments and flow visualisation in the case of a Couette geometry. We consider a series of experiments performed at various rotational velocities Ω. We recall that the shear stress distribution $\tau\ (r, \Omega)$ within the gap of a Couette geometry is given by $\tau\ (r, \Omega) = T(\Omega)/2\pi r^2 H$, where $T(\Omega)$ is the torque exerted on the inner cylinder. The local shear rate $\dot{\gamma}(r, \Omega)$ is inferred from the velocity profiles and computed as:

$$\dot{\gamma}(r, \Omega) = \frac{V\ (r, \Omega)}{r} - \frac{\mathrm{d}V\ (r, \Omega)}{\mathrm{d}r} \tag{20}$$

If the material is homogeneous, i.e. if it is the same at any position r in the gap, then all the $\tau\ (r, \Omega)$ and $\dot{\gamma}(r, \Omega)$ data can be combined into a series of data of the constitutive law $\tau = f(\dot{\gamma})$ measured locally. An example of such constitutive law determination is shown and commented on below for a model yield stress fluid (a concentrated emulsion, see Fig. 18b).

In the case where the material is inhomogeneous, the local volume fraction $\phi\ (r, \Omega)$ has to be measured and taken into account, which is also possible with some of the methods described above (Ovarlez et al. 2006). When the inhomogeneous density profile $\phi\ (r)$ is independent of Ω, a change of variables can be performed between the radial position r and $\phi\ (r)$. A local stress/strain-rate curve $\tau(\dot{\gamma}, \phi)$—at fixed and well-defined density $\phi\ (r)$—is then obtained by collecting all measurements of local stress $\tau\ (r, \Omega)$ and shear rate $\dot{\gamma}(r, \Omega)$ for a fixed r and varying Ω (note that the analysis is also possible but more complex when $\phi\ (r, \Omega)$ depends on Ω). An example of such constitutive law determination is shown and commented later for a suspension of particles in a Newtonian fluid.

Finally, it is worth noting that, in contrast with purely macroscopic experiments, this combination of techniques allows the derivation of the constitutive law of complex suspensions without any hypothesis.

Flow localisation. We now discuss the observation of flow inhomogeneities in yield stress fluids; more details can be found in Ovarlez et al. (2009, 2013).

Shear localisation. In all geometries, the shear stress distribution is inhomogeneous (even if the inhomogeneity is slight in cone and plate and thin gap Couette geometries). In yield stress fluids, when the applied macroscopic stress is close to the yield stress τ_Y, this implies that the stress can be locally lower than τ_Y and that the material might not flow in some regions of the gap of the geometry. This phenomenon is called 'shear localisation' as flow occurs only in some regions.

This is illustrated in Fig. 18a where local velocities measured in a concentrated emulsion flowing in a wide gap Couette geometry are depicted. It is observed that the material does not flow in the whole gap, the velocity being zero in a region near the outer cylinder. Moreover, the fraction of the material that is sheared decreases when the rotational velocity Ω of the inner cylinder decreases. Shear localisation is here attributed to the $1/r^2$ decrease of the shear stress $\tau\,(r, \Omega)$, which passes below τ_Y at the interface $R_c(\Omega)$ between the flowing and the non flowing regions. When Ω tends to 0, $R_c(\Omega)$ tends to R_i, i.e. the thickness of the flowing region tends to zero and the stress $\tau\,(R_i, \Omega)$ at the inner cylinder tends to τ_Y.

In standard macroscopic experiments, this phenomenon of shear localisation prevents one from calculating the true (local) shear rate within the gap with the standard equations of rheometry (see Sect. 2) and may thus lead to wrong evaluation of the rheological properties of the material. It is nevertheless possible to take this artefact into account thanks to Eq. 11, which is supposed to be valid without any hypothesis about the flow. When shear is localised, the shear rate at the outer cylinder is zero and this equation reduces to Eq. 12, which is easy to handle. From a whole set of macroscopic measurements $T(\Omega)$, the material constitutive law can then be computed. Figure 18b shows a comparison of data obtained using this approach and data measured locally from the velocity profiles thanks to the approach presented above, in the concentrated emulsion of Fig. 18a. It is observed that the data obtained with Eq. 11 match perfectly the local behaviour, which is a strong validation of this equation (Ovarlez et al. 2008; Coussot et al. 2009).

In conclusion, one has to be careful when studying yield stress fluids. Shear localisation forbids the use of simple standard equations of rheometry, which lead to underestimation of the true shear rate. However, shear localisation can be properly taken into account using more elaborated equations such as Eq. 11 in a Couette geometry.

Shear banding. All yield stress fluids show shear localisation in an inhomogeneous shear stress field. However, very different features are observed in simple and thixotropic yield stress fluids. In simple yield stress fluids, the velocity profiles tend smoothly towards zero at the interface between the sheared and the unsheared regions; by contrast, in thixotropic yield stress fluids flow stops abruptly: there is a

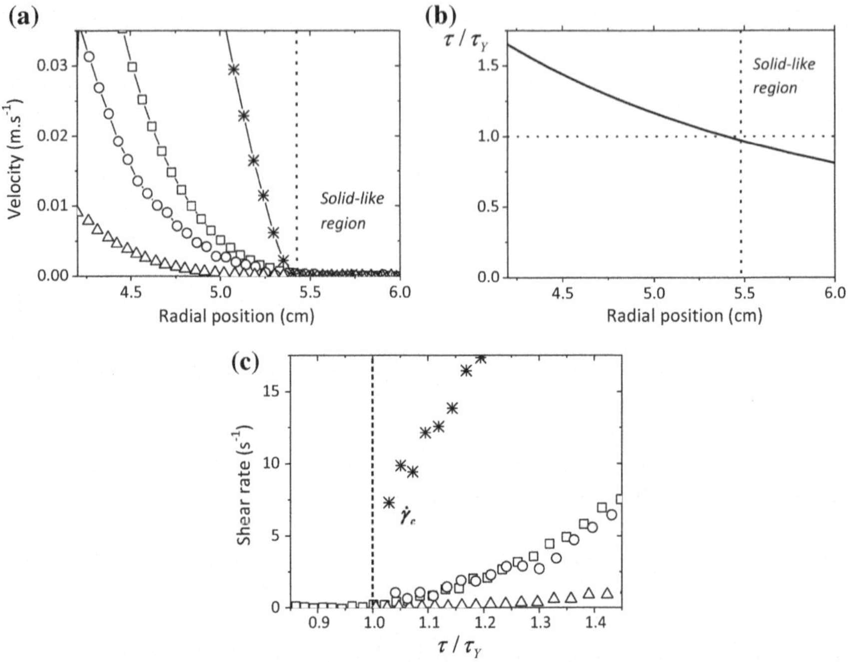

Fig. 19 **a** Azimuthal velocity profiles of simple yield stress fluids (open symbols) and a thixotropic fluid (stars) in a wide gap Couette. **b** Shear stress profiles in the geometry. **c** Local flow curves extracted from Fig. 19a. Replotted from Ovarlez et al. (2013)

discontinuity of the slope of the velocity profiles at the interface between the sheared and the unsheared regions. Figure 19 illustrates these features by comparing velocity profiles in both situations.

To understand these observations, it should be recalled that the shear rate is $\dot{\gamma}(r) = V(r)/r - \mathrm{d}V(r)/\mathrm{d}r$ in a Couette geometry, which reduces to $\dot{\gamma}(r) = -\mathrm{d}V(r)/\mathrm{d}r$ at the interface between the sheared and the unsheared regions. A non-zero slope of the velocity profile at flow stoppage is thus the signature of the phenomenon of viscosity bifurcation (see Sect. 4): thixotropic materials cannot flow steadily for a shear rate below a value $\dot{\gamma}_c$. This critical shear rate $\dot{\gamma}_c$ is thus an intrinsic property of thixotropic yield stress fluids, in addition to their yield stress. In contrast, the smooth transition between flow and rest observed in simple yield stress fluids means that $\dot{\gamma}(\tau) \overset{\tau \to \tau_Y}{\to} 0$, consistently with standard constitutive laws such as the Bingham and Herschel–Bulkley laws. For more details on this issue, see Ovarlez et al. (2009, 2013).

An important consequence of this specific behaviour of thixotropic materials is that flow localisation also occurs in homogeneous stress fields for these materials, in contrast with simple yield stress fluids (see also Sect. 4 for a discussion). This coexistence of two material states (here, sheared and unsheared) in a homogeneous

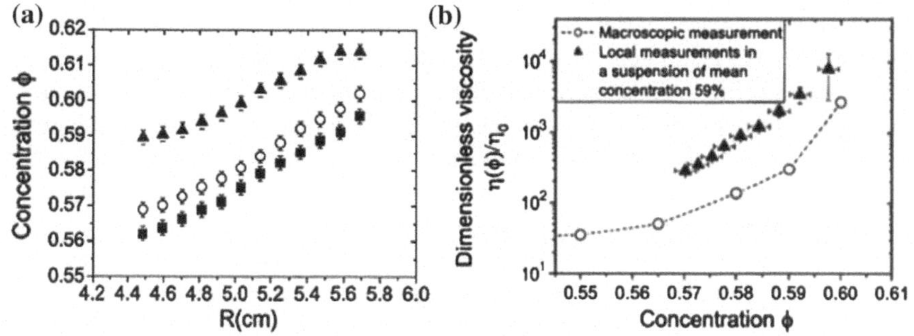

Fig. 20 **a** Radial concentration profiles in a Couette cell for suspensions of monodisperse spherical nonBrownian particles at various initial volume fraction (58, 59 and 60%) **b** Dimensionless viscosity of nonBrownian suspensions versus particle volume fraction: macroscopic measurement (triangles) and local measurements (circles). Reprinted with permission from Ovarlez et al. (2006). Copyright (2006) by the Society of Rheology

stress field is called 'shear banding'. In any slow flow, these materials will experience shear-banding.

Material inhomogeneities. This section discusses the observations of flow-induced material inhomogeneities in suspensions with a focus on their consequences.

Shear induced migration. Volume fraction measurements have led to show the development of concentration inhomogeneities in noncolloidal suspensions in wide-gap Couette flows (Phillips et al. 1992; Ovarlez et al. 2006), parallel-plate flows (Chow et al. 1994; Barentin et al. 2004; Merhi et al. 2005), pipe flows (Sinton and Chow 1991) and extrusion experiments (Altobelli et al. 1997). It has been observed in neutrally buoyant suspensions and is thus a shear-induced migration phenomenon; other phenomena occur in the presence of gravity. It was observed mostly in suspensions of noncolloidal particles in Newtonian media. However, some experiments showed that the same phenomena occur when particles are suspended in viscoelastic fluids (Tehrani 1996) and in yield stress fluids (Hafid et al. 2015; Ovarlez et al. 2011, 2015). In polydisperse systems, these inhomogeneities can be accompanied by size segregation, which finds its origin in the particle size dependence of the kinetics of migration.

In wide-gap Couette flows, the consequence of migration is an excess of particles near the outer cylinder, i.e. particles tend to migrate from the regions of high shear to the regions of low shear (Fig. 20a). Migration is actually a direct consequence of shear inhomogeneities. The basic mechanism is as follows: due to their interactions, particles experience shear-induced diffusion (Leighton and Acrivos 1987). The shear-induced diffusion coefficient being an increasing function of the shear rate, shear inhomogeneities naturally lead to a flux of particles towards the zones of low shear. For more details and theoretical developments, the reader is referred to Phillips et al. (1992), Morris and Boulay (1999), Ovarlez et al. (2006) and Lhuillier (2009).

When dealing with macroscopic rheometric measurements, the main problem is to avoid this phenomenon. If migration occurs during an experiment, then macroscopic data cannot be analysed as the response of a well-defined system. Data obtained on an inhomogeneous system generally lead to underestimation of the apparent viscosity of the material (Fig. 20b, Ovarlez et al. 2006). It is often considered in the literature on complex suspensions that migration is a slow phenomenon and that it is therefore possible to perform rheological measurements on a homogeneous material before the development of any significant inhomogeneity. It was indeed observed in Newtonian suspensions that it is usually a slow diffusive phenomenon. The typical strainscale for migration to occur is expected to be rate independent and to scale as $\propto \text{gap}^2/a^2$ where a is the particle size, i.e. the use of large gap to particle size ratio is likely to prevent migration. This would, for example, lead to an expected strain of the order of 10 000 for migration to occur when the gap is 100 particles wide.

However, it has been recently shown that, at high particle volume fraction, migration may be very rapid in some cases (depending on the flow characteristics) and thus unavoidable: the strain to induce migration can be smaller than 50 with 40 μm particles in a 2 cm gap, which is more than 1000 times faster than expected (Fall et al. 2010, 2015). Consequently, this phenomenon can basically not be avoided when performing rheological measurements on such systems and it may occur in most dense suspensions studies.

The impact of migration on the rheological measurements can be quantified by comparing the true (local) behaviour $\tau(\dot{\gamma}, \phi_0)$ of the material at a well-defined constant volume fraction ϕ_0, obtained thanks to the technique detailed earlier, and the behaviour obtained from macroscopic experiments by assuming (wrongly) that the material is homogeneous (see Fig. 20b). It is clearly observed that the macroscopic measurements tend to underestimate the viscosity of the suspensions. Note that the discrepancy between local/macroscopic measurements increases with the volume fraction (Ovarlez et al. 2006; Hafid et al. 2015). This phenomenon is most likely the main source of error in dense suspensions rheology (Ovarlez et al. 2006; Hafid et al. 2015).

Migration is part of the behaviour of the material and cannot be prevented. The only thing one can do in a rheological experiment is to evaluate the possible occurrence of migration and its impact. As migration is an irreversible process, the classical experiment consists of applying exactly the same macroscopic shear rate (or shear stress) at the beginning and at the end of a series of experiments: if the same response is obtained in both cases, it should mean that the material—supposed to be homogeneous just after loading—has remained homogeneous and that all measurements can be trusted. However, much care has to be taken with this technique: remember that migration can be very fast. Preshear thus has to be avoided; if a steady inhomogeneous state is reached during the preshear, then the above procedure may lead one to wrongly conclude that the material is homogeneous. Ideally, the response at the very beginning of shear (typically, for a strain lower than ten) should be taken as a reference. It thus seems that clear conclusions can be reached from such approach in non-thixotropic systems only. In thixotropic materials, the time needed to reach steady state during the first measurement can be sufficient to generate steady inhomogeneities, in which case the above experiment may again lead one to wrongly conclude that the material is homogeneous.

Shear-induced settling. Particle volume fraction measurements coupled with rheometric experiments also allow the coupling between flow and settling to be studied. It has indeed been shown that particles that are stable at rest in a yield stress fluid tend to settle when the material flows (Ovarlez et al. 2010, 2012). To study this phenomenon properly, the effect of shear has to be decoupled as much as possible from the effect of gravity. A Couette geometry is then the best choice that can be made as it imposes shear in the plane perpendicular to sedimentation.

This phenomenon is illustrated in Fig. 21 for a suspension of noncolloidal particles in a yield stress fluid. After introduction in the Couette cell, the sample is homogeneous. When left at rest, the sample remains homogeneous and the particles do not settle. When the sample is sheared, particles move downwards. This is evidenced by the volume fraction profiles, which show classical features of sedimentation. The upper part is at 0% concentration, the middle part remains at the initial 5% concentration, and the particles accumulate at the bottom of the geometry, below the inner cylinder. The transition zone between the first two parts defines a sedimentation front that moves regularly towards the bottom as the flow duration increases. This sedimentation in a suspension that is stable at rest means that, under shear in the parallel plane, there is no more yield resistance to flow in the vertical direction. More details and a quantitative analysis can be found in Ovarlez et al. (2012). The main result is that the sedimentation velocity of particles in a yield stress fluid sheared at a shear rate $\dot{\gamma}$ is conveniently approximated by the Stokes velocity $V \propto \Delta\rho g R^2 / \eta_{app}$ in a fluid of viscosity equal to the apparent viscosity $\eta_{app} = \tau(\dot{\gamma})/\dot{\gamma}$. This simple equation allows evaluation of the sedimentation velocity and thus of the possible occurrence of this phenomenon when studying suspensions of particles in yield stress fluids such as muddy materials and mortars. Note that this settling velocity decreases when the particle volume fraction increases.

Finally, let us also mention the phenomenon of shear-induced resuspension, which was evidenced using the same methods (Acrivos et al. 1993): in a Couette cell, particles that tend to settle at rest are resuspended by shear. The origin of this phenomenon is basically the same as that of shear-induced migration. The efficiency of this phenomenon is enhanced at high shear rate and high volume fraction.

Flow in a vane-in-cup geometry. Flow visualisations are also crucial for studying flows in geometries for which no theoretical solution exists, as a complement to simulations. For example, for a vane-in-cup geometry, theoretical equations exist only for linear materials in an infinite cup. Below we illustrate the contribution of flow visualisation to the study of flows with a vane tool with two examples. For more details, the reader is referred to Ovarlez et al. (2011).

Flows near the yield stress. A practical method used to study the flow properties of complex suspensions with a vane tool, known as the Couette analogy (see Sect. 2), consists of analysing macroscopic data as if the material was sheared in a Couette geometry of inner cylinder radius $R_{i,eq}$. While the streamlines go into the virtual cylinder delimited by the blades in Newtonian media (Baravian et al. 2002), yielding a $R_{i,eq}$ value lower than the vane radius (Sherwood and Meeten 1991), it was found that the streamlines are nearly cylindrical (Barnes and Carnali 1990; Savarmand et al. 2007) and thus that $R_{i,eq} = R_i$ for shear-thinning fluids of index n lower than 0.5.

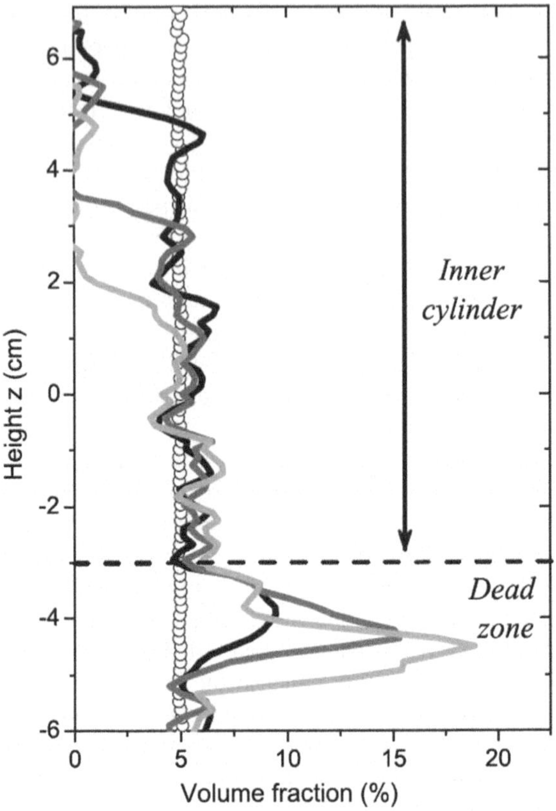

Fig. 21 Vertical volume fraction profiles observed in the gap of a Couette geometry in a suspension of 5% of glass beads in a concentrated emulsion, after a 24 h rest (circles) and after 15 min (black line), 30 min (dark grey line), and 45 min (light grey line) of shear at 4 s^{-1}. Reprinted from Ovarlez et al. (2012) with permission from Elsevier

For yield stress fluids, several works (Keentok et al. 1985; Yan and James 1997; Savarmand et al. 2007) show that at yield (i.e. at low shear rates), the material flows uniformly in a thin cylindrical layer near the blades. This is now widely accepted (Barnes and Nguyen 2001) and used to perform a Couette analogy with $R_{i,eq} = R_i$; the yield stress τ_Y is then simply extracted using Eq. 17 (Nguyen and Boger 1992).

Velocity measurements in a Herschel–Bulkley fluid of index $n=0.5$ show that the picture may be more complex (Ovarlez et al. 2011). Figure 22a shows ten azimuthal velocity profiles measured between two adjacent blades of a six-bladed vane rotated at low velocity (0.1 rpm). There is a strong θ-dependence of the velocity profiles. The velocity profile that starts near from a blade tip (at $\theta=0°$) has a much steeper slope than the profile measured midway between adjacent blades (at $\theta=30°$), which shows that the blade tip neighbourhoods are regions of high shear. Meanwhile the flow stops at a radius R_c which is larger at $\theta=30°$ than at $\theta=0°$. Figure 22b shows a two-dimensional map of the flow field, indicating both the boundary between the region of rigid body rotation (between the blades) and the sheared region, and the boundary between the sheared region and the outer region of fluid at rest (i.e. the

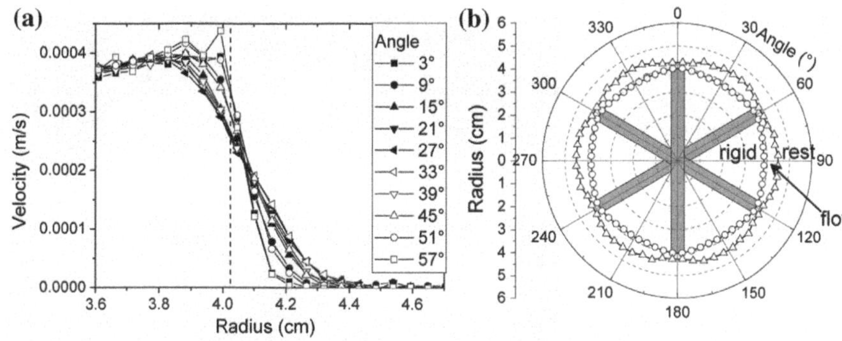

Fig. 22 **a** Azimuthal velocity profile of a yield stress fluid (concentrated emulsion) sheared at 0.1 rpm in a six-bladed vane-in-cup geometry, for various angular positions θ between two adjacent blades (from θ = 0° to θ = 60°). The vertical dashed line shows the radius of the vane. **b** Two-dimensional plot of the limit between rigid motion and shear (circles) and between shear and rest (triangles). The grey rectangles correspond to the blades. Reprinted with permission from Ovarlez et al. (2011). Copyright (2011) by the Society of Rheology

position where the yield criterion is satisfied). Flow is found to occur in a layer of complex shape which is far from being cylindrical even at this very low velocity.

These observations contradict the usually accepted picture for yield stress fluid flows at low rates, namely that the material flows uniformly in a thin cylindrical layer near the blades. This means that further investigation on the exact conditions under which this symmetry can be recovered is needed. It is particularly striking and counterintuitive that R_c is largest at the angular position ($\theta = 30°$) where shear at R_i is smallest. It thus seems that the link between the yield stress τ_Y and the torque T measured at yield with a vane-in-cup geometry is still an open question, although the classical formula given in Eq. 17 probably provides a sufficiently accurate determination of τ_Y in practice.

Slip with a vane tool. Figure 23 shows an image of a suspension of large particles in a yield stress fluid after a flow in a vane-in-cup geometry. In this image obtained by magnetic resonance imaging (MRI), brighter zones contain less particles than the bulk material. Quantitative analysis (Ovarlez et al. 2011) shows that there are actually no more particles in a thin layer (a few particle diameters wide), the position of which marks the transition between the unsheared material (which rotates as a rigid body between the blades) and the sheared material.

This observation of particle depletion near the blades when the yield stress fluid contains noncolloidal particles has important consequences. It has to be noted that the phenomenon is very rapid (Ovarlez et al. 2011) and irreversible, and thus probably unavoidable when studying flows of suspensions of large particles. It results in the creation of a pure interstitial yield stress fluid layer and thus in a kind of wall slip near the blades. It contrasts with the classical assumption that is made in the field of concentrated suspension rheology where the vane tool is mainly used to avoid this phenomenon (Saak et al. 2001). As the flow of the suspension is localised near the

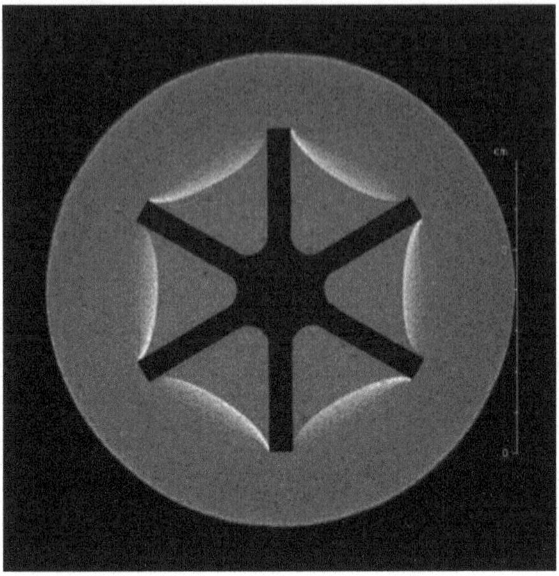

Fig. 23 Two-dimensional magnetic resonance image of a suspension of particles in a yield stress fluid after shear in a vane-in-cup geometry. Reprinted with permission from Ovarlez et al. (2011). Copyright (2011) by the Society of Rheology

inner tool at low shear rate, only the pure interstitial fluid created by migration near the blades remains in the sheared layer at a sufficiently low rotational velocity of the vane tool. This explains why the same apparent value of the yield stress can be found in the suspension as in the pure interstitial fluid (Mahaut et al. 2008a; Ovarlez et al. 2011). Consequently, in the case of pasty materials with large particles, the vane tool may finally be suitable only for the study of the solid (elastic) properties of materials and for the static yield stress measurements (Alderman et al. 1991; Mahaut et al. 2008a; Mahaut et al. 2008b). As the yield stress measurement may induce irreversible particle depletion near the blades, any new measurement then requires a new sample preparation. In order to study accurately the flows of pasty materials with large particles, these results suggest that when possible a coaxial cylinders geometry with properly roughened surfaces is preferable. If the use of a vane tool cannot be avoided, one should keep in mind these observations in order to carefully interpret any result.

References

Acrivos, A., Mauri, R., & Fan, X. (1993). Shear-induced resuspension in a Couette device. *International Journal of Multiphase Flow, 19,* 797–802.

Alderman, N. J., Meeten, G. H., & Sherwood, J. D. (1991). Vane rheometry of bentonite gels. *Journal of Non-Newtonian Fluid Mechanics, 39,* 291–310.

Altobelli, S. A., Fukushima, E., & Mondy, L. A. (1997). Nuclear magnetic resonance imaging of particle migration in suspensions undergoing extrusion. *Journal of Rheology, 41,* 1105–1115.

Balmforth, N., Frigaard, I., & Ovarlez, G. (2014). Yielding to stress: Recent developments in viscoplastic fluid mechanics. *Annual Review of Fluid Mechanics, 46,* 121–146.
Baravian, C., Lalante, A., & Parker, A. (2002). Vane rheometry with a large, finite gap. *Applied Rheology, 12,* 81–87.
Barentin, C., Azanza, E., & Pouligny, B. (2004). Flow and segregation in sheared granular slurries. *Europhysics Letters, 66,* 139–145.
Barnes, H. A., & Carnali, J. O. (1990). The vane-in-cup as a novel rheometer geometry for shear thinning and thixotropic materials. *Journal of Rheology, 34,* 841–866.
Barnes, H. A., & Nguyen, Q. D. (2001). Rotating vane rheometry—a review. *Journal of Non-Newtonian Fluid Mechanics, 98,* 1–14.
Bousmina, M., Aït-Kadi, A., & Faisant, J. B. (1999). Determination of shear rate and viscosity from batch mixer data. *Journal of Rheology, 43,* 415–433.
Boyer, F., Pouliquen, O., & Guazzelli, É. (2011). Dense suspensions in rotating-rod flows: Normal stresses and particle migration. *Journal of Fluid Mechanics, 686,* 5–25.
Chow, A. W., Sinton, S. W., Iwamiya, J. H., & Stephens, T. S. (1994). Shear-induced particle migration in Couette and parallel-plate viscometers: NMR imaging and stress measurements. *Physics of Fluids, 6,* 2561–2576.
Coussot, P. (2005). *Rheometry of pastes, suspensions and granular materials*. New York: John Wiley & Sons.
Coussot, P., Nguyen, Q. D., Huynh, H. T., & Bonn, D. (2002). Viscosity bifurcation in thixotropic, yielding fluids. *Journal of Rheology, 46,* 573–589.
Coussot, P., Tabuteau, H., Chateau, X., Tocquer, L., & Ovarlez, G. (2006). Aging and solid or liquid behavior in pastes. *Journal of Rheology, 50,* 975–994.
Coussot, P., Tocquer, L., Lanos, C., & Ovarlez, G. (2009). Macroscopic versus local rheology of yield stress fluids. *Journal of Non-Newtonian Fluid Mechanics, 158,* 85–90.
Deboeuf, A., Gauthier, G., Martin, J., Yurkovetsky, Y., & Morris, J. F. (2009). Particle pressure in a sheared suspension: A bridge from osmosis to granular dilatancy. *Physical Review Letters, 102,* 108301.
Derec, C., Ducouret, G., Ajdari, A., & Lequeux, F. (2003). Aging and nonlinear rheology in suspensions of polyethylene oxide–protected silica particles. *Physical Review E, 67,* 061403.
Dullaert, K., & Mewis, J. (2005). Thixotropy: Build-up and breakdown curves during flow. *Journal of Rheology, 49,* 1213–1230.
Estellé, P., Lanos, C., Perrot, A., & Amziane, S. (2008). Processing the vane shear flow data from Couette analogy. *Applied Rheology, 18,* 34037–34481.
Ewoldt, R. H., Hosoi, A. E., & McKinley, G. H. (2008). New measures for characterizing nonlinear viscoelasticity in large amplitude oscillatory shear. *Journal of Rheology, 52,* 1427–1458.
Fall, A., Lemaître, A., Bertrand, F., Bonn, D., & Ovarlez, G. (2010). Shear thickening and migration in granular suspensions. *Physical Review Letters, 105,* 268303.
Fall, A., Bertrand, F., Hautemayou, D., Mezière, C., Moucheront, P., Lemaître, A., et al. (2015). Macroscopic discontinuous shear thickening versus local shear jamming in cornstarch. *Physical Review Letters, 114,* 098301.
Ferraris, C. F., & Brower, L. E. (2001). *Comparison of concrete rheometers: International tests at LCPC (Nantes, France) in October, 2000*. National Institute of Standards and Technology Interagency Report (NISTIR) 6819.
Ferraris, C. F., & Brower, L. E. (2004). *Comparison of concrete rheometers: International tests at MB (Cleveland OH, USA) in May, 2003*. National Institute of Standards and Technology Interagency Report (NISTIR) 7154.
Hafid, H., Ovarlez, G., Toussaint, F., Jezequel, P. H., & Roussel, N. (2015). Assessment of potential concrete and mortar rheometry artifacts using magnetic resonance imaging. *Cement and Concrete Research, 71,* 29–35.
Jarny, S., Roussel, N., Rodts, S., Le Roy, R., & Coussot, P. (2005). Rheological behavior of cement pastes from MRI velocimetry. *Cement and Concrete Research, 35,* 1873–1881.

Jarny, S., Roussel, N., Le Roy, R., & Coussot, P. (2008). Modelling thixotropic behavior of fresh cement pastes from MRI measurements. *Cement and Concrete Research, 38,* 616–623.

Jop, P., Forterre, Y., & Pouliquen, O. (2006). A constitutive law for dense granular flows. *Nature, 441,* 727–730.

Keentok, M., Milthorpe J. F., & ODonovan, E. (1985). On the shearing zone around rotating vanes in plastic liquids: Theory and experiment. *Journal of Non-Newtonian Fluid Mechanics*, 17:23–35.

Koehler, E. P., Fowler, D. W., Ferraris, C. F., & Amziane, S. A. (2006). New, portable rheometer for fresh self-consolidating concrete. *ACI Materials Journal, 233,* 97–116.

Leighton, D., & Acrivos, A. (1987). Measurement of shear-induced self-diffusion in concentrated suspensions of spheres. *Journal of Fluid Mechanics, 177,* 109–131.

Lhuillier, D. (2009). Migration of rigid particles in non-Brownian viscous suspensions. *Physics of Fluids, 21,* 023302.

Macosko, C. W. (1994). *Rheology: Principles, measurements, and applications*. New York: Wiley.

Mahaut, F., Chateau, X., Coussot, P., & Ovarlez, G. (2008a). Yield stress and elastic modulus of suspensions of noncolloidal particles in yield stress fluids. *Journal of Rheology, 52,* 287–313.

Mahaut, F., Mokéddem, S., Chateau, X., Roussel, N., & Ovarlez, G. (2008b). Effect of coarse particle volume fraction on the yield stress and thixotropy of cementitious materials. *Cement and Concrete Research, 38,* 1276–1285.

Manneville, S., Becu, L., & Colin, A. (2004). High-frequency ultrasonic speckle velocimetry in sheared complex fluids. *The European Physical Journal Applied Physics, 28,* 361–373.

Merhi, D., Lemaire, E., Bossis, G., & Moukalled, F. (2005). Particle migration in a concentrated suspension flowing between rotating parallel plates: Investigation of diffusion flux coefficients. *Journal of Rheology, 49,* 1429–1448.

Morris, J. F., & Boulay, F. (1999). Curvilinear flows of noncolloidal suspensions: The role of normal stresses. *Journal of Rheology, 43,* 1213–1237.

Mujumdar, A., Beris, A. N., & Metzner, A. B. (2002). Transient phenomena in thixotropic systems. *Journal of Non-Newtonian Fluid Mechanics, 102,* 157–178.

Nguyen, Q. D., & Boger, D. V. (1992). Measuring the flow properties of yield stress fluids. *Annual Review of Fluid Mechanics, 24,* 47–88.

Ovarlez, G., & Chateau, X. (2008). Influence of shear stress applied during flow stoppage and rest period on the mechanical properties of thixotropic suspensions. *Physical Review E, 77,* 061403.

Ovarlez, G., Bertrand, F., & Rodts, S. (2006). Local determination of the constitutive law of a dense suspension of noncolloidal particles through magnetic resonance imaging. *Journal of Rheology, 50,* 259–292.

Ovarlez, G., Rodts, S., Ragouilliaux, A., Coussot, P., Goyon, J., & Colin, A. (2008). Wide-gap Couette flows of dense emulsions: Local concentration measurements, and comparison between macroscopic and local constitutive law measurements through magnetic resonance imaging. *Physical Review E, 78,* 036307.

Ovarlez, G., Rodts, S., Chateau, X., & Coussot, P. (2009). Phenomenology and physical origin of shear localization and shear banding in complex fluids. *Rheologica Acta, 48,* 831–844.

Ovarlez, G., Barral, Q., & Coussot, P. (2010). Three-dimensional jamming and flows of soft glassy materials. *Nature Materials, 9,* 115–119.

Ovarlez, G., Mahaut, F., Bertrand, F., & Chateau, X. (2011). Flows and heterogeneities with a vane tool: MRI measurements. *Journal of Rheology, 55,* 197–223.

Ovarlez, G. (2011). Introduction to the rheometry of complex suspensions. In N. Roussel (Ed.), *Understanding the rheology of concrete*. Cambridge: Woodhead Publishing.

Ovarlez, G., Bertrand, F., Coussot, P., & Chateau, X. (2012). Shear-induced sedimentation in yield stress fluids. *Journal of Non-Newtonian Fluid Mechanics, 177,* 19–28.

Ovarlez, G., Cohen-Addad, S., Krishan, K., Goyon, J., & Coussot, P. (2013). On the existence of a simple yield stress fluid behavior. *Journal of Non-Newtonian Fluid Mechanics, 193,* 68–79.

Ovarlez, G., Mahaut, F., Deboeuf, S., Lenoir, N., Hormozi, S., & Chateau, X. (2015). Flows of suspensions of particles in yield stress fluids. *Journal of Rheology, 59,* 1449–1486.

Phillips, R. J., Armstrong, R. C., Brown, R. A., Graham, A. L., & Abbott, J. R. (1992). A constitutive equation for concentrated suspensions that accounts for shear-induced particle migration. *Physics of Fluids, 4,* 30–40.
Raynaud, J. S., Moucheront, P., Baudez, J. C., Bertrand, F., Guilbaud, J. P., & Coussot, P. (2002). Direct determination by NMR of the thixotropic and yielding behavior of suspensions. *Journal of Rheology, 46,* 709–732.
Rodts, S., Bertrand, F., Jarny, S., Poullain, P., & Moucheront, P. (2004). Développements récents dans lapplication de lIRM à la rhéologie et à la mécanique des fluides. *Comptes Rendus Chimie, 7,* 275–282.
Roussel, N., Le Roy, R., & Coussot, P. (2004). Thixotropy modelling at local and macroscopic scales. *Journal of Non-Newtonian Fluid Mechanics, 117,* 85–95.
Roussel, N., Ovarlez, G., Garrault, S., & Brumaud, C. (2012). The origins of thixotropy of fresh cement pastes. *Cement and Concrete Research, 42,* 148–157.
Rueb, C. J., & Zukoski, C. F. (1997). Viscoelastic properties of colloidal gels. *Journal of Rheology, 41,* 197–218.
Saak, A. W., Jennings, H. M., & Shah, S. P. (2001). The influence of wall slip on yield stress and viscoelastic measurements of cement paste. *Cement and Concrete Research, 31,* 205–212.
Salmon, J. B., Manneville, S., Colin, A., & Pouligny, B. (2003). An optical fiber based interferometer to measure velocity profiles in sheared complex fluids. *The European Physical Journal Applied Physics, 22,* 1436154.
Savarmand, S., Heniche, M., Bechard, V., Bertrand, F., & Carreau, P. J. (2007). Analysis of the vane rheometer using 3D finite element simulation. *Journal of Rheology, 51,* 161–177.
Sherwood, J. D., & Meeten, G. H. (1991). The use of the vane to measure the shear modulus of linear elastic solids. *Journal of Non-Newtonian Fluid Mechanics, 41,* 101–118.
Sinton, S. W., & Chow, A. W. (1991). NMR flow imaging of fluids and solid suspensions in Poiseuille flow. *Journal of Rheology, 35,* 735–772.
de Souza Mendes, P. R. (2009). Modeling the thixotropic behavior of structured fluids. *Journal of Non-Newtonian Fluid Mechanics, 164,* 66–75.
Tehrani, M. A. (1996). An experimental study of particle migration in pipe flow of viscoelastic fluids. *Journal of Rheology, 40,* 1057–1077.
Toorman, E. A. (1997). Modelling the thixotropic behaviour of dense cohesive sediment suspensions. *Rheologica Acta, 36,* 56–65.
Viasnoff, V., & Lequeux, F. (2002). Rejuvenation and overaging in a colloidal glass under shear. *Physical Review Letters, 89,* 065701.
Wallevik, J. E. (2008). Minimizing end-effects in the coaxial cylinders viscometer: Viscoplastic flow inside the ConTec BML Viscometer 3. *Journal of Non-Newtonian Fluid Mechanics, 155,* 116–123.
Wilhelm, M., Maring, D., & Spiess, H. W. (1998). Fourier-transform rheology. *Rheologica Acta, 37,* 399–405.
Yan, J., & James, A. E. (1997). The yield surface of viscoelastic and plastic fluids in a vane viscometer. *Journal of Non-Newtonian Fluid Mechanics, 70,* 237–253.

Rheology of Visco-Plastic Suspensions

Guillaume Ovarlez

1 Introduction

This chapter deals with the rheological properties of viscoplastic suspensions. It is largely based on Ovarlez et al. (2015).

Dense suspensions involved in industrial processes (concrete casting, drilling muds, foodstuff transport...) and natural phenomena (debris-flows, lava flows...) are usually very polydisperse systems. Many of these systems are yield stress fluids, in which the yield stress primarily arises from the colloidal forces between the smallest suspended particles[1] (Mewis and Wagner 2012). When scale separation is possible between these last particles and the largest—nonBrownian—particles in suspension, these materials can be considered in a first step as viscoplastic suspensions, i.e., suspensions of nonBrownian particles embedded in a yield stress fluid [e.g. fresh concrete $\approx$ sand and granulate in cement paste (Geiker et al. 2002; Toutou and Roussel 2006; Erdogan et al. 2008; Joumana et al. 2008; Mahaut et al. 2008b); debris-flows $\approx$ rocks in mud (Coussot 1997; Ancey and Jorrot 2001)]. Substantial progress in the understanding of the behavior of such materials can thus be made by studying the impact of adding nonBrownian particles to a yield stress fluid of known properties (Mahaut et al. 2008a; Chateau et al. 2008; Vu et al. 2010). From a more fundamental point of view, these systems, viewed as rigid inclusions in a nonlinear material, may provide crucial tests for micromechanical approaches developed to describe the behavior of composite materials. Model materials with a simple behavior (e.g., close to the ideal elastoplastic behavior) can indeed be designed experimentally (Mahaut et al. 2008a).

[1] We do not consider here the case of frictional materials.

G. Ovarlez (✉)
Université de Bordeaux, CNRS, Solvay, LOF - UMR 5258, 33608 Pessac, France
e-mail: guillaume.ovarlez@u-bordeaux.fr

G. Ovarlez and S. Hormozi (eds.), *Lectures on Visco-Plastic Fluid Mechanics*,
CISM International Centre for Mechanical Sciences 583,
https://doi.org/10.1007/978-3-319-89438-6_5

Most studies of suspensions of nonBrownian particles have focused on the case of a Newtonian suspending fluid (Stickel and Powell 2005; Denn and Morris 2014). Similar to the interstitial fluid, such suspensions have a viscous behavior. Their shear viscosity $\eta(\phi)$ is an increasing function of the particle volume fraction ϕ that diverges at the approach of the maximum flowing packing fraction $\phi_{\rm div}$. This divergence comes from the divergence of the local shear rate and thus of viscous dissipation in the interstices between the particles, which get infinitely small as ϕ approaches $\phi_{\rm div}$ (Frankel and Acrivos 1967). The value of $\phi_{\rm div}$ depends on the particle shape and on the polydispersity. Even for monodisperse spheres, there is no consensus on its value (e.g. Boyer et al. (2011b) find 58% whereas Ovarlez et al. (2006) find 60.5%) which may indicate strong dependence on the details of particle interaction at close distance (Sierou and Brady 2002).

Suspensions in Newtonian fluids also exhibit non-Newtonian behavior, in particular normal stress differences (Zarraga et al. 2000; Boyer et al. 2011a; Couturier et al. 2011; Dbouk et al. 2013; Dai et al. 2013) proportional to the shear rate, which become significant at high shear rates. The $\dot\gamma$ dependence of normal stresses originates from hydrodynamic interactions, whereas the emergence of normal stress differences may be related to a shear-induced anisotropy of the microstructure (the particle spatial distribution) (Parsi and Gadala-Maria 1987; Morris 2009; Blanc et al. 2013) that is generated by hard sphere repulsions (Brady and Morris 1997; Blanc et al. 2011a). At a macroscopic scale, these materials are often found to develop volume fraction inhomogeneities (Abbott et al. 1991; Phillips et al. 1992; Tetlow et al. 1998; Shapley et al. 2004; Ovarlez et al. 2006; Altobelli et al. 1991, 1997).

Viscoplastic suspensions have been the subject of few studies and less is known about their behavior. The present chapter reviews their main features. In Sect. 2, we discuss the theoretical behavior of suspensions of rigid particles in linear and nonlinear media. In Sect. 3, we present appropriate model systems, experimental setups and methods: a key problem one has to face is indeed the design of experimental systems that can be described theoretically as rigid particles in a viscoplastic fluid interacting only through hydrodynamical interactions mediated by the fluid viewed as a continuum. The main experimental observations are presented in Sect. 4: we present the evolution of the elastic, plastic, and flow properties with the particle volume fraction, and we discuss the emergence of a shear-dependent microstructure; throughout this part, we compare the experimental results and the theoretical predictions. Finally, shear-induced migration and its link with normal stress differences are briefly discussed in Sect. 5.

2 Theory

In the following, the stress tensor is denoted σ_{ij}, the deviatoric stress tensor is denoted τ_{ij}, the strain tensor is denoted ε_{ij}, and the strain rate tensor is denoted d_{ij}. The shear stress is $\tau = \sqrt{\frac{1}{2}\tau_{ij}\tau_{ij}}$, the shear strain is $\gamma = \sqrt{2\varepsilon_{ij}\varepsilon_{ij}}$ and the shear rate is

$\dot{\gamma} = \sqrt{2d_{ij}d_{ij}}$. In the case of a simple shear flow, the flow is described only by γ and $\dot{\gamma}$, the stress is described by τ and by two normal stress differences $N_1 = \sigma_{11} - \sigma_{22} = \tau_{11} - \tau_{22}$ and $N_2 = \sigma_{22} - \sigma_{33} = \tau_{22} - \tau_{33}$; indexes 1, 2, and 3, correspond, respectively, to the velocity direction, to the velocity gradient direction, and to the vorticity direction.

We consider monodisperse[2] rigid particles of diameter d dispersed in a fluid at a volume fraction ϕ. The approach presented below describes the general case of a nonlinear suspending fluid (which includes yield stress fluids). As will be discussed in the following, the theoretical approach works strictly for an *isotropic* suspending fluid only. We thus assume that it can be described by a generalized Newtonian behavior

$$\tau_{ij} = 2\eta(\dot{\gamma})d_{ij} \tag{1}$$

i.e., $\tau = \eta(\dot{\gamma})\dot{\gamma}$ in a simple shear flow, where $\eta(\dot{\gamma})$ is the shear-rate-dependent apparent viscosity of the material. In the case of a suspending yield stress fluid, we will consider that the material flow behavior is described by the Herschel-Bulkley behavior for applied shear stresses τ above a yield stress $\tau_y(0)$ (which correspond to a von Mises criterion):

$$\tau_{ij} = 2\frac{\tau_y(0) + \eta_{HB}(0)\dot{\gamma}^{n(0)}}{\dot{\gamma}}d_{ij} \text{ when } \tau \geq \tau_y(0) \tag{2}$$

where $\eta_{HB}(0)$ is the consistency, and $n(0)$ the index; in simple shear, this yields:

$$\tau = \tau_y(0) + \eta_{HB}(0)\dot{\gamma}^{n(0)} \text{ when } \tau \geq \tau_y(0) \tag{3}$$

This corresponds to the behavior described by Eq. 1 with $\eta(\dot{\gamma}) = (\tau_y(0) + \eta_{HB}(0)\dot{\gamma}^{n(0)})/\dot{\gamma}$.

Homogenization approach

In order to predict the behavior of a composite material such as a suspension, one usually computes the stresses at the particle scale, which requires the evaluation of the local strain rate field $d_{ij}^{\text{local}}(\vec{x})$ in the fluid around the particles when a homogeneous strain rate d_{ij} is applied to the suspension; here $\vec{x}$ denote the 3D position vector in the suspension. Instead of trying to infer all properties from the (difficult) computation of $d_{ij}^{\text{local}}(\vec{x})$, the method used by Chateau et al. (2008) consists, first, in assuming that the overall properties of the suspension can be accurately estimated from an average estimate $\bar{\dot{\gamma}}_{\text{local}}$ of the local shear rate $\dot{\gamma}_{\text{local}}(\vec{x})$ over the suspending fluid domain, second, in using one of the material properties determined experimentally

[2]Some aspects of the linear and nonlinear behavior of polydisperse suspensions are discussed in Vu et al. (2010).

to estimate the value of $\bar{\dot{\gamma}}_{\mathrm{local}}$, and finally, in using this last value to predict the other properties. These various steps are described more precisely in the subsections below.

Choice of a local strain rate estimate It is shown in Chateau et al. (2008) that accurate estimates of the overall properties of the suspension are obtained by using

$$\bar{\dot{\gamma}}_{\mathrm{local}}(\phi) = \sqrt{< \dot{\gamma}^2_{\mathrm{local}}(\vec{x}) >} \tag{4}$$

as the local strain rate estimate in the interstitial material. Here, $< A >$ denotes the average of A over the suspending fluid domain. This nontrivial choice is not arbitrary and is an essential step in the derivation of the suspension behavior. Indeed, as discussed in Chateau et al. (2008), this choice has been shown to be optimal in the framework of a variational approach to the solution of the nonlinear homogenization problem under consideration. A simpler and a priori more natural choice would have been to use $\bar{\dot{\gamma}}_{\mathrm{local}}(\phi) =< \dot{\gamma}_{\mathrm{local}}(\vec{x}) >= \dot{\gamma}/(1-\phi)$, but Suquet (1997) has shown that estimates of the overall properties of the heterogeneous material obtained using this effective liquid strain rate are less accurate than those obtained using Eq. 4. It should be noted that the fact that Eq. 4 is an optimal choice has been shown theoretically for isotropic composite materials; in the absence of such a result for anisotropic materials, we will consider in the following that this remains the best possible estimate of the local strain rate in the interstitial fluid in anisotropic suspensions.

Computation of the local strain rate estimate The second step consists in assuming that, for a given microstructure for the suspension, the solution of the linear problem is known, and to use this information to compute $\bar{\dot{\gamma}}_{\mathrm{local}}(\phi)$. The linear problem is the estimation of the overall properties of a dispersion of particles in a Hookean material, or, equivalently, of a suspension of particles in a Newtonian material.

For a suspension of particles in a Newtonian fluid of viscosity $\eta(0)$ in a simple shear flow of shear rate $\dot{\gamma}$, the suspension is a priori characterized by stress components that vary linearly with $\dot{\gamma}$:

$$\tau = g(\phi)\,\eta(0)\dot{\gamma} \tag{5}$$

$$N_1 = h_1(\phi)\,\eta(0)\dot{\gamma} = \frac{h_1(\phi)}{g(\phi)}\,\tau \tag{6}$$

$$N_2 = h_2(\phi)\,\eta(0)\dot{\gamma} = \frac{h_2(\phi)}{g(\phi)}\,\tau = \frac{h_2(\phi)}{h_1(\phi)}\,N_1 \tag{7}$$

The functions $g(\phi)$, $h_1(\phi)$, and $h_2(\phi)$, depend on the microstructure. They are known for isotropic suspensions: in this case, $h_1(\phi) = h_2(\phi) = 0$, and $g(\phi)$ is the so-called dimensionless high frequency dynamic viscosity (Sierou and Brady 2002); $g(\phi)$ for isotropic suspensions has also been characterized experimentally by Mahaut et al. (2008a). For a steadily sheared suspension, data for $g(\phi)$, $h_1(\phi)$, and $h_2(\phi)$ can also be found in the literature (Boyer et al. 2011b; Dbouk et al. 2013; Dai et al. 2013; Denn and Morris 2014); in this last case, the (anisotropic) suspension microstructure is fixed by shear.

One can a priori imagine any microstructure for the suspension: the 'instantaneous' behavior of the suspension would then be characterized by different functions $g(\phi)$, $h_1(\phi)$, and $h_2(\phi)$, which could be measured by applying infinitesimal nonperturbative oscillations to the material. The multiplicity of possible $g(\phi)$, $h_1(\phi)$, and $h_2(\phi)$ values is seen, e.g., in shear reversal experiments where suspensions see their microstructure changing continuously (Blanc et al. 2013) together with their macroscopic properties (Narumi et al. 2002; Blanc et al. 2011b). At this stage, we will make no hypothesis on their value for the considered suspensions; we simply note that $g(\phi)$, $h_1(\phi)$, and $h_2(\phi)$ bear some information on the suspension microstructure that can be used to predict the behavior of suspensions of particles in non-Newtonian fluids with the same microstructure.

With this knowledge, the local strain rate estimate $\bar{\dot{\gamma}}_{local}(\phi)$ in a suspension of given microstructure can be computed in some cases by matching the density of energy e dissipated at the macroscopic scale in the suspension in a Newtonian fluid, and that dissipated at the local scale in the Newtonian interstitial fluid, e_{local}. It is thus assumed here that there is negligible dissipation from particle direct contacts (sliding friction or collisions). It seems to be a reasonable assumption for particle volume fractions $\phi \lesssim 0.4$; at larger volume fraction, one has to be cautious since, as shown by Gallier et al. (2014), contact forces provide the main contribution to the suspension shear stress; with a nonzero interparticle coefficient of friction, there is thus likely non-negligible frictional dissipation. In dense suspensions, close to jamming, the presented approach can thus probably not be conducted as is.

The density of energy dissipated at the local scale in the interstitial fluid, e_{local}, is

$$e_{local} = (1-\phi) < \tau_{ij}^{local}(\vec{x})\, d_{ij}^{local}(\vec{x}) > = (1-\phi) < \eta(0)\dot{\gamma}_{local}^2(\vec{x}) > \quad (8)$$

$$= (1-\phi)\eta(0)\bar{\dot{\gamma}}_{local}^2(\phi) \quad (9)$$

This quantity is here computed easily due to the fact that the interstitial material has an isotropic behavior (Newtonian), which is thus an important assumption. For a linear anisotropic interstitial fluid, due to the complexity of the strain rate field in the interstitial fluid, $\tau_{ij}^{local}\, d_{ij}^{local}$ would not be simply proportional to $\dot{\gamma}_{local}^2$ but would be also a function of another invariant of the strain rate tensor.

The density of energy e dissipated at the macroscopic scale can also be computed as a function of $\dot{\gamma}$ only, in two situations: (i) for an isotropic suspension, in which case $\tau_{ij}d_{ij} = 2\eta(0)\, g(\phi)\, d_{ij}d_{ij} = \eta(0)\, g(\phi)\, \dot{\gamma}^2$ for any prescribed strain rate d_{ij}; (ii) for an anisotropic suspension in a *simple shear flow* only, in which case normal stress differences do not produce any mechanical work, and $\tau_{ij}d_{ij} = \eta(0)\, g(\phi)\, \dot{\gamma}^2$ is independent of $h_1(\phi)$ and $h_2(\phi)$. In both cases, one finally gets

$$e = \eta(0)\, g(\phi)\, \dot{\gamma}^2 \quad (10)$$

By matching the density of energy dissipated at the macroscopic scale, and its microscopic origin, that is, the density of energy dissipated in the interstitial fluid, a simple relationship between the macroscopic shear rate $\dot{\gamma}$ and the estimate of the local strain rate is finally obtained:

$$\bar{\dot{\gamma}}_{\text{local}}(\phi) = \dot{\gamma}\sqrt{g(\phi)/(1-\phi)} \tag{11}$$

All the useful information on the material microstructure and on the way the material is sheared locally is thus supposed to be encoded in $g(\phi)$.

From linear to nonlinear properties The next step consists in using the information obtained on $\bar{\dot{\gamma}}_{\text{local}}$ to predict the value of the nonlinear properties. Chateau et al. (2008) have linearized the suspending fluid behavior (Eq. 1) at each prescribed macroscopic shear rate $\dot{\gamma}$, and have used Eq. 4 to estimate its apparent viscosity which is thus $\eta(\bar{\dot{\gamma}}_{\text{local}})$. Using Eq. 11, the suspending fluid apparent viscosity is then estimated as $\eta\left(\dot{\gamma}\sqrt{g(\phi)/(1-\phi)}\right)$, which is used to predict the behavior of the linearized suspension behavior associated to a shear rate $\dot{\gamma}$.

Two situations shall then be distinguished, as above. For an isotropic suspension, the macroscopic apparent viscosity is simply $g(\phi)$ times the suspending fluid apparent viscosity, and the full tensorial behavior of the suspension can be estimated as:

$$\tau_{ij} = 2\,\eta\left(\dot{\gamma}\sqrt{g(\phi)/(1-\phi)}\right) g(\phi)\, d_{ij} \tag{12}$$

For an anisotropic suspension, the behavior in simple shear can be estimated as

$$\tau = g(\phi)\,\eta\left(\dot{\gamma}\sqrt{g(\phi)/(1-\phi)}\right)\dot{\gamma} \tag{13}$$

$$N_1 = h_1(\phi)\,\eta\left(\dot{\gamma}\sqrt{g(\phi)/(1-\phi)}\right)\dot{\gamma} \tag{14}$$

$$N_2 = h_2(\phi)\,\eta\left(\dot{\gamma}\sqrt{g(\phi)/(1-\phi)}\right)\dot{\gamma} \tag{15}$$

Within this approach, the ratios of shear to normal stresses remain naturally unchanged when changing the suspending fluid, which comes from their independence on the interstitial fluid viscosity, i.e.:

$$N_1 = \frac{h_1(\phi)}{g(\phi)}\tau \quad \text{and} \quad N_2 = \frac{h_2(\phi)}{g(\phi)}\tau = \frac{h_2(\phi)}{h_1(\phi)}N_1 \tag{16}$$

This will be shown to have important consequences when modeling migration.

An important assumption that has been made above is that the same functions $g(\phi)$, $h_1(\phi)$ and $h_2(\phi)$ can be used to estimate the linearized suspension characteristics independently of $\dot{\gamma}$. In other words, it is assumed that the microstructure does not vary significantly with $\dot{\gamma}$. Although it has no reason to be true for nonlinear suspending fluids, this has been somehow validated for particles in power-law fluids by Liard et al. (2014), who have shown that distributions of shear rates in the interstitial fluid are similar for power-law and Newtonian interstitial fluids. This might not be true for other complex fluids, as will be shown below for a suspending yield stress fluid (see Sect. 4.2); changes in the microstructure as a function of the shear rate specific to nonlinear interstitial fluids might then have a significant impact on the value of the rheological properties. Nevertheless, even in such cases, it is possible to use the above estimates in ranges of shear rates characterized by a similar microstructure; different $g(\phi)$, $h_1(\phi)$, and $h_2(\phi)$ functions shall then be used in these different ranges.

Application to yield stress fluids Within this approach, in the case of a suspending fluid described by a Herschel-Bulkley behavior (Eq. 2), for both isotropic and anisotropic suspensions, the shear stress τ in the suspension is found to be:

$$\tau = \sqrt{(1-\phi)g(\phi)}\left[\tau_y(0) + \eta_{HB}(0)\left(\frac{g(\phi)}{1-\phi}\right)^{n(0)/2} \dot{\gamma}^{n(0)}\right] \tag{17}$$

$$= \tau_y(\phi) + \eta_{HB}(\phi)\dot{\gamma}^{n(\phi)} \tag{18}$$

It is then first seen that

$$n(\phi) = n(0) \tag{19}$$

i.e., the suspension shear stress is that of a Herschel-Bulkley material of same index $n \equiv n(0)$ as the interstitial fluid.[3] The dimensionless yield stress is

$$\tau_y(\phi)/\tau_y(0) = \sqrt{(1-\phi)g(\phi)} \tag{20}$$

and the dimensionless consistency is

$$\eta_{HB}(\phi)/\eta_{HB}(0) = \sqrt{(g(\phi))^{n+1}(1-\phi)^{1-n}} \tag{21}$$

which can also be rewritten as

$$\eta_{HB}(\phi)/\eta_{HB}(0) = (\tau_y(\phi)/\tau_y(0))^{n+1}(1-\phi)^{-n} \tag{22}$$

It is worth noting that, within this approach, the knowledge of a single mechanical property is sufficient to help predicting the value of all other properties. Equation 22 shows in particular that the validity of the homogenization approach can be tested without the knowledge of the linear properties of the material.

For an isotropic suspension, the material tensorial behavior is an isotropic Herschel-Bulkley behavior (Eq. 2) of index n, and yield stress and consistency given by Eqs. 20 and 21. For an anisotropic suspension in simple shear, the shear stress is given by Eq. 17, and the material is also predicted to be characterized by normal stress differences

[3]For simplicity, we will use n to denote the index of both the suspending fluid and the suspension in the sequel.

$$N_1 = h_1(\phi)\frac{\tau_y(0) + \eta_{HB}(0)\bar{\dot{\gamma}}_{\mathrm{local}}(\phi)^n}{\bar{\dot{\gamma}}_{\mathrm{local}}(\phi)}\dot{\gamma} \tag{23}$$

$$= \frac{h_1(\phi)}{\sqrt{g(\phi)/(1-\phi)}}\left[\tau_y(0) + \eta_{HB}(0)\big(g(\phi)/(1-\phi)\big)^{n/2}\dot{\gamma}^n\right] \tag{24}$$

and

$$N_2 = \frac{h_2(\phi)}{\sqrt{g(\phi)/(1-\phi)}}\left[\tau_y(0) + \eta_{HB}(0)\big(g(\phi)/(1-\phi)\big)^{n/2}\dot{\gamma}^n\right] \tag{25}$$

It is predicted in particular that these suspensions exhibit 'yield normal stress differences'.

Domain of validity It is worth reminding the main assumptions made in all the above derivations, in order to better understand the domain of validity of the developed equations:

- Equation 4 has been chosen as a local strain rate estimate. It has been shown to be an optimal choice for isotropic suspensions only. The same result remains to be shown rigorously for anisotropic suspensions.
- The interstitial fluid has been assumed to have an isotropic behavior. This assumption is necessary to compute the suspension stresses by using Eqs. 5, 6, and 7 when linearizing the fluid and suspension behaviors. It is also used when calculating the density of energy dissipated in the complex flow at the local scale (Eqs. 8 and 9) to compute the local strain rate estimate. In the absence of baseline results for suspensions of particles in linear anisotropic fluids, the case of a nonlinear anisotropic interstitial fluid cannot yet be treated with the same approach.
- In computing the energy dissipated at the local scale, it was also assumed that there is negligible dissipation from particle direct contacts (sliding friction or collision). A consequence is that the presented approach may not work at high volume fraction. A finer micromechanical description of the suspensions would probably be necessary to describe cases where interparticle dissipation is not negligible.
- The case of anisotropic suspensions has been handled here only in the case of simple shear flows. In this case, normal stresses do not produce any mechanical work, which makes it possible to derive equations depending on a single function ($g(\phi)$). It would a priori be possible to develop the same approach for other flows, provided the behavior of suspensions in Newtonian fluids is well known for these flows; this would, however, probably yield more complex and less tractable expressions.
- It should also be noted that the prediction for $\tau_y(\phi)$ concerns the limit shear stress of the material in simple shear, and not its 'true' yield stress (as would be defined from a yield criterion based on the full stress tensor). Indeed, the yield criterion of the suspension should probably take into account its normal stress differences (which are different from zero at $\dot{\gamma} = 0$); this might have important consequences on the flow properties of the material at the solid/liquid transition (Cheddadi et al. 2012).

Building a 3D yield criterion for yield stress suspensions remains a challenging question.

- $g(\phi)$ has been assumed to be uniquely defined for anisotropic suspensions, independently of the fluid and of $\dot{\gamma}$, which seems to be a good approximation for power-law fluids (Liard et al. 2014). This is useful to get simple equations such as Eq. 22 and to derive all quantities from the sole knowledge of $g(\phi)$; this is useful also because $g(\phi)$ is well characterized for sheared viscous suspensions, which allows deriving phenomenological expressions for nonlinear interstitial fluids (see Sect. 4). $\dot{\gamma}$-dependent $g(\phi)$ can nevertheless be handled in Eqs. 13, 14, and 15. Their determination remains however challenging.

3 Experiments: Methodology

3.1 Materials

A first problem one has to face when studying viscoplastic suspensions is the design of model systems that can be described theoretically as rigid particles in a yield stress fluid interacting only through hydrodynamical interactions mediated by the fluid viewed as a continuum. Mahaut et al. (2008a) showed that, provided scale separation is ensured between the particles and the suspending fluid microstructure, and that special care is taken to avoid physicochemical interactions and experimental artefacts, such model materials can indeed be designed and characterized. They have studied the behavior of several suspensions (monodisperse glass or polystyrene beads suspended in concentrated emulsions, Carbopol gels or bentonite suspensions) and found similar rheological behaviors in all systems, thus showing the generic nature of the observed behavior. The results, obtained on model systems, have also been shown to be applicable to more complex systems such as model mortars made of rigid spherical particles in cement pastes (Mahaut et al. 2008b).

3.2 Experimental Procedures

As detailed in the theory section, the suspension microstructure is a key experimental parameter. We thus make the distinction here between isotropic suspensions, which can be studied at rest, and flowing suspensions, which are anisotropic.

Specific procedure for isotropic suspensions As shown by Mahaut et al. (2008a), an appropriate procedure to study the elastoplastic behavior of isotropic suspensions is the following:

- a suspension of beads in a yield stress fluid is prepared at a given volume fraction ϕ of beads in a cup. There is no preshear after preparation; however, in order to

perform all the measurements in the same conditions in the case of thixotropic materials, i.e. to start from a same state of destructuration of the material with and without particles, it is necessary to strongly stir the material by hand after loading. This procedure allows to avoid shear induced migration, and importantly also to perform measurements on an isotropic microstructure that allows for comparison with micromechanical models (Chateau et al. (2008)): any controlled shear of the material would induce an anisotropic microstructure as observed in suspensions of particles in Newtonian fluids (Parsi and Gadala-Maria (1987); Sierou and Brady (2002)).
- then, a vane tool is slowly inserted in the cup and oscillations are performed at rest in the linear regime during 10 min in order to get the elastic modulus; note that it is important to measure the elastic modulus evolution in time (and not its mean value) even for non thixotropic materials: this allows evidencing unwanted physicochemical interactions between the particles and the paste with some materials as shown by Mahaut et al. (2008a). The use of a vane-in-cup geometry, inappropriate for flows (see below), is the best possible choice for studying the elastic and plastic properties of isotropic suspensions: it does not disturb the material structure during loading, which contrasts with all other standard geometries.
- afterwards, a small rotational velocity is imposed to the vane tool in order to get the yield stress; it can be checked that the elasticity measurement is nonperturbative: the same yield stress is measured if a zero stress is imposed instead of oscillations before the yield stress measurement.

Finally, as the yield stress measurement may induce migration and a microstructure anisotropy, any new measurement requires a new sample preparation. Note that the manual preparation does not strictly ensure that one always gets the same initial state of destructuration in the case of thixotropic materials; that is why the yield stress measurement is performed after 10 minutes of rest: any slight initial irreproducibility have a negligible influence on the structure after such a time of rest.

Homogeneity is confirmed by MRI measurements of volume fractions profiles. Isotropy is confirmed by X-ray microtomography determination of the pair distribution functions (see Sect. 4.2).

Rheometry of suspensions When dealing with suspensions of large nonBrownian particles, it is needed to use a geometry with a wide gap to ensure that the properties of a continuous medium (the suspension) are measured. A geometry classically used to study the flows of such suspensions is the vane-in-cup geometry (Koehler et al. 2005; Martínez-Padilla and Rivera-Vargas 2006; Jau and Yang 2010), which is supposed to avoid wall slip (Keentok 1982; Dzuy and Boger 1983; Saak et al. 2001): it is assumed that the material in the gap of the geometry is sheared by the (same) material that is trapped between the blades. However, Ovarlez et al. (2011) have shown that a thin layer made of the pure interstitial yield stress fluid appears quickly at the interface between the sheared material and that trapped between the blades (Fig. 1). This layer acts as a slip layer, leading to shear rate overestimations; moreover, flows are localized near the tool (and thus, in the depleted layer) at low shear rates, which makes one underestimate the suspension yield stress.

Fig. 1 2D Magnetic Resonance Image of a suspension of particles in a yield stress fluid in a vane-in-cup geometry after a 1 h shear at $\Omega = 100$ rpm. The cross symbols correspond to the limit between rigid motion and shear. The image is coded in grey scales; a brighter zone contains less particles. Reprinted with permission from Ovarlez et al. (2011). Copyright (2011) by the Society of Rheology

It is thus more relevant to use a wide gap concentric-cylinder Couette. In order to avoid wall slip, sandpaper can be glued on the walls. Since the shear stress distribution is inhomogeneous in this geometry, it can be expected that flow causes particle migration towards the outer cylinder as in viscous suspensions (Leighton and Acrivos 1987b; Phillips et al. 1992). In rheological studies, migration is usually considered as an artefact that has to be avoided when possible. As it may not always be avoided, in particular at high particle volume fractions, one has to find a way to take it into account; a possible way consists in measuring the local velocity and volume fraction fields (e.g., with rheo-NMR methods, Callaghan (1999)) and in using this information to infer the rheological properties values as a function of the locally measured volume fraction (Ovarlez et al. 2006).

Inhomogeneities and Rheo-NMR Proton NMR (Callaghan 1991, 1999) can be used as a non-intrusive technique in order to get measurements of the local material velocity and of the local particle concentration inside the gap of a Couette cell. The cell thus has to be designed to be inserted in a MRI scanner.

If the material is homogeneous in the radial direction, local data $\left(\tau(r, \Omega), \dot{\gamma}(r, \Omega)\right)$ measured at various r and various Ω can finally be combined to obtain the constitutive law $\tau(\dot{\gamma})$ consistent with the observed flows (more details about this reconstruction technique can be found in Ovarlez et al. (2008)). Note that this analysis is not affected by a possible wall slip.

Of course, for a homogeneous simple yield stress fluid, this analysis is not necessary and may seem uselessly complex since a simple rheometer with appropriate procedures provides the same result (Ovarlez et al. 2013). However, in the case of suspensions, this methodology can become indispensable. Indeed, as already mentioned, particles tend to migrate toward the outer cylinder in a concentric-cylinder Couette flow, which can lead to strong particle volume fraction inhomogeneities (Phillips et al. 1992; Morris and Boulay 1999; Ovarlez et al. 2006, 2015). The kinetics of this migration phenomenon is slow at low volume fraction, but gets more and more rapid as the volume fraction increases. This is also at high particle volume fraction that the impact of inhomogeneities is the strongest since rheological properties diverge at the approach of ϕ_{div}; data obtained with a rheometer only may then lead to gross underestimations of these rheological properties (Ovarlez et al. 2006; Hafid et al. 2015). It can take as little as a few units of strains for the material to become inhomogeneous at high volume fraction (Ovarlez et al. 2006). It may thus be impossible to characterize the flow properties of the homogeneous material. E.g., for 50% of particles in a yield stress fluid, in a gap that is 100 particles wide, it is found that a steady inhomogeneous volume fraction profile is reached after less than 1000 units of strain.

In such case, the above analysis to obtain the constitutive behavior from local measurements can be conducted again. Indeed, since $\phi(r)$ is independent of Ω (this will be shown in the section dealing with shear-induced inhomogeneities), a change of variables can be performed between radial position r and ϕ. A local stress/strain-rate relationship $\tau(\dot{\gamma}, \phi)$ at fixed and well-defined volume fraction ϕ is then obtained by collecting all measurements of local stress $\tau(r)$ and shear rate $\dot{\gamma}(r)$ as above, for varying Ω, but for a single fixed r. More details on this methodology can be found in Ovarlez et al. (2006), Fall et al. (2010, 2015). The application to viscoplastic suspensions is found in Hafid et al. (2015), Ovarlez et al. (2015).

Microstructure and Rheo-X-ray microtomography The suspensions' microstructure, i.e., the spatial distribution of particles in the suspensions can be determined with the help of 3D tomography (Deboeuf et al. 2018). Indeed, a rheometer can be coupled to a X-ray microtomograph (Maire et al. 2001), which allows one to get 3D images of suspensions and to determine the particles' position. To characterize the suspension microstructure, a given shear history is imposed to the suspension in the rheometric cell, and then stopped as long as needed for the complete scanning of the sample (about 1 h). The suspension microstructure is then assumed to be the same just before and after the interruption of shear thanks to the fluid yield stress, although slight relaxation of the interstitial fluid can be expected below the yield stress (Mohan et al. 2015). The absence of blur in the images is used to confirm that the particles do not move at rest during the 1 h scan.

X-ray absorption has to be different between the particles and the fluid, so that the particles are easily identified; the spherical shape of the particles then allows one to determine the positions of particle centers with a subvoxel resolution (Deboeuf et al. 2018). Various quantities characterizing the particle distribution in space can then be computed to describe the suspension microstructure. Pair distribution functions will be discussed in Sect. 4.2.

4 Phenomenological Expressions and Experimental Results

For both isotropic and anisotropic suspensions, the development of phenomenological expressions for the yield stress and the consistency relies on the existence of phenomenological expressions for the linear response $g(\phi)$. Model yield stress fluids have a linear elastic behavior below their yield stress: $\tau_{ij} = G'(0)\gamma_{ij}$ when $\tau < \tau_y$. The best way to get $g(\phi)$ thus seems to determine it directly on the studied materials as $g(\phi) = G'(\phi)/G'(0)$, $G'(\phi)$ being the elastic modulus of the suspension at particle volume fraction ϕ. $g(\phi)$ can also a priori be obtained from the response of any suspension of particles in a linear (Hookean or Newtonian) material with the same structure as the studied yield stress suspension.

From now, we will distinguish $g_{\text{iso}}(\phi)$ and $g_{\text{aniso}}(\phi)$ for, respectively, isotropic and anisotropic suspensions.

4.1 *Isotropic Suspensions*

Mahaut et al. (2008a) have measured both the elastic modulus (Fig. 2a) and the yield stress (Fig. 2b) of isotropic suspensions of particles in yield stress fluids; measuring their consistency is not possible as the isotropic structure is destroyed by flow. The dimensionless elastic modulus of the suspensions is found to be well fitted to the Krieger-Dougherty equation up to $\phi = 50\%$ (see Fig. 2a), as classically found for the dimensionless viscosity of suspensions in Newtonian fluids [this is due to the similarity of the elastic and viscous problems (Torquato 2002)]:

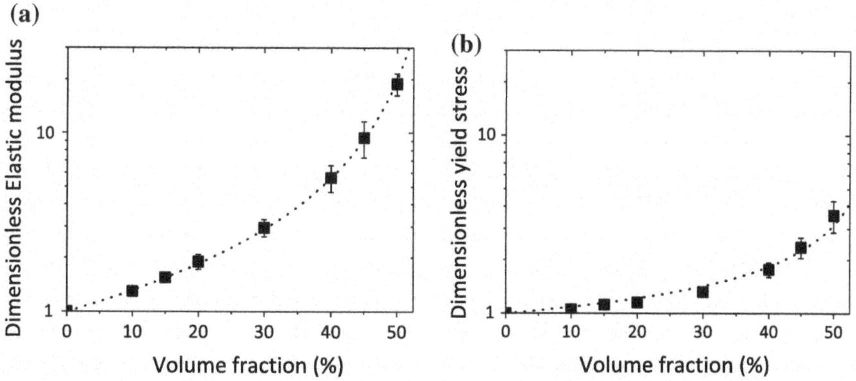

Fig. 2 Dimensionless elastic modulus $G'(\phi)/G'(0)$ (**a**) and dimensionless static yield stress $\tau_y(\phi)/\tau_y(0)$ (**b**) measured in isotropic suspensions of monodisperse spherical particles in yield stress fluids by Mahaut et al. (2008a) (data are averages over 13 different suspensions). The dotted line in Fig. 2a is a Krieger-Dougherty equation (Eq. 26), the dotted line in Fig. 2b is the equation derived by Chateau et al. (2008) (Eq. 27). Replotted from Ovarlez et al. (2015)

$$g_{\text{iso}}(\phi) = G'(\phi)/G'(0) = (1 - \phi/\phi_{\text{div}})^{-2.5\phi_{\text{div}}} \quad \text{with} \quad \phi_{\text{div}} = 0.57 \tag{26}$$

Combining Eqs. 20 and 26 then yields a simple phenomenological law for the yield stress:

$$\tau_y(\phi)/\tau_y(0) = \sqrt{(1 - \phi)\,(1 - \phi/\phi_{\text{div}})^{-2.5\phi_{\text{div}}}} \quad \text{with} \quad \phi_{\text{div}} = 0.57 \tag{27}$$

This expression is in very good agreement with the observations of Mahaut et al. (2008a, b) for the static yield stress of isotropic yield stress suspensions, for $\phi \leq 50\%$ (Fig. 2b). This validates Eqs. 11 and 20, and shows that the yield stress of these suspensions can be derived from the sole knowledge of their elastic modulus.

4.2 *Anisotropic Suspensions*

Expressions determined from the literature In principle, the same approach as above could be used to determine directly $g_{\text{aniso}}(\phi)$ on the studied suspensions from the measurement of the suspension elastic modulus after flow cessation or by using parallel superposition (Vermant et al. 1998). This cannot be done easily, however, at volume fractions $\phi \gtrsim 40\%$, due to volume fraction inhomogeneities; parallel superposition poses other problems in yield stress fluids, such as accounting for possible flow inhomogeneities.

Another possibility consists in studying the same particles in a Newtonian fluid of viscosity $\eta(0)$ and to assume that the suspension microstructure is the same as in the yield stress fluid in order to get $g_{\text{aniso}}(\phi)$ as $g_{\text{aniso}}(\phi) = \eta(\phi)/\eta(0)$. This is the approach chosen by Liard et al. (2014), Dagois-Bohy et al. (2015). This approach poses several problems. First, it will be shown below that the suspension microstructure depends on the shear rate in a yield stress fluid, which contrasts with the case of a Newtonian interstitial fluid. Second, for given particles, the function $g_{\text{aniso}}(\phi)$ depends on the (Newtonian) interstitial fluid that is chosen, as can be seen from a survey of the literature. E.g., the same PS particles (same size: 40 µm, same provider: Dynoseeds) lead to $(1 - \phi/\phi_{\text{div}})^{-2}$ scaling for the viscosity divergence in a silicone oil (Bonnoit et al. 2010) whereas Fall et al. (2010) report a $(1 - \phi/\phi_{\text{div}})^{-1}$ scaling in water. In addition, for similar PS beads, the viscosity was observed to diverge at $\phi_{\text{div}} = 60.5\%$ for particles dispersed in a silicone oil (Ovarlez et al. 2006) and in water (Fall et al. 2010), whereas $\phi_{\text{div}} = 58\%$ was obtained by Boyer et al. (2011b) in polyethylene glycol. These differences likely come from the strong dependence of the suspension behavior on the short range interactions between the particles (Brady and Morris 1997; Sierou and Brady 2002), which affects the cut-off on lubrication forces, and on the interparticle friction coefficient (Sierou and Brady 2002; Gallier et al. 2014). Friction and interparticle interactions most probably depend on the physicochemical properties of the interstitial fluid.

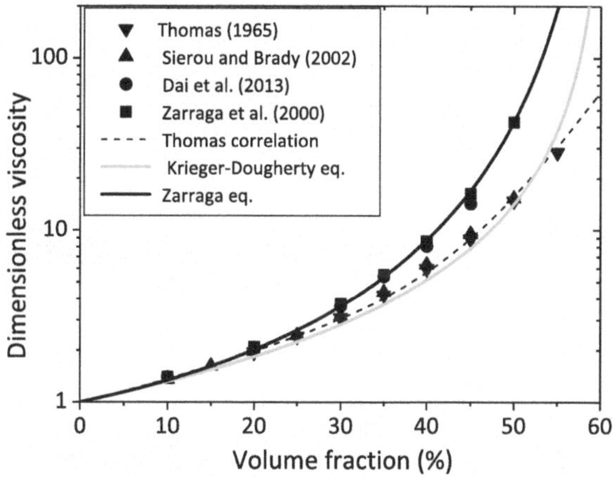

Fig. 3 Dimensionless viscosity $g_{\text{aniso}}(\phi) = \eta(\phi)/\eta(0)$ obtained in steady-state simple shear flows of monodisperse spherical particles in Newtonian fluids: data replotted from the literature (see legend). The grey line is Eq. 28; the dotted line is a correlation proposed by Thomas (1965). The black line is Eq. 29. Replotted from Ovarlez et al. (2015)

This implies that one cannot rely on measurements performed with beads in a given Newtonian fluid to predict quantitatively the behavior of the same beads in another material. Instead of trying to study the range of possible viscosities for our beads in Newtonian fluids by using many different fluids, it is thus proposed below to rely on results of the literature.

To describe the possible linear response of sheared viscous suspensions, one can thus use the sets of data shown in Fig. 3, which seem to provide lower and upper limits for values observed in the literature; e.g., at $\phi = 50\%$, the data of Zarraga et al. (2000) and those of Thomas (1965) differ by a factor 3. The data reported by Thomas (1965) come from his analysis of many works of the literature; they are consistent with the simulations of Sierou and Brady (2002), and may thus correspond to cases where interparticle friction does not contribute much to dissipation. The correlation proposed by Thomas (1965), plotted in Fig. 3, does not diverge at high ϕ; his data are thus described here with the Krieger-Dougherty equation (Eq. 28), with $\phi_{\text{div}} = 0.605$ consistent with the finding of Ovarlez et al. (2006), in fair agreement with the Thomas (1965), Sierou and Brady (2002) data. Zarraga et al. (2000) have proposed another equation (Eq. 29) which provides a good fit to their data and to those of Dai et al. (2013).

In the following, a comparison is made between experimental results and two different expressions for g_{aniso}:

$$g_{\text{aniso},1}(\phi) = (1 - \phi/\phi_{\text{div}})^{-2.5\phi_{\text{div}}} \quad \text{with} \quad \phi_{\text{div}} = 0.605 \tag{28}$$

and

$$g_{\text{aniso},2}(\phi) = (1 - \phi/\phi_{\text{div}})^{-3}\, e^{-2.34\phi} \quad \text{with} \quad \phi_{\text{div}} = 0.62 \tag{29}$$

These expressions yield, respectively

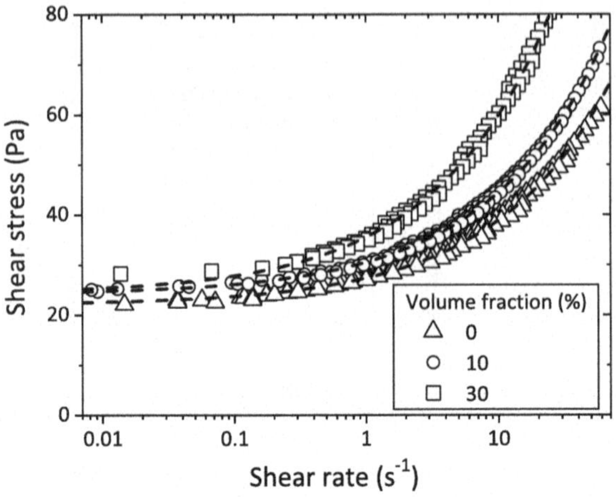

Fig. 4 Local constitutive behavior $\tau(\dot{\gamma})$ measured using MRI techniques in a concentrated emulsion, and in suspensions of particles in the same emulsion at volume fractions $\phi = 10\%$ and $30 \pm 0.5\%$. The dashed lines are fits of a Herschel-Bulkley behavior $\tau_y + \eta_{HB}\dot{\gamma}^{0.5}$ to the data with, from bottom to top, $\tau_y = 22.1$ Pa and $\eta_{HB} = 5.3$ Pa.s$^{0.5}$, $\tau_y = 24.1$ Pa and $\eta_{HB} = 6.4$ Pa.s$^{0.5}$, $\tau_y = 24.3$ Pa and $\eta_{HB} = 11.25$ Pa.s$^{0.5}$. Replotted from Ovarlez et al. (2015)

$$\tau_y(\phi)/\tau_y(0) = \sqrt{(1-\phi)\,(1-\phi/\phi_{\text{div}})^{-2.5\phi_{\text{div}}}} \quad \text{with} \quad \phi_{\text{div}} = 0.605 \tag{30}$$

$$\eta_{HB}(\phi)/\eta_{HB}(0) = \sqrt{(1-\phi/\phi_{\text{div}})^{-2.5(n+1)\phi_{\text{div}}}\,(1-\phi)^{1-n}} \quad \text{with} \quad \phi_{\text{div}} = 0.605 \tag{31}$$

and

$$\tau_y(\phi)/\tau_y(0) = \sqrt{(1-\phi)\,(1-\phi/\phi_{\text{div}})^{-3}\,e^{-2.34\phi}} \quad \text{with} \quad \phi_{\text{div}} = 0.62 \tag{32}$$

$$\eta_{HB}(\phi)/\eta_{HB}(0) = \sqrt{(1-\phi/\phi_{\text{div}})^{-3(n+1)}\,e^{-2.34(n+1)\phi}\,(1-\phi)^{1-n}} \quad \text{with} \quad \phi_{\text{div}} = 0.62 \tag{33}$$

Experimental flow curves In Fig. 4, we present the local constitutive behavior $\tau(\dot{\gamma})$ obtained from the analysis of MRI data for a concentrated emulsion, and for suspensions of 10 and 30% particles in the same emulsion (from Ovarlez et al. (2015)), together with fits of Herschel-Bulkley laws to the data.

It is first observed that the yield stress increases only slightly when adding particles, whereas a significant increase of the viscous effects is visible. To analyze quantitatively the data, a reference to a model is necessary. Since the Herschel-Bulkley law provides a good fit to the emulsion behavior, all of the data are analyzed in the framework of this model. It is observed in Fig. 4 that the three materials can be fairly

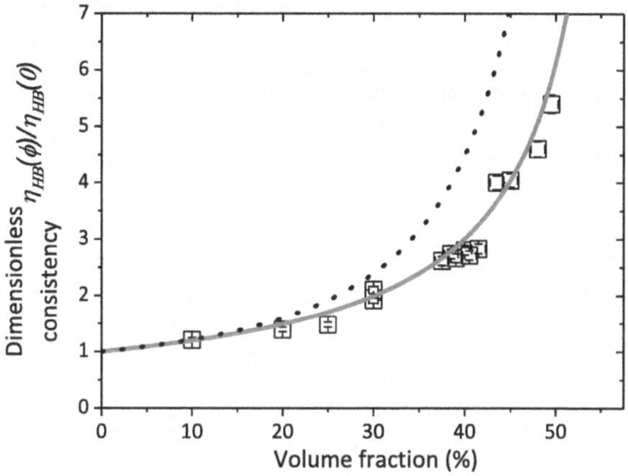

Fig. 5 Dimensionless consistency $\eta_{HB}(\phi)/\eta_{HB}(0)$ versus volume fraction ϕ for all the studied suspensions of particles in concentrated emulsions, extracted from a fit of the $\tau(\dot{\gamma})$ data to a Herschel-Bulkley law of index $n = 0.5$. The grey line is the model (Eq. 31) based on the linear behavior observed by Thomas (1965), Sierou and Brady (2002); the dotted line is the model (Eq. 33) based on the linear behavior observed by Zarraga et al. (2000), Dai et al. (2013). Replotted from Ovarlez et al. (2015)

modeled by a Herschel-Bulkley law of same index n (here, $n \simeq 0.5$) in all cases, as predicted by the theory. This is a general feature, observed for all concentrations.

In the following, we discuss the evolution with the particle volume fraction ϕ of the dimensionless consistency $\eta_{HB}(\phi)/\eta_{HB}(0)$ and of the dimensionless yield stress $\tau_y(\phi)/\tau_y(0)$ extracted from the fit to the data of a Herschel-Bulkley behavior with $n = 0.5$.

Consistency In Fig. 5, the consistency is observed to increase significantly with the volume fraction, this increase being of a factor of order 5 for $\phi = 50\%$. The observed evolution is in very good agreement with the prediction of Eq. 31, which is based on the linear response of suspensions as observed by Thomas (1965), Sierou and Brady (2002). By contrast, there is a strong mismatch between the data and the predictions of Eq. 33, which is based on the linear behavior observed by Zarraga et al. (2000), Dai et al. (2013). This suggests that the studied suspension microstructure is similar to that of suspensions of low friction particles in Newtonian media, as this corresponds to the situation of the Sierou and Brady (2002) simulations; indeed, as discussed previously, it seems that the lowest possible $g_{\text{aniso}}(\phi)$ are obtained when interparticle friction is negligible (Sierou and Brady 2002; Gallier et al. 2014).

Note that the equation used here for $g_{\text{aniso}}(\phi)$ (Eq. 28) was validated only up to 50% in Fig. 3; it is known, however, that the asymptotic behavior of $g_{\text{aniso}}(\phi)$, at the vicinity of jamming, is different from Eq. 26 (Ovarlez et al. 2006; Boyer et al. 2011b). The model for the consistency (Eq. 31) is thus expected to be valid only in the range of volume fraction investigated here, from 0 to 50%. The behavior for $\phi_{\text{div}} > \phi > 45\%$ has been investigated by Dagois-Bohy et al. (2015). The authors

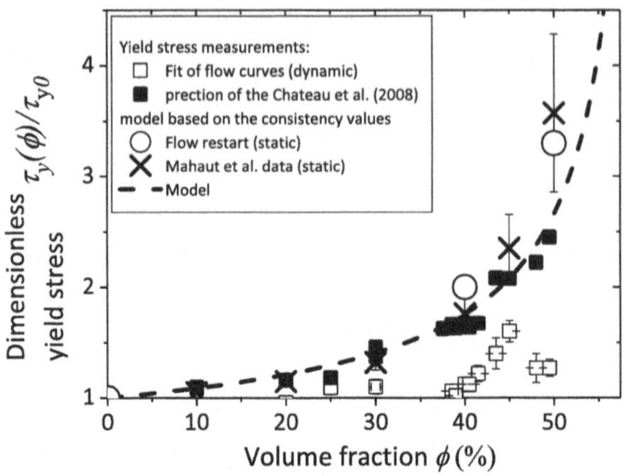

Fig. 6 Open squares: dimensionless yield stress $\tau_y(\phi)/\tau_y(0)$ versus volume fraction ϕ for suspensions of particles in concentrated emulsions, extracted from a fit to the $\tau(\dot{\gamma})$ data of a Herschel-Bulkley law of index $n = 0.5$. Open circles: dimensionless static yield stresses measured after a preshear at high shear rate. Crosses: dimensionless static yield stress data from Mahaut et al. (2008a). Dashed line: model (Eq. 30) based on the linear behavior observed by Thomas (1965), Sierou and Brady (2002). Filled squares: dimensionless consistency data plotted as $[\eta_{HB}(\phi)/\eta_{HB}(0)]^{1/(n+1)}(1-\phi)^{n/(n+1)}$ with $n = 0.5$. Replotted from Ovarlez et al. (2015)

have added a free parameter to the relationship (Eq. 11) between the local shear rate and the macroscopic shear rate in order to fit their data to the model presented here; they report that, if the $g_{\text{aniso}}(\phi)$ they measure in a viscous suspension is correct, then $\bar{\dot{\gamma}}_{\text{local}}(\phi) \simeq 2 \times \dot{\gamma}\sqrt{g_{\text{aniso}}(\phi)/(1-\phi)}$, which is not at all justified in the theoretical approach derived in Sect. 2. As already pointed out, this might be due to the choice of PEG as an interstitial fluid to measure $g_{\text{aniso}}(\phi)$; another choice would have provided a different $g_{\text{aniso}}(\phi)$. This might also come from two limitations of the theoretical approach: direct interparticle dissipation was neglected, and the estimate chosen for $\bar{\dot{\gamma}}_{\text{local}}(\phi)$ (Eq. 4) might not be relevant in the very dense regime, where most of viscous dissipation occurs in the small volume occupied by the fluid between close particles. The behavior at the approach of ϕ_{div} thus remains to be investigated theoretically; it will not be discussed further here.

Yield stress We now turn to the analysis of the yield stress data, as obtained from a fit of steady-state flow curves to a Herschel-Bulkley model. In Fig. 6, $\tau_y(\phi)$ is observed to increase only moderately with the volume fraction ϕ, this increase being of less than 10% for $\phi \leq 40\%$, and the maximum observed increase being of a factor 1.6. The scattered data at the highest volume fraction investigated, where the yield stress at 50% seems to be lower than the yield stress at 45%, is likely due to the fact that yield stresses shown here were obtained from a fit to $\tau(\dot{\gamma})$ data with a limited amount of data at low shear rates, insufficient to describe accurately the yield stress plateau, as pointed out by Ovarlez et al. (2015).

By contrast with the consistency data, the yield stress values are much lower than expected from the model (Eq. 30). The model can actually be tested independently of the previous results. Indeed, one of the predictions is that there exists an equation (Eq. 22) without any fitting parameter linking the dimensionless consistency and the dimensionless yield stress. To test this prediction, the dimensionless consistency data are plotted in Fig. 6 in the form $[\eta_{HB}(\phi)/\eta_{HB}(0)]^{1/(n+1)}(1-\phi)^{n/(n+1)}$ with $n=0.5$, which should be equal to $(\tau_y(\phi)/\tau_y(0))$ according to Eq. 22. There is no agreement at all, whereas the $[\eta_{HB}(\phi)/\eta_{HB}(0)]^{1/(n+1)}(1-\phi)^{n/(n+1)}$ data match quite well the model for the yield stress (Eq. 30).

As mentioned in Sect. 2, a key assumption to get Eq. 22 is that the material microstructure does not depend on $\dot{\gamma}$, i.e. that it is similar at low and high shear rate, which is far from being obvious for a nonlinear interstitial fluid. To go one step further, it is necessary to question this assumption, which is done in the next section.

Microstructure: history dependence The spatial distribution of particles is known to depend on flow history in suspensions of nonBrownian particles in Newtonian fluids, which has an impact on the material properties (Morris 2009; Blanc et al. 2011b, 2013). In the case of a nonlinear interstitial fluid, the steady-state particle microstructure under flow is moreover a function of the shear rate, which contrasts with the case of a Newtonian interstitial fluid.

To illustrate this point, as described in Sect. 3.2, X-ray microtomography can be used to determine the positions of all particle centers for different shear histories in a simple shear flow. From these data, one can compute the pair distribution function $g_2(\vec{r})$, which is a relevant quantity to describe the suspension microstructure (Morris 2009). $g_2(\vec{r})$ is the probability of finding a pair of particles whose centers are separated by the vector $\vec{r}$ normalized by the mean particle density, so that the asymptotic value of g_2 for large values of $\|\vec{r}\|$ is $g_2(\|\vec{r}\| \to \infty) = 1$. We focus here on the microstructure of the suspension in the velocity-velocity gradient plane of the simple shear flow imposed locally by the geometry, through the computation of the statistical pair distribution function $g_{2,r}(\rho, \phi)$, with ρ and ϕ the polar coordinates of a pair of particles located in the velocity-velocity gradient plane.

In Fig. 7, steady-state pair distribution functions (pdf) obtained in a suspension of 37% beads in a concentrated emulsion are displayed; they correspond to three different conditions: (a) after shear at low shear rate (10^{-2} s^{-1}); (b) after shear at high shear rate (10 s^{-1}); and (c) after loading into the rheometer. The three pdfs are clearly different: the material is isotropic after loading and anisotropic in the plane of shear under flow, and the details of the anisotropy depend on the applied shear rate.

More precisely, after loading, the pdf $g_{2,r}(\rho, \phi)$ depends only on the distance ρ, and not on the angular direction ϕ. It can be hypothesized that obtaining such isotropy requires to achieve a mixing close to chaotic mixing when preparing the suspension, and to avoid imposing a significant unidirectional flow to the material before any measurement.

Under flow, $g_{2,r}(\rho, \phi)$ now depends on both ρ and ϕ. For distances ρ close to the value of the particle diameter, which correspond to near neighbours and possibly to

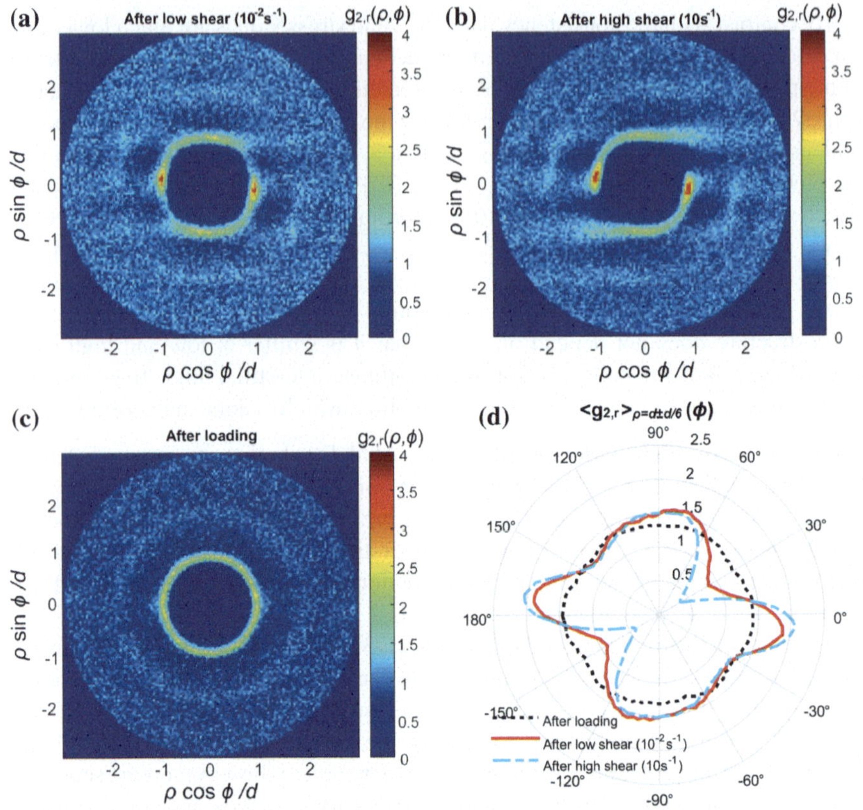

Fig. 7 **a–b–c** Color maps of pair distribution functions $g_{2,r}(\rho, \phi)$ in the velocity-velocity gradient plane of the simple shear flow imposed locally by a parallel plate geometry for a suspension of 37% particles in a concentrated emulsion, in three different conditions: **a** after shear at low shear rate (10^{-2} s^{-1}); **b** after shear at high shear rate (10 s^{-1}); and **c** after loading the sample into the geometry. (ρ, ϕ) are the polar coordinates of a particle pair in the velocity (horizontal axis)-velocity gradient (vertical axis) plane. **d** Polar plot of the pair distribution function as a function of the pair angle ϕ for close neighbors: $g_{2,r}$ is here averaged over pairs of distance ρ equal to $d \pm d/6$, d being the particle diameter. The pdfs are computed from the particle positions determined using X-ray microtomography for the different shear histories. Reprinted with permission from Ovarlez et al. (2015). Copyright (2015) by the Society of Rheology

particles in contact, it is observed (Fig. 7d) that $g_{2,r}(\rho, \phi)$ has maximal and minimal values corresponding, respectively, to an increase of the number of particle pairs roughly aligned with the flow, in the compressional region of simple shear, and to a decrease of the number of pairs in the extensional region. These features are qualitatively similar to what is observed for particles in Newtonian fluids (Morris 2009; Blanc et al. 2013).

Importantly, the shape of the pdf $g_{2,r}(\rho, \phi)$ depends on the shear rate, which contrasts with the case of interstitial Newtonian fluids: the depletion of particle pairs in close contact in the extensional region is indeed enhanced by high shear rates (Fig. 7d), which is counter-balanced by a growing tail (Fig. 7b). The failure of Eq. 22 observed in Fig. 6, i.e. the apparent absence of a link between the dynamic yield stress value and the consistency, can thus be explained qualitatively from the differences in the particle microstructures. Indeed, the consistency value is obtained from the suspension behavior at high shear rate whereas the dynamic yield stress is obtained from the behavior at low shear rate. These two properties thus reflect two different microstructures.

Static yield stress vs dynamic yield stress: history dependence To test further this idea, additional macroscopic experiments can be performed by preshearing the suspension at high shear rate, stopping abruptly the rotation and leaving the material at rest to relax (Ovarlez et al. 2015). With this procedure, it is expected that the particle microstructure formed under shear is gelled at rest; any measurement performed subsequently should then bear the signature of this microstructure. In particular, if a shear flow at low shear rate is subsequently imposed on the material, a transient stress response is expected before the material reaches a steady-state stress corresponding to the steady-state microstructure characterizing a slow flow. In the following, to account for such possible transient shear-history-dependent effects, the distinction is made between the *static* yield stress $\tau_{y,s}$, which is the stress above which flows start, and the *dynamic* yield stress $\tau_{y,d}$, which is the stress below which flows stop.

Whereas the behavior of the pure emulsion does not show any significant dependence on flow history, as classically observed for such simple yield stress fluids (Ovarlez et al. 2013), it can be shown that the suspensions do display a static yield stress $\tau_{y,s}$ much higher than their dynamic yield stress $\tau_{y,d}$ (Ovarlez et al. 2015). It is worth noting that the flow curve $\tau(\dot{\gamma})$ is followed reversibly if the shear stress is ramped-up just after a down-ramp: in this case, the static yield stress is equal to the dynamic yield stress because the microstructure gelled at rest is that formed under low shear rate.

It can thus be concluded that the static yield stress is the yield stress of the material which has the microstructure formed during its preparation, whereas the dynamic yield stress is the yield stress of the material whose microstructure is the steady-state microstructure formed under low shear.

Values of the dimensionless static yield stresses measured in such experiments are reported in Fig. 6, together with the dynamic yield stress, previous data of Mahaut et al. (2008a), and the predictions of the model. Of course, these static yield stresses are macroscopic values, possibly affected by migration effect and heterogeneities of the suspension microstructure; they provide nevertheless relevant estimates of the impact of microstructure on the behavior.

It is remarkable that the observed static yield stresses for materials presheared at high shear rate are now in much better agreement with the model (Eq. 30), and with the consistency data replotted as $[\eta_{HB}(\phi)/\eta_{HB}(0)]^{1/(n+1)} \times (1-\phi)^{n/(n+1)}$ with $n = 0.5$. In other words, there seems to be a link between the consistency values and the yield stress values as predicted by Eq. 22, provided the yield stress is measured for the microstructure formed at high shear, as is the consistency, which makes sense.

This allows one to provide a full consistent picture of the behavior of the materials with the model presented in Sect. 2. It also confirms that the shear-dependent function $g(\phi)$ bears sufficient information on the microstructure for each shear history as a first approximation.

It remains, however, that no prediction yet exists for the value of the dynamic yield stress. It suggests that the properties of shear in the interstitial fluid are specific at the onset of plasticity. To characterize $g(\phi)$ for any shear history, a solution could consist in using parallel superposition (Vermant et al. 1998) or to stop abruptly the flow and to apply small oscillations to measure the suspension linear properties for the gelled microstructure; this cannot be done simply, however, in the case of inhomogeneous materials.

We also note in Fig. 6 that there is a good agreement between the model and the data of Mahaut et al. (2008a). This is surprising and seems to be fortuitous. Indeed, the static yield stress measurements of Mahaut et al. (2008a) where performed after loading into the rheometer, on isotropic (unsheared) materials (see Fig. 7c), and there is a priori no reason why materials with an isotropic microstructure and materials with the microstructure formed at high shear rate should have the same properties. The link between the microstructure and the rheological properties of yield stress suspensions has to be built to understand this puzzling observation.

5 Shear-Induced Migration

Suspensions in Newtonian fluids are often found to develop volume fraction inhomogeneities (Abbott et al. 1991; Phillips et al. 1992; Tetlow et al. 1998; Shapley et al. 2004; Ovarlez et al. 2006; Altobelli et al. 1991, 1997). E.g., in wide-gap concentric-cylinder Couette flows, particles migrate from the inner to the outer cylinder. In the Suspension Balance Model (SBM), this comes from particle fluxes counterbalancing the gradients in particle normal stresses (Nott and Brady 1994; Mills and Snabre 1995; Lhuillier 2009; Nott et al. 2011). Although migration is usually slow at low concentration, it can be very rapid and unavoidable at the approach of ϕ_{div} (Ovarlez et al. 2006; Fall et al. 2010, 2015); this poses experimental problems, since the determination of rheological properties with classical (macroscopic) means of investigation requires that the material is homogeneous (Ovarlez et al. 2006).

Less is known about the case of particles in yield stress fluids. From the theory presented in the chapter dealing with their rheological properties, it is seen that suspensions of particles in yield stress fluids display normal stress differences similar to those found in suspensions in Newtonian fluids; the expressions derived theoretically can then be used to model migration.

Theory

Normal stress differences are particularly important when dealing with shear-induced migration. Shear-induced migration is indeed driven by a gradient in particle normal stresses (Nott and Brady 1994; Mills and Snabre 1995; Morris and Boulay 1999;

Lhuillier 2009; Nott et al. 2011). Noting Σ^p_{ij} the particle stresses, the particle normal stresses are

$$\Sigma^p_{11} = \alpha_1(\phi)\eta(0)\,\dot\gamma, \qquad \Sigma^p_{22} = \alpha_2(\phi)\eta(0)\,\dot\gamma, \qquad \Sigma^p_{33} = \alpha_3(\phi)\eta(0)\,\dot\gamma \tag{34}$$

in a Newtonian fluid of viscosity $\eta(0)$, where $\alpha_i(\phi)$ are the dimensionless particle normal stress. Recent experimental values for particle normal stresses can be found in Boyer et al. (2011b), Dbouk et al. (2013), Garland et al. (2013). We recall that indexes 1, 2, and 3, correspond, respectively, to the velocity direction, to the velocity gradient direction, and to the vorticity direction. This can be rewritten as:

$$\Sigma^p_{11} = \frac{\alpha_1(\phi)}{g(\phi)}\tau, \qquad \Sigma^p_{22} = \frac{\alpha_2(\phi)}{g(\phi)}\tau, \qquad \Sigma^p_{33} = \frac{\alpha_3(\phi)}{g(\phi)}\tau \tag{35}$$

In the linearized yield stress fluid, Eq. 35 should remain valid with τ given by Eq. 17.

Fully modeling shear-induced migration and its kinetics is far out of the scope of the present chapter. The focus is here on steady flow properties of suspensions of particles in yield stress fluids. We only discuss the steady volume fraction profiles generated by migration in these suspensions.

Steady-state volume fraction profiles can easily be predicted by ensuring stress balance at both the suspension and the particle phase scale: $\phi(\vec r)$ profiles are found from solving $\partial_i\sigma_{ij} = 0$ and $\partial_i\Sigma^p_{ij} = 0$, with $\phi(\vec r)$ a free function (Morris and Boulay 1999; Lhuillier 2009; Nott et al. 2011).

For concentric-cylinder Couette flows, the relevant equations are (Morris and Boulay 1999)

$$\partial_r(r^2\tau) = 0 \tag{36}$$

$$\partial_r\Sigma^p_{rr} = \frac{\Sigma^p_{\theta\theta} - \Sigma^p_{rr}}{r} \tag{37}$$

Thanks to Eq. 35, Eq. 37 can be changed into

$$\tau\,\partial_r\left[\frac{\alpha_2(\phi(r))}{g(\phi(r))}\right] + \left[\frac{\alpha_2(\phi(r))}{g(\phi(r))}\right]\partial_r\tau = \frac{\alpha_1(\phi(r)) - \alpha_2(\phi(r))}{g(\phi(r))r}\tau \tag{38}$$

When combined with Eq. 36 this finally yields

$$\partial_r\left[\frac{\alpha_2(\phi(r))}{g(\phi(r))}\right] = \frac{\alpha_1(\phi(r)) + \alpha_2(\phi(r))}{g(\phi(r))r} \tag{39}$$

which, once the material properties $\alpha_1(\phi)$, $\alpha_2(\phi)$ and $g(\phi)$ are known, allows one to determine the steady-state inhomogeneous volume fraction profile $\phi(r)$.

It is striking here is that it is the exact same equation as for particles suspended in a Newtonian medium (Morris and Boulay 1999). In other words, it is predicted that steady volume fraction profiles in a concentric-cylinder Couette geometry should be the same in yield stress fluids and in Newtonian materials, under two conditions: (i) if $\alpha_1(\phi)$, $\alpha_2(\phi)$ and $g(\phi)$ are the same in the yield stress fluid as in a Newtonian material, i.e., if the microstructure is similar in both cases; and (ii), if the yield stress suspension is sheared in the whole gap of the Couette geometry, which cannot be the case near the yielding transition, where shear localization occurs: in such case, the volume fraction in the unsheared region cannot change.

Note that this result cannot be generalized to other flows. Indeed, it is due to (i) the proportionality between the particle normal stresses and the suspension shear stress for both suspending fluids, which implies that the particle stress balance equation has the same form when expressed as a function of the suspension shear stress for both suspending fluids, and (ii) the independence of the shear stress profile on the material behavior in a concentric-cylinder Couette flow. Both equations being independent of the fluid behavior finally implies that the particle volume fraction at equilibrium is the same in both fluids (and in any nonlinear fluid in which the same approach can be conducted). One would thus expect the same features in a Poiseuille flow (as long as the plug region is of negligible extent in the yield stress suspension), where the shear stress profile is also independent of the fluid behavior, but not, e.g., in a parallel plate geometry.

Experimental results: Volume fraction profiles and normal stresses

One of the predictions of the theory is that steady volume fraction profiles in a yield stress fluid, in the absence of shear localization, should be similar to those observed in a Newtonian fluid. To test this prediction, experimental data obtained by Ovarlez et al. (2015) for particles in a viscoplastic fluid (a concentrated emulsion) are compared with data from experiments performed with suspensions of monodisperse spherical particles in a Newtonian oil (polyethylene glycol, as in Boyer et al. (2011b)), in which steady volume fraction profiles were measured; note that the dimensionless viscosity $g(\phi)$ of this system is better fitted by the Zarraga equation (Eq. 29) than by the Krieger-Dougherty equation (Boyer et al. 2011b), which implies that perfect match between experiments performed with the Newtonian interstitial fluid and with the yield stress fluid cannot be expected; moreover, the dimensions of the Couette geometries are slightly different.

The steady volume fraction profiles obtained in the two fluids are shown in Fig. 8 for a particle volume fractions of 50%. It is observed that these profiles have the same shape in both fluids, and are characterized by the same amplitude of volume fraction inhomogeneities. There is thus no noticeable difference between steady-state migration in a yield stress fluid and in a Newtonian fluid, as long as there is no plug zone in the case of the yield stress suspensions; of course, things should be different for the kinetics of the phenomenon, as it can be expected to have a complex dependence on the shear rate in a nonlinear fluid.

As explained above, this independence of the steady volume fraction profiles on the fluid properties can be predicted from the assumption that the ratio of the particle normal stress to the shear stress should be a function of ϕ only, as in a Newtonian fluid, and that this function should not be affected by the nature of the fluid. To illustrate this point further, we show the combination of normal stress differences

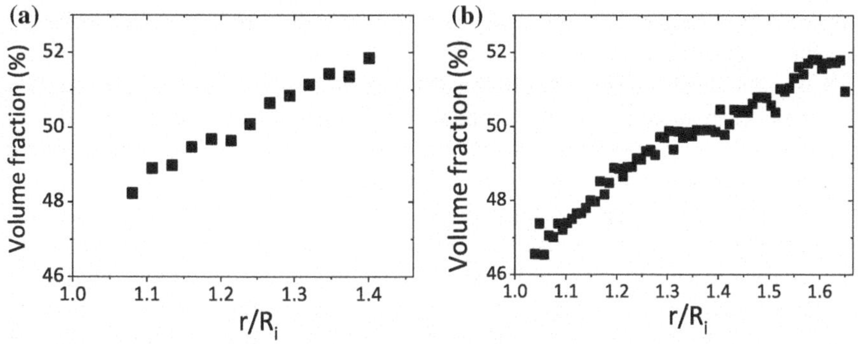

Fig. 8 Steady volume fraction profiles versus dimensionless radial position r/R_i in suspensions 50% particles in **a** a concentrated emulsion, and **b** a Newtonian oil. For the yield stress suspensions, the inner cylinder radius is $R_i = 4.1$ cm, and the gap is 1.9 cm wide; for the viscous suspensions, $R_i = 3$ cm, and the gap is 2 cm wide. Replotted from Ovarlez et al. (2015)

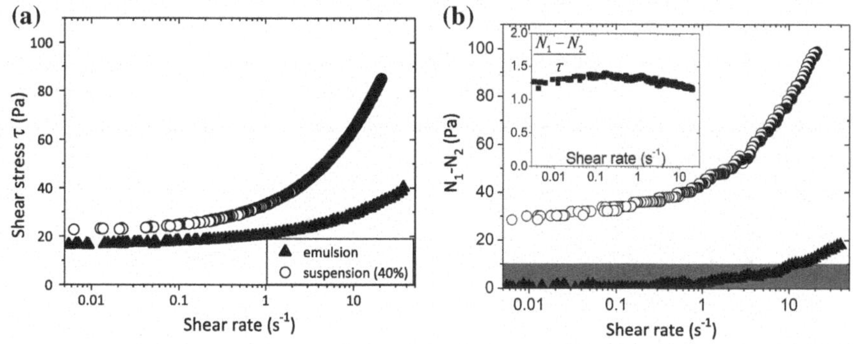

Fig. 9 Shear stress τ (**a**) and combination of normal stress differences $N_1 - N_2$ (**b**) versus shear rate $\dot{\gamma}$, measured in a parallel plate geometry in an emulsion (triangles) and in the same emulsion filled with 40% of particles (open circles). The grey zone shows the range of possible tension surface effects. The inset of Fig. 9b shows the ratio $(N_1 - N_2)/\tau$ versus $\dot{\gamma}$ in the suspension. Replotted from Ovarlez et al. (2015)

$N_1 - N_2$ measured in a parallel plate geometry for a pure emulsion, and for the same emulsion filled with 40% of particles. The data are shown in Fig. 9.

It is first observed that the flows of the concentrated emulsion seem to be characterized themselves by a combination of normal stress differences $N_1 - N_2$ which tends towards zero at low shear rate as expected theoretically (Seth et al. 2011). Nevertheless, in the range of shear rates investigated, it is hard to say what is the true value of $N_1 - N_2$ because of the shear-rate-dependent contribution of surface tension due to free surface shape changes. The normal stress differences can nevertheless be expected to be relatively important at high shear rates ($\geq 100\ s^{-1}$). They can thus be expected to play a role on the rheology of the suspensions at high shear rate and at high ϕ (where the shear rate in the interstitial fluid is highly magnified), which remains to be investigated theoretically (we recall that the theory presented in Sect. 2 is valid for an isotropic interstitial fluid).

It is particularly striking that, in the suspension, $N_1 - N_2$ is characterized by a nonzero critical value $[N_1 - N_2]_y$, which is its limit value as $\dot{\gamma}$ tends toward zero (given its value, it cannot be an artefact due to surface tension). This feature, absent from the interstitial fluid, is in agreement with the theory (see Eq. 24), and comes from the fact that normal stress differences are assumed to be proportional to the apparent interstitial viscosity, as in viscous suspensions. To further check this point, we plot the ratio of $N_1 - N_2$ to the shear stress τ in Fig. 9c. As predicted by the theory, and in line with the observations on shear-induced migration, it is observed that this ratio does not depend much on the shear rate. It is here of order 1.2. For suspensions of 40% particles in a Newtonian fluid, it was observed to be of order 0.3 by Dai et al. (2013) and of order 0.5 by Dbouk et al. (2013). We cannot comment further this difference, as a detailed study of normal stress differences in viscoplastic suspensions has not yet be performed in the literature.

Finally, note that some conclusions of this part, namely the validity of Eq. 35 and the existence of a yield particle normal stress, are supported by the paper by Dagois-Bohy et al. (2015) where shear stress and particle normal stresses have been studied in detail for $\phi_{\text{div}} > \phi > 45\%$. A notable result from this work is that the apparent friction coefficient of the yield stress suspensions follows the same law as a function of ϕ as in viscous suspensions.

6 Conclusion

The main features of the behavior of viscoplastic suspensions can be summarized as follows:

- The steady flow behavior of the suspensions can be modeled by a Herschel-Bulkley behavior of same index as the interstitial fluid, which is predicted by the theory, in which the only source of nonlinearity of the behavior is in the interstitial fluid. This should be strictly true only if the particle microstructure does not show significant variations with the shear rate at high shear rate; the yield stress has no importance for this property: this should hold for power-law materials.
- There is good agreement between the increase of the consistency with the particle volume fraction in the 0–50% range and the theoretical prediction (Eq. 31) based on the works of Chateau et al. (2008), Mahaut et al. (2008a). The similarity between the function $g(\phi)$ used to model the behavior of the isotropic suspensions studied by Mahaut et al. (2008a) and that used to model the suspension behavior under steady shear at high shear rate remains to be understood, the microstructures being significantly different. The fact that this function $g(\phi)$ is the same as for some viscous suspensions is however consistent with the conclusions of Liard et al. (2014) for power-law fluids. The similarity with the Sierou and Brady (2002) simulations suggests that the particles behave as low friction particles; this might

be due to the large size of the constitutive elements of the yield stress fluid as compared to the particle roughness and to their high concentration, which may hinder frictional interactions.

- The steady-state suspension microstructure (the particle spatial distribution) in a yield stress fluid depends on the shear rate, which contrasts with the case of a Newtonian interstitial fluid. This leads to an original shear-dependent macroscopic behavior. In particular, whereas the emulsion is a simple yield stress fluid, the yield stress of the suspensions at flow start can differ significantly from the yield stress at flow stoppage, depending on flow history. The static yield stress is the yield stress of the material which has the microstructure formed during its preparation, whereas the dynamic yield stress is the yield stress of the material whose microstructure is the steady-state microstructure formed under low shear. The existence of a shear-rate dependent microstructure also explains the discrepancy between the theory prediction for the consistency and that for the yield stress at flow stoppage, since a unique microstructure is assumed in the theory. Further progresses in the modeling of suspensions of particles in yield stress fluids would imply to model the dependence of microstructure on the shear rate, and to take it into account in the modeling of the macroscopic properties. It remains that, at this stage, we are unable to predict the value of the dynamic yield stress of the suspensions.
- It can also be shown that particle normal stresses are proportional to the shear stress as for suspensions of particles in Newtonian fluids, which is a strong validation of the theoretical approach, which consists in linearizing the interstitial fluid behavior: in this approach, a change of the apparent viscosity of the interstitial material (due to a change of the shear rate) impacts similarly all components of the suspension stress tensor. In line with this last observation, one can also observe the emergence of a nonzero normal stress difference at the yielding transition, which is absent of the interstitial fluid behavior. This is a signature of the fact that the rheology is governed by the apparent viscosity of the sheared interstitial material.

References

Abbott, J. R., Tetlow, N., Graham, A. L., Altobelli, S. A., Fukushima, E., Mondy, L. A., et al. (1991). Experimental observations of particle migration in concentrated suspensions: Couette flow. *Journal of Rheology*, *35*, 773–795.

Acrivos, A. (1995). Shear-induced particle diffusion in concentrated suspensions of non-colloidal particles? Bingham Award lecture, 1994. *Journal of Rheology*, *39*, 813.

Altobelli, S. A., Givler, R. C., & Fukushima, E. (1991). Velocity and concentration measurements of suspensions by nuclear magnetic resonance imaging. *Journal of Rheology*, *35*, 721–734.

Altobelli, S. A., Fukushima, E., & Mondy, L. A. (1997). Nuclear magnetic resonance imaging of particle migration in suspensions undergoing extrusion. *Journal of Rheology*, *41*, 1105–1115.

Ancey, C., & Jorrot, H. (2001). Yield stress for particle suspensions within a clay dispersion. *Journal of Rheology*, *45*, 297–319.

Blanc, F., Peters, F., & Lemaire, E. (2011a). Experimental signature of the pair trajectories of rough spheres in the shear-induced microstructure in noncolloidal suspensions. *Physical Review Letters*, *107*, 208302.

Blanc, F., Peters, F., & Lemaire, E. (2011b). Local transient rheological behavior of concentrated suspensions. *Journal of Rheology*, *55*, 835–854.
Blanc, F., Lemaire, E., Meunier, A., & Peters, F. (2013). Microstructure in sheared non-Brownian concentrated suspensions. *Journal of Rheology*, *57*, 273–292.
Bonnoit, C., Darnige, T., Clement, E., & Lindner, A. (2010). Inclined plane rheometry of a dense granular suspension. *Journal of Rheology*, *54*, 65–79.
Boyer, F., Pouliquen, O., & Guazzelli, É. (2011a). Dense suspensions in rotating-rod flows: Normal stresses and particle migration. *Journal of Fluid Mechanics*, *686*, 5–25.
Boyer, F., Guazzelli, É., & Pouliquen, O. (2011b). Unifying suspension and granular rheology. *Physical Review Letters*, *107*, 188301.
Brady, J. F., & Morris, J. F. (1997). Microstructure of strongly sheared suspensions and its impact on rheology and diffusion. *Journal of Fluid Mechanics*, *348*, 103–139.
Callaghan, P. T. (1991). *Principles of nuclear magnetic resonance spectroscopy*. Oxford: Clarendon Press.
Callaghan, P. T. (1999). Rheo-NMR: Nuclear magnetic resonance and the rheology of complex fluids. *Reports on Progress in Physics*, *62*, 599.
Chateau, X., Ovarlez, G., & Trung, K. L. (2008). Homogenization approach to the behavior of suspensions of noncolloidal particles in yield stress fluids. *Journal of Rheology*, *52*, 489–506.
Cheddadi, I., Saramito, P., & Graner, F. (2012). Steady Couette flows of elastoviscoplastic fluids are nonunique. *Journal of Rheology*, *56*, 213–239.
Coussot, P. (1997). *Mudflow rheology and dynamics*. Rotterdam: Balkema.
Couturier, É., Boyer, F., Pouliquen, O., & Guazzelli, É. (2011). Suspensions in a tilted trough: Second normal stress difference. *Journal of Fluid Mechanics*, *686*, 26–39.
Dagois-Bohy, S., Hormozi, S., Guazzelli, É., & Pouliquen, O. (2015). Rheology of dense suspensions of non-colloidal spheres in yield-stress fluids. *Journal of Fluid Mechanics*, *776*, R2.
Dai, S. C., Bertevas, E., Qi, F., & Tanner, R. I. (2013). Viscometric functions for noncolloidal sphere suspensions with Newtonian matrices. *Journal of Rheology*, *57*, 493–510.
Dbouk, T., Lobry, L., & Lemaire, E. (2013). Normal stresses in concentrated non-Brownian suspensions. *Journal of Fluid Mechanics*, *715*, 239–272.
Deboeuf, S., Lenoir, N., Hautemayou, D., Bornert, M., Blanc, F., & Ovarlez, G. (2018). Imaging non-Brownian particle suspensions with X-ray tomography: Application to the microstructure of Newtonian and visco-plastic suspensions. *Journal of Rheology*, *62*, 643–663.
Denn, M. M., & Morris, J. F. (2014). Rheology of non-Brownian suspensions. *Annual Review of Chemical and Biomolecular Engineering*, *5*, 203–228.
Dzuy, N. Q., & Boger, D. V. (1983). Yield stress measurement for concentrated suspensions. *Journal of Rheology*, *27*, 321–349.
Erdoğan, S. T., Martys, N. S., Ferraris, C. F., & Fowler, D. W. (2008). Influence of the shape and roughness of inclusions on the rheological properties of a cementitious suspension. *Cement and Concrete Composites*, *30*, 393–402.
Fall, A., Lemaitre, A., Bertrand, F., Bonn, D., & Ovarlez, G. (2010). Shear thickening and migration in granular suspensions. *Physical Review Letters*, *105*, 268303.
Fall, A., Bertrand, F., Hautemayou, D., Mézière, C., Moucheront, P., Lemaitre, A., et al. (2015). Macroscopic discontinuous shear thickening versus local shear jamming in cornstarch. *Physical Review Letters*, *114*, 098301.
Frankel, N. A., & Acrivos, A. (1967). On the viscosity of a concentrated suspension of solid spheres. *Chemical Engineering Science*, *22*, 847–853.
Gallier, S., Lemaire, E., Peters, F., & Lobry, L. (2014). Rheology of sheared suspensions of rough frictional particles. *Journal of Fluid Mechanics*, *757*, 514–549.
Garland, S., Gauthier, G., Martin, J., & Morris, J. F. (2013). Normal stress measurements in sheared non-Brownian suspensions. *Journal of Rheology*, *57*, 71–88.
Geiker, M. R., Brandl, M., Thrane, L. N., & Nielsen, L. F. (2002). On the effect of coarse aggregate fraction and shape on the rheological properties of self-compacting concrete. *Cement, Concrete and Aggregates*, *24*, 3–6.

Hafid, H., Ovarlez, G., Toussaint, F., Jezequel, P. H., & Roussel, N. (2015). Assessment of potential concrete and mortar rheometry artifacts using magnetic resonance imaging. *Cement and Concrete Research*, *71*, 29–35.
Hutton, J. F. (1972). Effect of changes of surface tension and contact angle on normal force measurement with the Weissenberg rheogoniometer. *Rheologica Acta*, *11*, 70–72.
Jau, W. C., & Yang, C. T. (2010). Development of a modified concrete rheometer to measure the rheological behavior of conventional and self-consolidating concretes. *Cement and Concrete Composites*, *32*, 450–460.
Keentok, M. (1982). The measurement of the yield stress of liquids. *Rheologica Acta*, *21*, 325–332.
Koehler, E. P., Fowler, D. W., Ferraris, C. F., & Amziane, S. (2005). A new, portable rheometer for fresh self-consolidating concrete. *ACI Special Publications*, *233*, 97.
Leighton, D., & Acrivos, A. (1987a). Measurement of shear-induced self-diffusion in concentrated suspensions of spheres. *Journal of Fluid Mechanics*, *177*, 109–131.
Leighton, D., & Acrivos, A. (1987b). The shear-induced migration of particles in concentrated suspensions. *Journal of Fluid Mechanics*, *181*, 415–439.
Lhuillier, D. (2009). Migration of rigid particles in non-Brownian viscous suspensions. *Physics of Fluids*, *21*, 023302.
Liard, M., Martys, N. S., George, W. L., Lootens, D., & Hebraud, P. (2014). Scaling laws for the flow of generalized Newtonian suspensions. *Journal of Rheology*, *58*, 1993–2015.
Maire, E., Buffiere, J. Y., Salvo, L., Blandin, J. J., Ludwig, W., & Letang, J. M. (2001). On the application of X-ray microtomography in the field of materials science. *Advanced Engineering Materials*, *3*, 539–546.
Mahaut, F., Chateau, X., Coussot, P., & Ovarlez, G. (2008a). Yield stress and elastic modulus of suspensions of noncolloidal particles in yield stress fluids. *Journal of Rheology*, *52*, 287–313.
Mahaut, F., Mokeddem, S., Chateau, X., Roussel, N., & Ovarlez, G. (2008b). Effect of coarse particle volume fraction on the yield stress and thixotropy of cementitious materials. *Cement and Concrete Research*, *38*, 1276–1285.
Martínez-Padilla, L. P., & Rivera-Vargas, C. (2006). Flow behavior of Mexican sauces using a vane-in-a-large cup rheometer. *Journal of Food Engineering*, *72*, 189–196.
Mewis, J., & Wagner, N. (2012). *Colloidal suspension rheology*. Cambridge: Cambridge University Press.
Mills, P., & Snabre, P. (1995). Rheology and structure of concentrated suspensions of hard spheres: Shear induced particle migration. *Journal de Physique II*, *5*, 1597–1608.
Mohan, L., Cloitre, M., & Bonnecaze, R. T. (2015). Build-up and two-step relaxation of internal stress in jammed suspensions. *Journal of Rheology*, *59*, 63–84.
Morris, J. F., & Boulay, F. (1999). Curvilinear flows of noncolloidal suspensions: The role of normal stresses. *Journal of Rheology*, *43*, 1213–1237.
Morris, J. F. (2009). A review of microstructure in concentrated suspensions and its implications for rheology and bulk flow. *Rheologica Acta*, *48*, 909–923.
Narumi, T., See, H., Honma, Y., Hasegawa, T., Takahashi, T., & Phan-Thien, N. (2002). Transient response of concentrated suspensions after shear reversal. *Journal of Rheology*, *46*, 295–305.
Nott, P. R., & Brady, J. F. (1994). Pressure-driven flow of suspensions: Simulation and theory. *Journal of Fluid Mechanics*, *275*, 157–199.
Nott, P. R., Guazzelli, E., & Pouliquen, O. (2011). The suspension balance model revisited. *Physics of Fluids*, *23*, 043304.
Ovarlez, G., Bertrand, F., & Rodts, S. (2006). Local determination of the constitutive law of a dense suspension of noncolloidal particles through magnetic resonance imaging. *Journal of Rheology*, *50*, 259–292.
Ovarlez, G., Rodts, S., Ragouilliaux, A., Coussot, P., Goyon, J., & Colin, A. (2008). Wide-gap Couette flows of dense emulsions: Local concentration measurements, and comparison between macroscopic and local constitutive law measurements through magnetic resonance imaging. *Physical Review E*, *78*, 036307.

Ovarlez, G., Mahaut, F., Bertrand, F., & Chateau, X. (2011). Flows and heterogeneities with a vane tool: Magnetic resonance imaging measurements. *Journal of Rheology*, *55*, 197–223.

Ovarlez, G., Cohen-Addad, S., Krishan, K., Goyon, J., & Coussot, P. (2013). On the existence of a simple yield stress fluid behavior. *Journal of Non-Newtonian Fluid Mechanics*, *193*, 68–79.

Ovarlez, G., Mahaut, F., Deboeuf, S., Lenoir, N., Hormozi, S., & Chateau, X. (2015). Flows of suspensions of particles in yield stress fluids. *Journal of Rheology*, *59*, 1449–1486.

Parsi, F., & Gadala-Maria, F. (1987). Fore-and-Aft asymmetry in a concentrated suspension of solid spheres. *Journal of Rheology*, *31*, 725–732.

Phillips, R. J., Armstrong, R. C., Brown, R. A., Graham, A. L., & Abbott, J. R. (1992). A constitutive equation for concentrated suspensions that accounts for shear-induced particle migration. *Physics of Fluids A: Fluid Dynamics*, *4*, 30–40.

Saak, A. W., Jennings, H. M., & Shah, S. P. (2001). The influence of wall slip on yield stress and viscoelastic measurements of cement paste. *Cement and Concrete Research*, *31*, 205–212.

Seth, J. R., Mohan, L., Locatelli-Champagne, C., Cloitre, M., & Bonnecaze, R. T. (2011). A micromechanical model to predict the flow of soft particle glasses. *Nature materials*, *10*, 838.

Shapley, N. C., Brown, R. A., & Armstrong, R. C. (2004). Evaluation of particle migration models based on laser doppler velocimetry measurements in concentrated suspensions. *Journal of Rheology*, *48*, 255–279.

Sierou, A., & Brady, J. F. (2002). Rheology and microstructure in concentrated noncolloidal suspensions. *Journal of Rheology*, *46*, 1031–1056.

Stickel, J. J., & Powell, R. L. (2005). Fluid mechanics and rheology of dense suspensions. *Annual Review of Fluid Mechanics*, *37*, 129–149.

Suquet, P. (Ed.) (1997). *Continuum micromechanics* Wien: Springer.

Tetlow, N., Graham, A. L., Ingber, M. S., Subia, S. R., Mondy, L. A., & Altobelli, S. A. (1998). Particle migration in a Couette apparatus: experiment and modeling. *Journal of Rheology*, *42*, 307–327.

Thomas, D. G. (1965). Transport characteristics of suspension: VIII A note on the viscosity of Newtonian suspensions of uniform spherical particles. *Journal of Colloid Science*, *20*, 267–277.

Torquato, S. (2002). *Random heterogeneous materials: Microstructure and macroscopic properties* Heidelberg: Springer Science & Business Media.

Toutou, Z., & Roussel, N. (2006). Multi scale experimental study of concrete rheology: From water scale to gravel scale. *Materials and Structures*, *39*, 189–199.

Vermant, J., Walker, L., Moldenaers, P., & Mewis, J. (1998). Orthogonal versus parallel superposition measurements. *Journal of Non-Newtonian Fluid Mechanics*, *79*, 173–189.

Vu, T. S., Ovarlez, G., & Chateau, X. (2010). Macroscopic behavior of bidisperse suspensions of noncolloidal particles in yield stress fluids. *Journal of Rheology*, *54*, 815–833.

Yammine, J., Chaouche, M., Guerinet, M., Moranville, M., & Roussel, N. (2008). From ordinary rhelogy concrete to self compacting concrete: A transition between frictional and hydrodynamic interactions. *Cement and Concrete Research*, *38*, 890–896.

Zarraga, I. E., Hill, D. A., & Leighton, D. T, Jr. (2000). The characterization of the total stress of concentrated suspensions of noncolloidal spheres in Newtonian fluids. *Journal of Rheology*, *44*, 185–220.

Industrial Applications of Yield Stress Fluids

D. Ian Wilson

1 Flow and Transport of YSFs

Introduction. Industry has been working with yield stress fluids for many years. In many cases they are known as viscoplastic fluids (VPFs).

VPFs are characterised by the existence of a characteristic yield stress below which they do not flow, and above which they flow readily (See Balmforth et al. 2014 for a recent review.).

Stress ≤ yield stress

No flow, but possibly creep

Does it yield elasto-plastically, or visco-elasto-plastically?

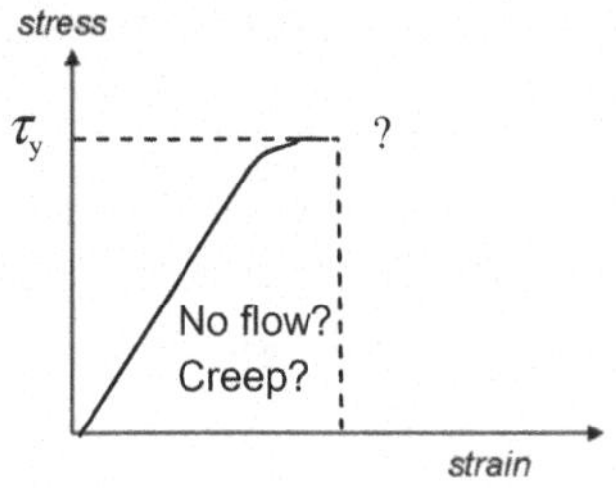

Stress ≥ yield stress

The material will flow

Is there a shear rate dependency?

Does it have memory (thixotropy)?

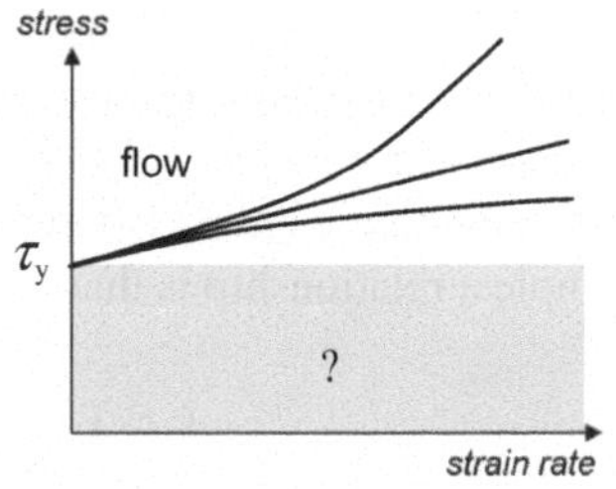

D. I. Wilson (✉)
Department of Chemical Engineering and Biotechnology,
Philippa Fawcett Drive, Cambridge CB3 0AS, UK
e-mail: diw11@cam.ac.uk

G. Ovarlez and S. Hormozi (eds.), *Lectures on Visco-Plastic Fluid Mechanics*,
CISM International Centre for Mechanical Sciences 583,
https://doi.org/10.1007/978-3-319-89438-6_6

YSFs represent a crossover between two branches of engineering:

(i) plastic materials, such as metals, which yield at a certain *strain*
(ii) viscous fluid behaviour, which is determined by *strain rate*

The detailed subtleties of YSF behaviour, including thixotropy, are usually only considered if they are important for the process under consideration.

The key consequence is that regions within a flow experiencing small shear stresses will not be deformed and will form a plug, whereas other regions will experience shear. Changes in geometry can also cause the formation of static zones at the wall.

The presence of non-deforming regions is particularly important in

Mixing—regions of low shear stress will not move and mix by convection.
Heat transfer—energy transfer through low shear regions occurs by conduction alone.
Pumping—the flow characteristics will differ noticeably from a Newtonian fluid.

Constitutive equations. Strictly speaking, yield occurs when the deviatoric stress in the material, s, exceeds a particular value. In 3-D we write

$$s = \sqrt{\frac{1}{2}\sigma_D : \sigma_D} > \tau_y \quad \text{flow} \tag{1}$$

Here, τ_y is the shear yield stress. Its relationship to the bulk yield stress, σ_y, depends on the material. A common assumption used in YSFs is that they obey von Mises' yield behaviour, which is expected for ductile materials, giving

$$\tau_y = \frac{\sigma_y}{\sqrt{3}} \tag{2}$$

and we will use this result in these lectures unless otherwise stated. Equation (2) should be used with care when dealing with highly filled suspensions, clays or soils, where more complex yielding behaviour is observed.

The simplest relationship is that of a *Bingham fluid*: (in 3-D)

$$\tau \le \tau_y \quad \dot{\gamma} = 0 \quad \left|s^2\right| \le \tau_y \tag{3a}$$

$$\tau > \tau_y \quad \tau = \tau_y + \mu_B\dot{\gamma} \quad \tau = \left(\mu_B + \frac{\tau_y}{2\left|s^2\right|^{1/2}}\right)\dot{\gamma} \tag{3b}$$

where τ is the shear stress, $\dot{\gamma}$ is the shear strain rate in simple shear (quite a mouthful—let's call it the shear rate) and μ_B is the Bingham viscosity (the measure of shear rate dependency). An example is tomato ketchup, with τ_y ~ 1 Pa and μ_B ~ 15 Pa s.

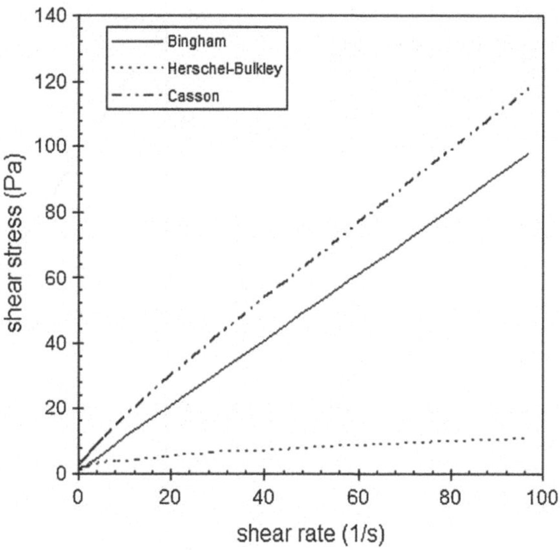

Fig. 1 Flow curves for Bingham, Herschel-Bulkley (with $n = ½$) and Casson fluids for the case when $\tau_y = 1$ Pa, and μ_B, k_{HB} and k_C all have magnitude 1 (but different units)

The Bingham fluid is a specific case of a *Herschel-Bulkley fluid*, which is a more general model:

$$\tau \le \tau_y \quad \dot{\gamma} = 0 \tag{4a}$$

$$\tau > \tau_y \quad \tau = \tau_y + k_{HB}\dot{\gamma}^n \tag{4b}$$

where the shear rate dependency is now that of a power law fluid with flow index n and flow consistency k_{HB}. Examples: toothpaste, Carbopol suspensions (these give hair 'gel' its consistency).

Another VPF model in regular use, particularly to describe inks, drilling muds and molten chocolate, is the Casson model:

$$\tau \le \tau_y \quad \dot{\gamma} = 0 \tag{5a}$$

$$\tau > \tau_y \quad \tau^{1/2} = \tau_y^{1/2} + k_C\dot{\gamma}^{1/2} \tag{5b}$$

This empirical model presumes a power law dependency of ½ and a different stress relationship.

Comparing constitutive models

I have written all three models with the same term for the yield stress, τ_y. If you **fit** data with the Casson and Herschel-Bulkley models you will get different values for the yield stress, so **beware** using the result from one model with another! This is illustrated in Fig. 1.

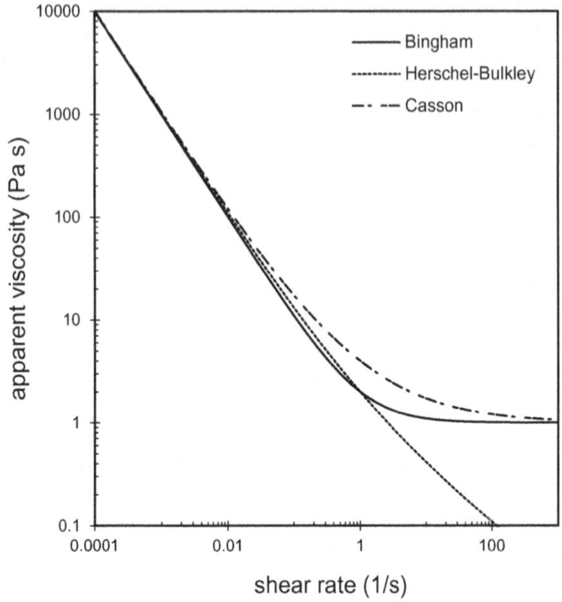

Fig. 2 Plots of apparent viscosity for the data sets in Fig. 1 (n Herschel-Bulkley = ½)

Apparent viscosity

Consider the Bingham model: the apparent viscosity is defined as

$$\begin{aligned}\eta_a &= \frac{\tau}{\dot{\gamma}} \\ &= \frac{\tau_y}{\dot{\gamma}} + \mu_B\end{aligned} \tag{6}$$

Some observations:

(i) This says that $\eta_a \to \infty$ as $\dot{\gamma} \to 0$, or that η_a decreases as $\dot{\gamma}$ increases, i.e. that VSFs are always *shear-thinning*. Note, however, that a shear-thinning fluid is *not* necessarily viscoplastic.

(ii) One test often used for VSFs is that the gradient of the flow curve on log-log axes should approach−1. Figure 2 considers more data and presents them in this form—which is the normal way to plot such data!

The standard approach is to make rheometrical measurements of flow (either flow rate or shear rate) for a range of imposed shear stresses in a known geometry. The yield stress is estimated and the viscous component identified from plots of $\tau - \tau_y$ versus $\dot{\gamma}$. There are several methods for measuring or (more often) estimating τ_y. Table 1 summarises those in common use.

Challenging questions. There are four challenges in the estimation of the yield stress. Some of these are covered in detail in other lectures in the workshop.

Table 1 Methods used to determine yield stress (based on Steffe[a] Table 1.4)

Method	Measurement technique	Typical tool(s)	Schematic of configuration
Extrapolation	Measure apparent viscosity and extrapolate to zero shear rate	Capillary or rotational rheometers	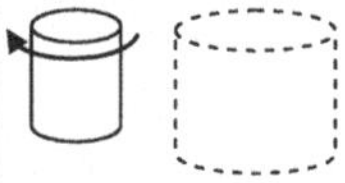
Stress decay	Measure shear stress after a flow is stopped	Capillary rheometer, also squeeze flow devices	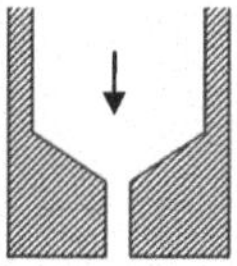
Stress to initiate flow	Increase shear stress and determine when finite shear rate is detected	Rotational rheometer with vane tool	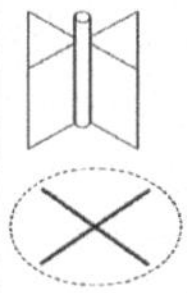
Velocimetry	Identify boundary of flow region and relate this to estimate of local shear stress	Flow imaging, e.g. PIV; tomography, e.g. MRI	
Dynamic testing	Interpretation of small amplitude oscillatory testing—end of linear region, or *G*', *G*" crossover	Rotational rheometers	
Vertical plate coating	Withdraw plate from fluid and measure thickness of film	Wilhemy plate device	

(continued)

(a) **Timescale**: If a material is just on the point of yielding, things move slowly. Phenomena such as spatial rearrangement caused by Brownian motion, which are not important at higher deformation rates, can then influence the rate of deformation. Creep is another example. The yield stress inferred from measurements near the yield point may not give a reliable description of the phenomena active at higher deformation rates. The measurements may be accurate, but their interpretation requires care. Timescales are revisited in Sect. 4.

(b) **Start-up or steady**: Measurements made under steady flow conditions often give different results to those made when a sample is taken from the static state

Table 1 (continued)

Method	Measurement technique	Typical tool(s)	Schematic of configuration
Squeeze flow testing	Interpretation of deformation between parallel circular disks, subject to constant force	Strain frame or texture analyser	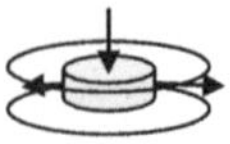
Cone penetrometer	Measure depth of penetration under given loading—interpret depth in terms of plastic deformation	Strain frame or texture analyzer with controlled force operation	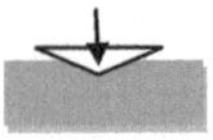

[a]https://sites.google.com/site/jfsteffe/freeman-press

through the yield point. For example, measurements of the yield stress using a vane tool at different times are likely to exhibit the following behaviour:

Sketch

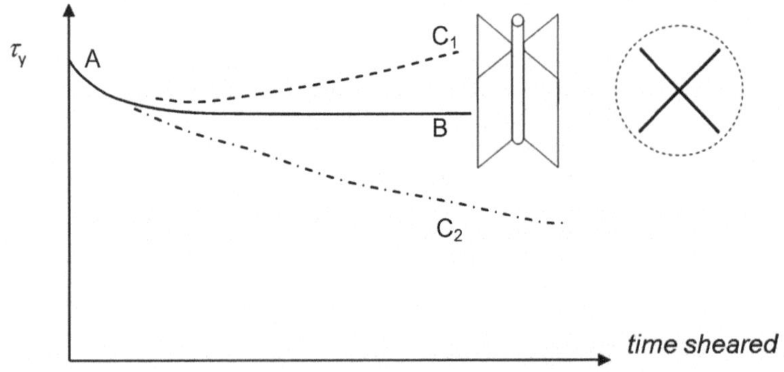

- A is the static shear yield stress and is the property of the unsheared structure.
- B is the dynamic shear yield stress and is that associated with flow at the test conditions. This is the value of τ_y to use in steady flow calculations.
- C_1 and C_2 are other possible scenarios and represent differing cases of thixotropy.

(c) VSFs are often **difficult to handle**, so that they cannot be loaded into the test device. The microstructure responsible for yield behaviour can be disturbed by loading, so care must be taken when a sample is to be transferred or loaded.

(d) **Different tests** often give different estimates for τ_y. Be prepared for uncertainty in parameters, and variation in the values reported by different workers.[1] Figure 3

[1]See Aeschlimann and Beckett (2000) Measurements of the Casson parameters for a standard chocolate across 23 laboratories reported a 7-fold range in yield value.

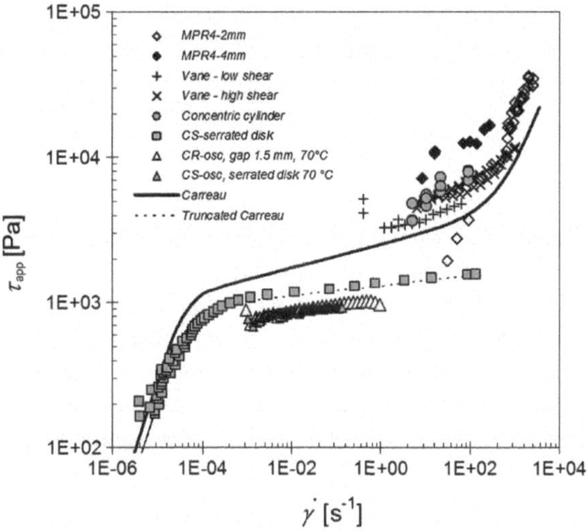

Fig. 3 Flow curve for a 50 vol.% ZnO/wax paste obtained using different rheometers at 70 °C. Note the wide range of shear rates. Reproduced from Barker and Wilson (2008)

shows data collected for a zinc oxide/wax paste using several rheometers. The behaviour at low shear shear stress is time-dependent and is caused by *creep* in response to a prolonged stress. From an engineering standpoint, the material can be classified as a Herschel-Bulkley material with $\tau_y \sim 2$ kPa.

Dimensionless groups

Bingham

The Bingham number has been defined in earlier chapters, but here it is again for a Bingham fluid flowing at mean velocity v_m:

$$Bn \equiv \frac{\tau_y}{\mu_B\,[v_m/D]} = \frac{\text{yield stress}}{\text{viscous stress}} \tag{7}$$

while for a Herschel-Bulkley fluid

$$Bn = \frac{\tau_y}{k_{HB}}\left(\frac{D}{v_m}\right)^n \tag{8}$$

where D is a characteristic dimension such as duct diameter. *Bn* quantifies the contribution from the plastic component to the flow: a low value of *Bn* indicates that viscous behaviour dominates.

Inspection of Eq. (7) indicates that under certain conditions flow behaviour will be dominated by the viscous contribution (where $\tau_y < K_{HB}\dot{\gamma}^n$) or the solid-like term ($\tau_y > K_{HB}\dot{\gamma}^n$). Since τ_y, K and p are determined by the paste formulation and $\dot{\gamma}$ is a characteristic shear strain rate for the process, this gives a basis for determining pastes as either 'soft' or 'stiff', with the former being materials whose behaviour is dominated by viscous contributions at the flow rate (and hence characteristic shear strain rate) of interest.

'*Stiffness*' (Wilson and Rough 2012).

A second group which is important in practice is the 'stiffness' of the YSF, Y:

$$Y = \frac{\tau_y}{\rho g H} = \frac{yield}{weight} \tag{9}$$

which compares τ_y as a measure of the bulk yield 'strength' with the self-weight of a sample of height H and density ρ. Y therefore characterizes the likelihood of a sample of YSF to deform under its own weight. Materials with large Y can be considered 'stiff' and are then suitable for **forming** or **shaping operations**, whereas those with small Y are better suited for transport (flow) applications.

A group similar to Y also appears in descriptions of sedimentation, where the denominator is then the buoyancy force associated with a body of length-scale H (e.g. the diameter of a chocolate chip in cookie dough).

$$Y = \frac{\tau_y}{g \Delta \rho D} = \frac{yield}{buoyancy} \tag{10}$$

Table 2 compares values of the product $YH \equiv \tau_y/\rho g$ for different YSFs.

The materials are organized in Table 2 in terms of function. Those primarily intended for use in transport applications have YH values of order mm or less, while those intended for forming have YH values in metres.

Time

The Deborah number, De, is defined as

$$De = \frac{t_{process}}{t_{material}} \tag{11}$$

where t_{process} is a characteristic time of the process, and t_{material} is the time taken for the material to change its nature or properties in response to the stress state induced by the process. For polymers, this is the relaxation time of the polymer chains.

With VSFs, the process timescale is set by the flow rate. Recall that the flow rate sets the characteristic strain rate, which has units of time^{-1}, so Eq. (11) can be written as

$$De^{-1} = \dot{\gamma} \times t_{material} \tag{12}$$

If De is small, the material does not have time to change and the rheology of the material can be treated as constant.

The mechanisms by which the material can change are dictated by the **nature** of the material. These can be broadly categorised as

(i) *Microscale*

Here, the microstructure of the material changes in response to the imposed strain or stress. An example is a Carbopol gel, where at low strain rates the interactions between individual polymer particles give rise to thixotropy and

Table 2 Values of $Y \times H$ for different yield stress fluids

Material	τ_y/Pa	ρ/kg m^{-3}	*YH*	Source
Flow materials				
Tomato ketchup	1	1050	97 μm	Steffe
0.12 wt% Carbopol 940	19	1000	2.0 mm	Ortega-Avila et al. (2016)
Carbopol solutions (different pH)	30–240	~1000	3.4–28 mm	Curran et al. (2002)
Portland cements				(ASTM C150 Type 1)
100 mm flow	660–780	2110	32–38 mm	Assaad et al. (2016)
200 mm flow	40–60	1825	2.2–3.4 mm	
Zinc oxide/wax paste	2000	1550	13 cm	Barker and Wilson (2008)
Forming materials				
Potato starch dough	0.12 MPa[a]	950	12.6 m	Rahman et al. (2001)
Microcrystalline cellulose/water dough	0.28 MPa[a]	990	29 m	Zhang et al. (2013)
Talc paste	0.15 MPa[a]	1900	8 m	Martin et al. (2004)
Ceramic oxide pastes	0.2–0.74 MPa[a]	3500–3800 (estimated)	6–20 m	Powell and Blackburn (2010)
Tungsten carbide paste	0.40 MPa[a]	6000	5 m	Ferstl et al. (2012)
Solid soap	0.56 MPa[a]	800	71 m	Amarasinghe and Wilson (1998)

[a]The value reported is the measured bulk yield stress rather than the shear yield stress. For a von Mises material, $\sigma_y = \sqrt{3}\tau_y$

creep. The local composition does not change, but the structural arrangemnt does.

(ii) *Macroscale*

The stress applied to make the material flow, and the associated dissipation of energy, can give rise to local changes in composition and this in turn changes the rheology of the material.

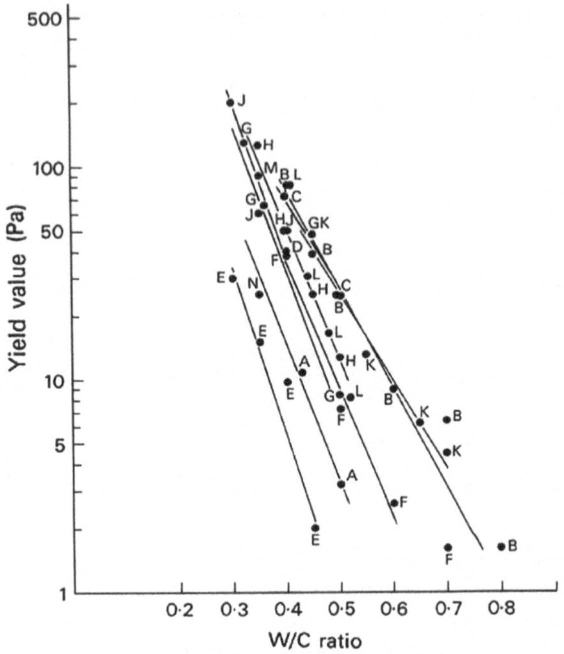

Fig. 4 Effect of water/cement (W/C) ratio on yield stress of cement pastes. Each letter denotes a different series of published data: reproduced from Banfill (2008)

Dense suspensions are particularly sensitive to macroscale phenomena. When the continuous phase is a mobile liquid, the stress imposed on the liquid (termed the 'pore pressure' in soil mechanics) can cause it to move relative to the solid phase. This gives rise to regions with high and low solids volume fraction. Since the yield stress of dense suspensions, like soils, are strongly sensitive to solids volume fraction (see Fig. 4 for data collected for cement pastes), the rheology of the material varies over space and time.

A second important example is local heating caused by viscous dissipation, which can promote softening or melting of a component, increasing the local volume fraction of liquid. As before, this will reduce the apparent viscosity in the affected region, *in addition to* any effect of increased temperature on the rheology of the liquid fraction. Since the dissipation rate per unit volume is given by the product $\tau \times \dot{\gamma}$, regions with high shear rate and shear stress are particularly sensitive to this phenomenon: for instance, any YSF-solid boundary. It is time to talk about *wall slip*.

Wall Slip. The standard assumption in flow analyses is that the relative velocity of material at a confining wall is zero with respect to the wall. In many cases there is a rapid change in velocity near the wall which is not consistent with the bulk behaviour and the material appears to change velocity over a small distance—called 'slip'.

Sketch

YSF pipe flow, no slip: $v_{wall} = 0$ YSF pipe flow, positive slip, $v_{wall} = v_s > 0$

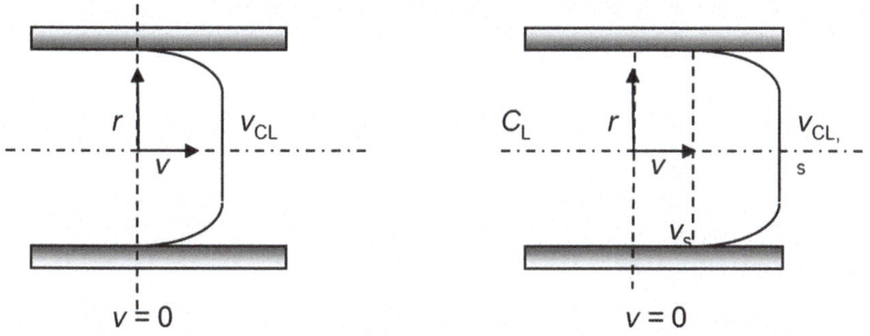

Wall slip is

(i) **Good** for pumping and extrusion. If the flow rates in the two scenarios above are the same, $v_{CL,s} < v_{CL}$. The material then undergoes less shearing in the bulk, so the pressure drop will be smaller. The *limiting case* is where the shear stress at the wall required to cause slip is less than the bulk yield stress, and plug flow occurs. In forming operations, slip often dominates behaviour at the wall.

(ii) **Bad** if it occurs in your rheometer and you don't notice it, as you are now measuring a mixed response—part wall behaviour, part bulk behaviour. In the limiting case, the rheometer measures slip behaviour and gives no information about bulk yield.

(iii) **Commonly observed** with YSFs. Roughened surfaces are therefore often used in rheological measurements of YSFs to promote bulk deformation.

Causes of slip

(i) *Wall-induced ordering*. A wall can induce local ordering and thus difference in material microstructure to the bulk, either due to steric/geometric effects, interactions arising from the walls different properties and composition, or its condition (charge, temperature, vibration etc.).

Consider a dense suspension: the bulk packing arrangement cannot be sustained in the wall region.

Sketch

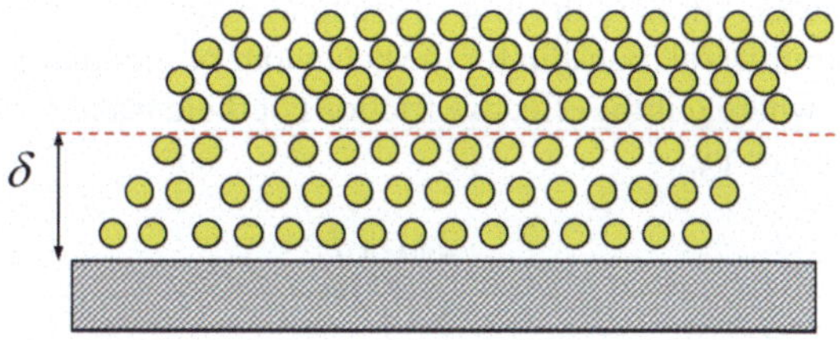

Shear of particles near a wall tends to (a) align elongated particles and (b) cause liquid to move towards the wall. A suspension with higher liquid content will have a lower solid content and have a lower viscosity. If the wall layer is thin, it can be considered as a slit of thickness δ undergoing simple shear with

$$\begin{aligned}\tau_w &= \eta_a \dot{\gamma}_{wall} \\ &= \eta_a \frac{v_s - 0}{\delta} = \eta_a \frac{v_s}{\delta}\end{aligned} \tag{13}$$

If $\eta_a \ll \eta_{bulk}$, v_s can be large and the material appears to 'slip' at the wall. In practice, $\delta \sim 5$–10 particles.

Rearranging (13) gives

$$\tau_w = \frac{\eta_a}{\delta} v_s = a \cdot v_s \tag{14}$$

which is an example of the Navier wall slip law:

$$v_s = \beta_{N,a} \tau_w^m \tag{15a}$$

with $m = 1$. At this point we note that some workers write the Navier slip law as

$$\tau_w = \beta_{N,b} v_s^m \tag{15b}$$

Always **check** which form is used!

The slip coefficient incorporates the effects of temperature and surface nature. For the form in (15b), we have

- No slip : $\beta_{N,b} \to \infty$
- Perfect slip: $\beta_{N,b} = 0$

Users of Eq. (15a) often assume that wall friction is independent of normal stress effects such as Coulombic friction. The simplest case including this is the Mohr-Coulomb law, widely used in granular materials:

$$\tau_w = \mu_w \sigma_N + c \tag{16}$$

where μ_w is the friction coefficient and σ_N is the normal stress acting on the wall.

Normal stress effects on wall friction have been studied in polymer flows, and arise widely in considerations of the flow of granular materials. Jop et al. (2006) have developed a constitutive relationship for the behaviour of granular materials, termed the $\mu(I)$ rheology, where μ is the effective friction coefficient and I is a dimensionless intertial number, given by

$$I = \frac{2\,\|\mathbf{D}\|\, d}{\sqrt{p/\rho^*}} \tag{17}$$

Here **D** is the second invariant of the strain rate tensor, d is the particle size, p is the hydrostatic component of the stress tensor and ρ^* is the solid density of the particles. The definition of I in (17) is taken from Gray and Edwards (2014), who provide a detailed account of the $Y(I)$ rheology and its application to a set of flows of granular materials.

(ii) *Viscous heating*. In simple shear the work done against viscous friction sees mechanical energy converted into heat. The local rate of viscous dissipation is given by $\tau_w \times \dot{\gamma}$ or $\eta_{app} \times \dot{\gamma}^2$ (units Wm^{-3}) and will therefore be greatest where the shear stress and the shear rate are largest, i.e. at the wall.

Viscous dissipation raises the temperature in the region near the wall and, for most fluids this will reduce the local viscosity: since

$$\tau_w = \eta_a \dot{\gamma}_{wall} = \eta_a \left(\frac{dv}{dr} \right)_{wall} \tag{18}$$

decreasing η_a increases dv/dr, giving rise to larger velocity gradients than expected and therefore ‘apparent slip’.

Viscous dissipation is important in viscous materials (e.g. molten polymers) and temperature sensitive materials (e.g. ice cream—see Elhweg et al. 2009). The influence of viscous heating is characterised by the Nahme-Griffiths (or Nahme) number, *Na*. For a pipe, *Na* is given by;

$$Na \equiv \frac{\beta_V \tau_w \dot{\gamma}_{wall} R^2}{4\lambda} \tag{19}$$

where β_v is the temperature sensitivity of viscosity, R is the pipe radius and λ is the fluid’s thermal conductivity. The material of the wall also plays a role, as it determines whether the heat liberated can be conducted away quickly.

In summary:

- Wall slip is often encountered with YSFs, particularly in forming operations.
- (Not discussed here): it can be difficult to model in detail in 2-D and 3-D.

Mooney’s method for quantification of slip in pipe flow. Mooney’s method is a general treatment which applies to any continuous fluid and is used to analyse data collected from capillary rheometers. Consider a fluid flowing along a pipe with wall shear stress τ_w and no slip:

Sketch

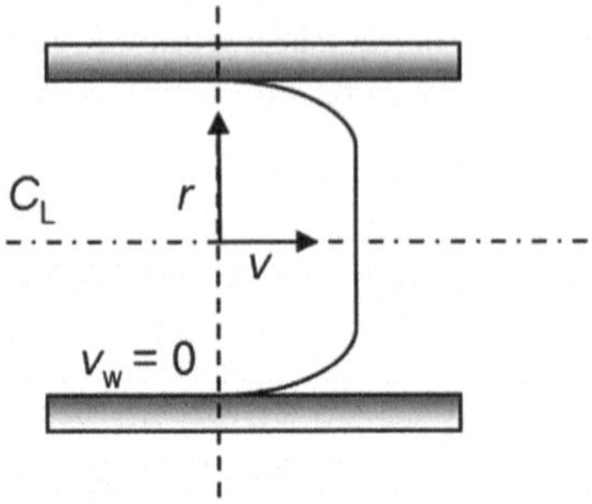

We can write the flow rate in the absence of slip, Q_{ns}, as

$$Q_{ns} = \frac{\pi R^3}{\tau_w^3} \int_0^{\tau_w} \tau^2 \dot{\gamma} d\tau \tag{20}$$

The local flow shear rate is a function of shear stress, so we can write

$$Q_{ns} = \frac{\pi R^3}{\tau_w^3} \int_0^{\tau_w} \tau^2 f(\tau) d\tau \tag{21}$$

If slip occurs, the total flow rate comes from two components:
Sketch

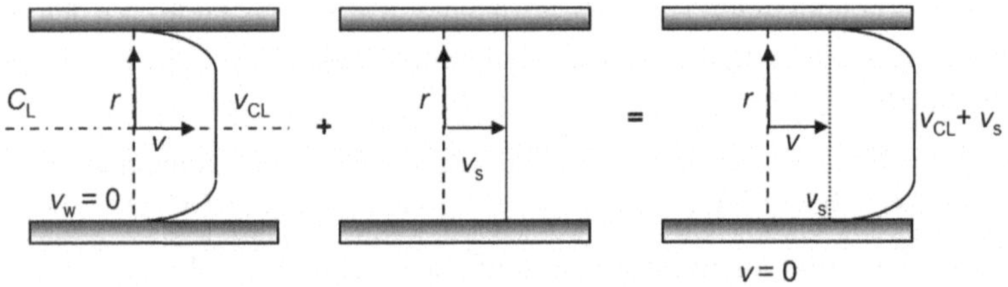

The flow caused by slip, Q_s, is (plug flow):

$$Q_s = \pi R^2 v_s \tag{22}$$

If the slip layer is thin so that the shear stress at its edge is ~τ_w (this is reasonable as $\tau_\delta = \tau_w \frac{R-\delta}{R}$ and δ is small), the total flow rate is

$$Q = Q_{ns} + Q_s = \pi R^2 v_s + \frac{\pi R^3}{\tau_w^3} \int_0^{\tau_w} \tau^2 f(\tau) d\tau \tag{23}$$

Assuming a Navier slip condition applies:

$$Q = \pi R^2 \beta_{N,a} \tau_w + \frac{\pi R^3}{\tau_w^3} \int_0^{\tau_w} \tau^2 f(\tau)\, d\tau \tag{24}$$

giving

$$\frac{Q}{\pi R^3 \tau_w} = \frac{\beta_{N,a}}{R} + \frac{1}{\tau_w^4} \int_0^{\tau_w} \tau^2 f(\tau)\, d\tau \tag{25}$$

The second term on the RHS is a function of τ_w alone: $\beta_{N,a}$ (or $\beta_{N,b}$) can be evaluated by performing experiments with capillaries of different size (at least three radii) and comparing values of $Q/\pi R^3 \tau_w$ at selected values of τ_w.

This sound straightforward, but in practice it is hard to control τ_w directly, so $Q/\pi R^3 \tau_w$ is plotted against τ_w and data at a given value of τ_w are obtained by interpolation. The latter step can introduce errors and uncertainty, so much so that some workers, particularly Jastrzebski (1967) proposed that $\beta_{N,a}$ be related to R by $\beta_{N,a} = b/R$. There is, however, no physical basis for this, and Martin and Wilson (2005) demonstrated that Jastrzebski's conclusions could be accommodated by Mooney's approach if Tikhonov regularisation was used for the interpolation.

It is unlikely that a simple mathematical form such as Eq. (14) will represent the behaviour of a real YSF—in much the same way that yield can be described as a simple yield stress.

Other methods for estimating wall exist for rotational devices exist, notably that presented by Yoshimura and Prud'homme (1988), which requires the use of different gap heights in parallel plate tests. There is a considerable body of literature on the topic, including the effect of alignment errors.

Slip in transport (pipe and annulus flows). Equation (23) can also be written as

$$\frac{Q}{\pi R^3} = \frac{\pi R^2 v_m}{\pi R^3} = \frac{v_s}{R} + \frac{1}{\tau_w^3} \int_0^{\tau_w} \tau^2 \dot{\gamma}\, d\tau$$

$$\Rightarrow \frac{v_m}{R} = \frac{v_s}{R} + \frac{1}{\tau_w^3} \int_0^{\tau_w} \tau^2 \dot{\gamma}\, d\tau \tag{26}$$

The LHS is related to the apparent shear rate for the flow, which is the shear rate at the wall for a Newtonian fluid with no wall slip (given by $4v_m/R$). The RHS has two components: the contribution from slip and that from shearing within the fluid.

- For a YSF, if $\tau_y > \tau_w$, then there is no deformation in the bulk and the second term on the RHS is zero: the apparent shear rate is all due to slip and $v_m = v_s$. Bulk deformation only starts when $v_m \geq \tau_y/\beta_{N,b}$, which can be seen to depend on the nature of the material and its interaction with the wall.

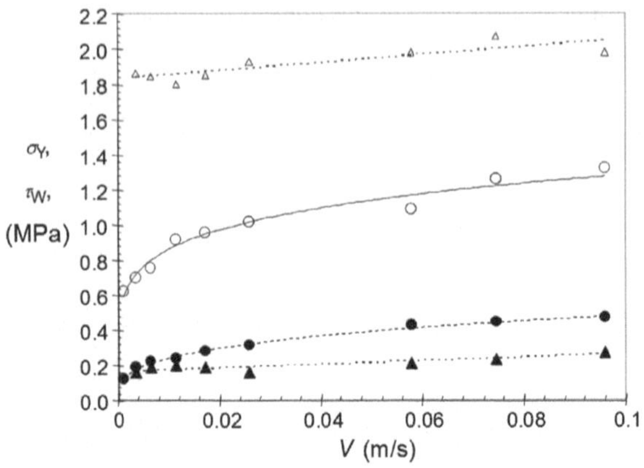

Fig. 5 Effect of extrusion velocity on highly filled tungsten carbide pastes with wax binder (triangles) and alcohol-aqueous binder (circles). Open symbols—bulk yield stress, σ_Y; solid symbols—die wall shear stress, τ_W. Reproduced from Wilmot et al. (2009)

- This gives the second method of applying Mooney's method. The apparent shear rate (or v_m/R is plotted against τ_w and the slip velocity estimated for a given value of τ_w: the v_s values are then compared with (26) and parameters extracted.
- This is the scenario desired for **forming**: nearly all the work is done at the wall. Figure 5 shows experimental results from extrusion of hardmetal pastes: the wall shear stress values are all smaller than the bulk yield strength/$\sqrt{3}$.

Examples—annular flows. Ortega-Avila et al. (2016) presented a study of the flow of a 0.12 wt% Carbopol solution along an annulus. They determined the local velocity of the YSF using PIV measurements and compared the results with the predictions for steady, fully developed flow. Similar behaviour for another Carbopol solution in a capillary was reported by Perez-Gonzalez et al. (2012).

Figure 6, reproduced from the paper, shows the configuration and distributions of shear stress and velocity. U_s is the slip velocity at either wall.

The slip velocity on the inner and outer wall is given by

$$U_s(\kappa R) = \beta_{N,a}(-\tau_{rz}(\kappa R))^m \tag{27a}$$

$$U_s(R) = \beta_{N,a}(\tau_{rz}(R))^m \tag{27b}$$

Shear rheology testing of the material indicated that it exhibited Herschel-Bulkley behaviour with parameters $\tau_y = 19.2$ Pa, $K_{HB} = 5.4$ Pa s^n and $n = 0.38$.

Figure 7 shows a series of velocity distributions for different overall pressure drops, plotted alongside predictions from their model. There is almost complete slip at the wall until the pressure drop ΔP exceeds 32.7 kPa. From this information they extracted a slip law of the form $U_s = 1.3 \times 10^{-5}\tau_w^{1.98}$.

Figure 8 shows an example of the (very good) agreement with model predictions. Details of the experimental and calculation methods are given in the original paper.

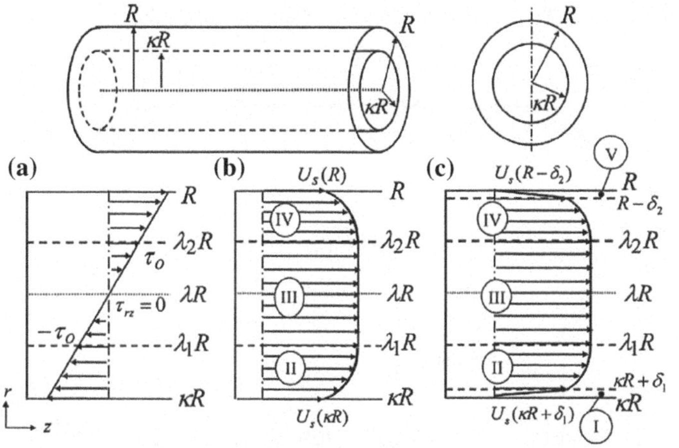

Fig. 6 Flow pattern for fully developed flow in a cylindrical annulus, with inner and outer radii κR and R, respectively, reproduced from Ortega-Avila et al. (2016). **a** shear stress distribution; axial velocity distribution with **b** true slip and **c** apparent slip at the wall. In regions II and IV the shear stress is greater than the yield stress τ_0 and there is shear flow: in region III the fluid flows as a plug

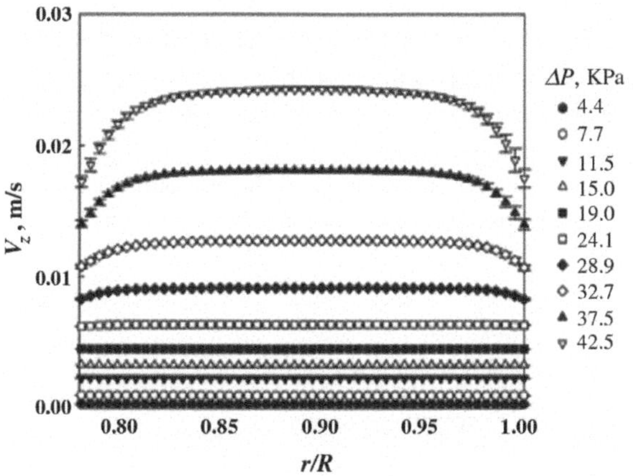

Fig. 7 PIV results for fully developed velocity distributions in an annular gap: ΔP is the overall pressure drop along the annulus. Reproduced from Ortega-Avila et al. (2016)

2 Forming of YSFs

In a forming operation the material changes *shape* and thus involves extensional flow. Many uses of viscoplastic materials are based on their ability to retain their shape after flow (hence large Y).

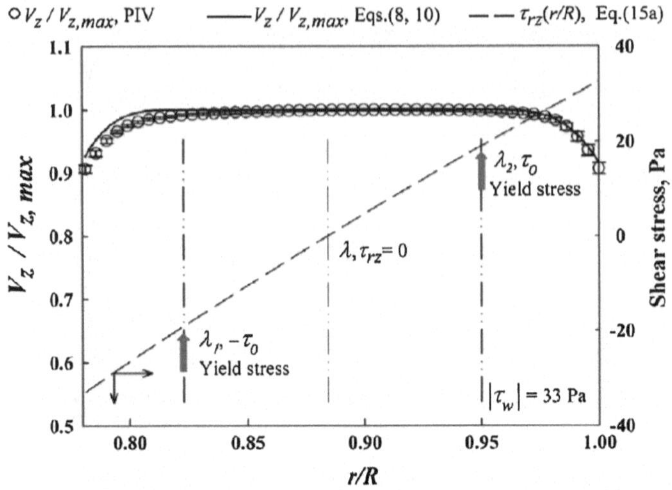

Fig. 8 Comparison of velocity distribution measured by PIV with model prediction for a wall shear stress of 33 Pa. Reproduced from Ortega-Avila et al. (2016)

Consider a plastic material, which is characterized by flow when the material is subject to a stress above its yield point. A *perfectly plastic* material requires no more work above the yield stress, i.e. it is independent of rate of strain and therefore velocity.

Consider a uniaxial deformation, such as the drawing of a cylindrical billet of initial diameter D_o to form a wire or rod of smaller diameter, D.

Sketch

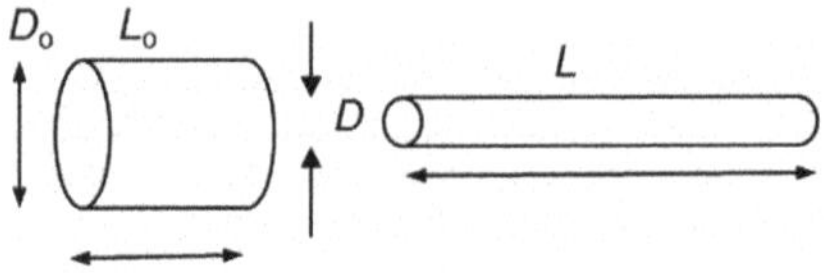

The force required to cause it to change shape is

$$F = A \cdot \sigma_p \tag{28}$$

where A is the cross-sectional area and σ_p is the plastic bulk yield strength. The work done, dW, to change its length by dL is then

$$dW = FdL = A\sigma_p dL \tag{29}$$

If the material is incompressible, the volume of an element is constant so

$$A_o L_o = AL \tag{30}$$

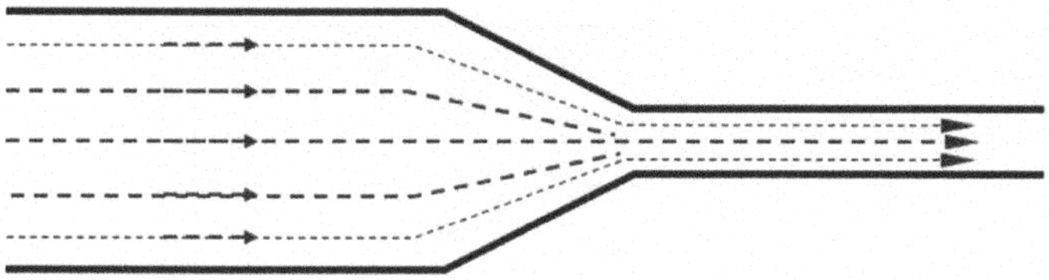

Fig. 9 Schematic of flow through a smooth contraction

$$dW = A_o L_o \sigma_p \frac{dL}{L} \tag{31}$$

giving

$$W = A_o L_o \sigma_p \int_{L_o}^{L} d \ln L = A_o L_o \sigma_p \ln \frac{L}{L_o} = A_o L_o \sigma_p \ln \varepsilon_L \tag{32}$$

where ε_L is the extensional strain. The work done per unit volume—which is the amount of energy to be provided by an external agent to make this change, or change in hydrostatic pressure, ΔP—is

$$\Delta P = \frac{W}{A_o L_o} = \sigma_p \ln \varepsilon_L = 2\sigma_p \ln \frac{D_o}{D} \tag{33}$$

Equation (33) gives the amount of work required to overcome the internal resistance of the material to its change in shape, which is termed the *homogeneous* work.

No work has been done against an external agent, such as wall friction, or dissipated via internal mechanisms such as viscous forces. These are described as *redundant work* and the energy is lost as heat.

Consider *extrusion* of a perfect plastic through a perfectly smooth planar or cylindrical die, so that there is no friction at the wall, shown schematically in Fig. 9. For the extensional flow of YSFs, we have to

(i) Select the appropriate form of the constitutive equation, relating σ_p and τ_y (as we usually measure the latter in shear—driven rheometers).
(ii) Calculate the redundant work associated with internal viscous contributions in extension (this introduces shear rate dependency), wall friction, etc.

We note that the die wall is a shallow cone, the flow is predominantly simple shear and there is little extension: the flow can be modelled using lubrication approaches as flow along a pipe with progressively narrower diameter.

Interpretation of capillary rheometry data: YSF flow through an orifice die
In capillary rheometry, material is forced from a reservoir into a long duct. Long die lands are used so that the contribution from the die entry term is small. The normal

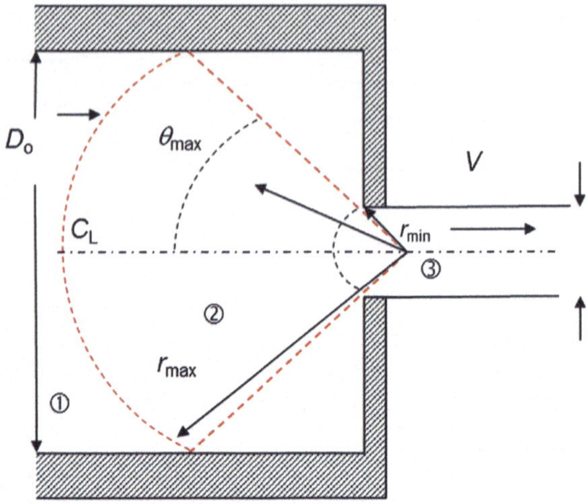

Fig. 10 Schematic of flow field in a square-ended axisymmetric cylindrical die

procedure is to use die lands of different length/diameter ratio so that the die entry contribution can be identified from a Bagley plot.[2]

Forming operations use short die lands and the die entry term contains the information about extension.

The standard treatment method is to assume a velocity profile, integrate the constitutive equation, and compare the result to the data in order to identify the model parameters.

In an orifice die, the die land is very short so the pressure drop across the orifice is that associated with the die entry. The following is a summary of the analysis presented by Basterfield et al. (2005). Figure 10 shows a YSF flowing from a cylindrical barrel through a concentric orifice die.

For square ended dies, as shown above, if there is **any** friction at the wall stagnant zones (labeled SR) are formed. The shape of the stagnant zone depends on the wall friction, which can range from *smooth* ($\tau_w = 0$) to *fully rough* ($\tau_w = \tau_y$).

Basterfield et al. (2005) assumed that the flow in the die entry was approximately conical, and the interface between the stagnant region and the flowing zone was smooth (perfectly lubricated), so their result gives the plastic and internal redundant work.

The flow is predominantly extensional. The Herschel-Bulkley model is then

$$\sigma = \sigma_y + k_u \dot{\varepsilon}^n \tag{34}$$

where

σ is the effective uniaxial stress
σ_y is the uniaxial yield stress
$\dot{\varepsilon}$ is the strain rate
k_u is the Herschel-Bulkley extensional consistency

---

[2]In a Bagley plot the overall pressure drop ΔP is plotted against L/D. A straight trend-line confirms that the wall shear stress is independent of pressure and the intercept gives the Bagley correction.

Assumptions

- The flow index, n, in extension is equal to that in shear, and
- The material does not exhibit flow anisotropy

  – this gives $\tau_y = \frac{\sigma_y}{\sqrt{3}}$ and $k_{HB} = \frac{k_u}{(\sqrt{3})^{n+1}}$

There are three flow regions

(1) Plug flow along the barrel up to the convergent zone.
(2) A radially convergent zone in the die entry between $0 \leq \theta \leq \theta_{max}$
Stagnant regions in corner, defined by $\theta_{max} \sim 45°$ in Fig. 10. The boundary of stagnant region is smooth so v_r does not depend on θ.
(3) Plug flow on exit from the convergent region.

Model
More assumptions

- The flow is incompressible.
- The flow is 'slow': slow enough that inertial effects can be ignored.
- Irrotational: elements stretch but do not rotate.
- Symmetrical: the flow involves spherical co-ordinates but we assume that v_r (the velocity in the radial direction) is the only non-zero component of velocity: $v_\theta = v_\varphi = 0$.

The volumetric flow rate is Q.

The flow across the surface of the conical section of radius r—see sketch is proportional to $r^2 v_r$: note that v_r is negative

This gives $Q = -r^2 v_r \Theta_{max} = -r_{min}^2 v_{r,min} \Theta_{max}$
where Θ_{max} is the solid angle subtended by the cone

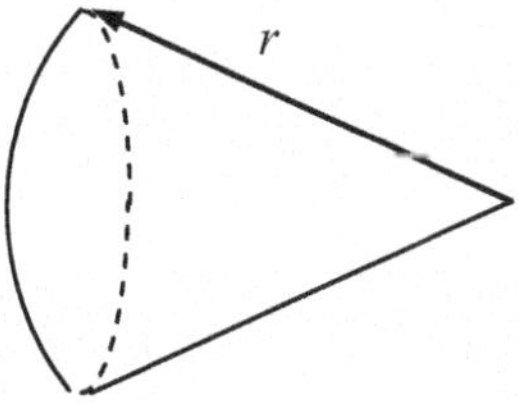

Q is also the flowrate in the plug flow region, of diameter D. A bit of geometry gives

$$v_r = -\frac{V \sin^2 \theta_{max}}{2(1 - \cos \theta_{max})} \frac{r_{min}^2}{r^2} = -a \frac{r_{min}^2}{r^2} \tag{35}$$

This result allows us to calculate the shear rate tensor.

The assumptions of a lubricated boundary and symmetry means that all the non-diagonal terms in the shear rate tensor are zero. For instance,

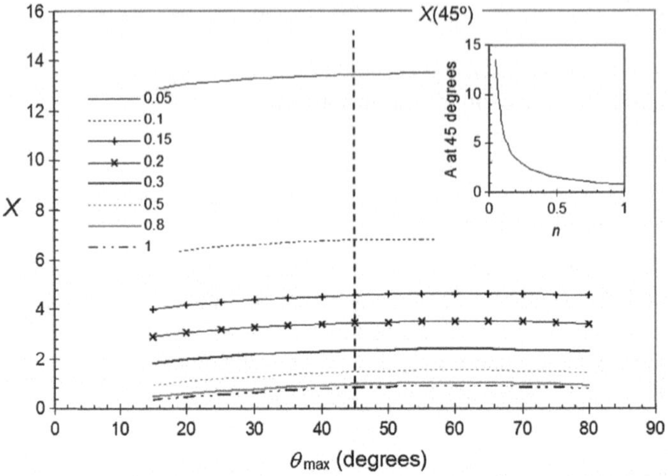

Fig. 11 Effect of θ_{max}, static zone shape, on geometrical factor X in Eq. (38). Inset shows effect of n on X when $\theta_{max} = 45°$

$$\varepsilon_{rr} = \frac{\partial v_r}{\partial r} = \frac{\partial}{\partial r}\left(-a\frac{r_{min}^2}{r^2}\right) = 2a\frac{r_{min}^2}{r^3} \quad (36)$$

Conservation of volume means that the extensional strain rate tensor is

$$\dot{\underline{\underline{\varepsilon}}} = \begin{bmatrix} \varepsilon_{rr} & 0 & 0 \\ 0 & \varepsilon_{\theta\theta} & 0 \\ 0 & 0 & \varepsilon_{\phi\phi} \end{bmatrix} = a\frac{r_{min}^2}{r^3}\begin{bmatrix} 2 & 0 & 0 \\ 0 & -1 & 0 \\ 0 & 0 & -1 \end{bmatrix} \quad (37)$$

This is then combined with (34) to give the local stress, which is then integrated to give the extrusion pressure, P_{ex}. The result (the interested reader can consult the original paper) is

$$P_{ex} = 2\sigma_y \ln\left(\frac{D_o}{D}\right) + \frac{2}{3n}k_u\left(\frac{2V}{D}\right)^n\left(1-\left(\frac{D}{D_o}\right)^{3n}\right)\underbrace{\left\{\frac{2}{3n}\left[\sin\theta_{max}\left(1+\cos\theta_{max}\right)\right]^n\right\}}_{X}$$

$$P_{ex} = \textit{plastic work} + \textit{internal redundant work} \quad (38)$$

This looks complicated but we note that

- V/D is the apparent shear rate in the die land or defined at the orifice
- All the other terms are geometrical or paste properties
- The main dependency is on the flow index, n: see Fig. 11.

Figure 11 shows that the result is relatively insensitive to the angle formed by the static zones—but note that this assumes that the boundary is fully lubricated and the convergent zone is conical.

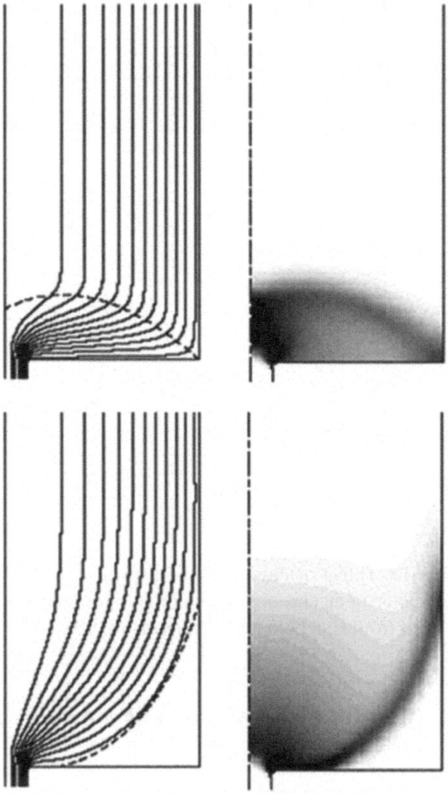

Fig. 12 Streamlines (left) and strain rate (right) distributions for a perfectly plastic von Mises material being extruded through cylindrical, square ended dies with smooth (top) and rough (bottom) walls. Reproduced from Horrobin (1999)

In most cases the static zone boundary will exert some friction and the boundary will then be curved: at this point we require numerical solutions. It also means that any model parameters extracted from data using Eq. (38) are subject to systematic error.

Effect of wall friction on extrusion. In most practical cases there will be some wall friction, introducing further dissipative losses and giving rise to curved yield surfaces. The shape of the yield surface is then determined by minimization of redundant work: identification of the yield surface is the subject of many other lectures in the workshop.

Figure 12 shows the importance of wall friction for the flow of a perfectly plastic (no shear rate effects) in a concentric cylindrical square ended die. With a smooth die there are no static zones and a region of high extensional strain exists above the die entry. With fully rough walls regions of high strain rates are also located at the static zone boundaries.

The assumption that stiff pastes can be modeled as plastic materials in extrusion flows forms the basis of the Benbow-Bridgwater (1993) characterization technique for quantifying the rheology of stiff pastes.

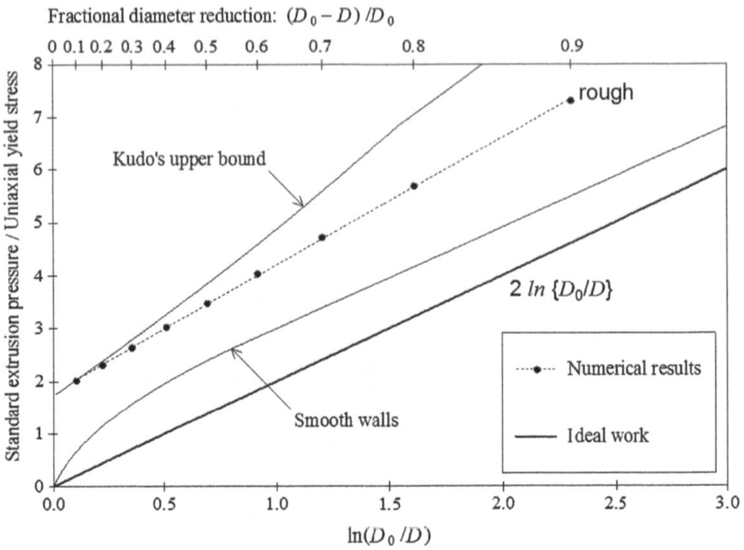

Fig. 13 Effect of die reduction, expressed as $\ln\{D_0/D\}$, on extrusion pressure for axisymmetric extrusion of an elastoplastic material through perfectly rough, cylindrical square entry dies. Reproduced from Horrobin and Nedderman (1998). The ideal (homogeneous) work is given by (33)

Modelling extrusion of YSFs. An accurate model of YSF extrusion must be able to account for

(i) Extensional work incurred in changing shape (small if the die has a long tapered entry)
(ii) Viscous (shear rate dependant) work
(iii) Work against wall friction

It should also be able to predict the shape of zones where the material is not deforming, including regions of plug flow and static zones formed in corners and behind die features (e.g. locating arms in spider dies).

Both plasticity-based and fluid flow approaches have been used. In the former, non-deforming regions are identified explicitly, while the latter incorporates velocity dependencies more readily.

Plasticity approach
Horrobin (1999) used the finite element tool ABAQUS to compute the flow of an elasto-plastic material through axisymmetric, cylindrical, square entry dies as well as dies of different shape and cross-section. The calculations featured extensive re-meshing in order to track the evolution of the large strains developed in extrusion. Horrobin modelled the wall interaction using a so-called *elastic stick* formulation. The results in Figs. 13 and 14 are presented here for comparison with those from a regularised YSF.

Fluid mechanics approach

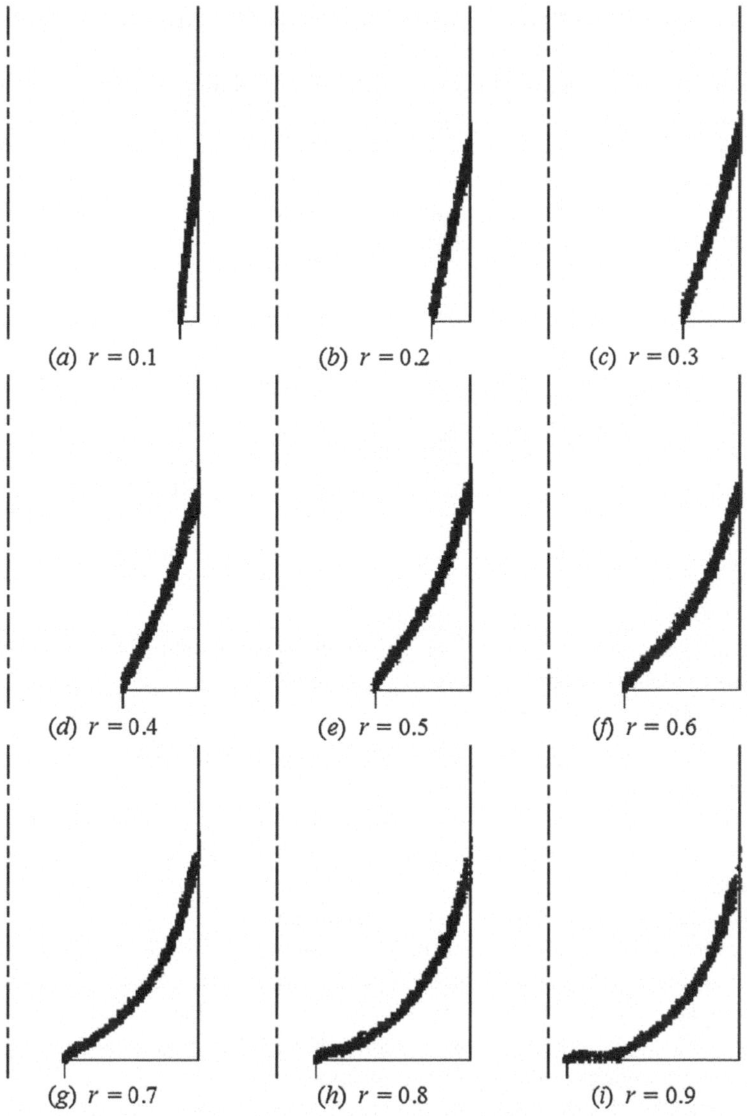

Fig. 14 Prediction of static zone boundary shape for axisymmetric extrusion of an elasto-plastic material through rough, cylindrical square entry dies. r is the reduction ratio, $r = (1 - D/D_o)$. Reproduced from Horrobin (1999)

There have been several studies of flows of YSFs through convergent dies but few have considered wall slip. Mitsoulis and Hatzikiriakos (2009) considered the extrusion of a compressible PTFE paste, using their own code. Bryan et al. (2015) used the commercial CFD code Polyflow™ to calculate the flow pattern and extrusion pressure associated with the flow of a Bingham fluid through an axisymmetric, cylindrical square entry die with a 25:3.5 reduction. They compared their results with flow pattern information obtained by extruding micro-crystalline cellulose water pastes through a similar, hemicylindrical die, with a glass wall as the diametrical plane which allowed the motion of the material to be tracked in situ. We'll look at some of their results.

Inertial terms in the Navier-Stokes equations were ignored so that the problem involved solution of the Stokes equations. The rheology of the material was modelled using the Papanastasiou regularisation, viz.

$$\eta_a = \mu_B + \tau_y \left(\frac{1 - \exp(-c\dot{\gamma})}{\dot{\gamma}} \right) \tag{39}$$

where c is the regularisation parameter.

Wall slip was modelled using Eq. (15b) with $m = 1$ (i.e. $\tau_w \propto v_s$ and $\beta_{N,a}$ values ranging from 1 to 10^{15}, to capture wall behaviours ranging from nearly perfect slip to rough. They also investigate the effect of c, starting with $c = 10{,}000$ s (see Mitsoulis, 2007).

They presented results for four Bingham numbers (where the length scale was the diameter of the barrel upstream of the die as this is the region of interest) of 0.1, 1, 10 and 100.

Figure 15 shows boundaries of non-deforming zones, identified by inspecting the solutions to identify locations where $s = \tau_y$ (Eq. (1)).

Points to note include

- For a small yield stress contribution (Bn =0.1), a small static zone is predicted and this is difficult to identify reliably.
- For larger Bn, the boundaries are mostly curved, in agreement with Fig. 12. For a given Bn value the slip function determines the size and shape of the static zone formed in the corner (see $Bn = 1$ and 10), but information about the static zone shape alone does not allow Bn and β to be quantified.
- For a very stiff paste (e.g. $Bn = 100$), the plug of material in the barrel merges with the static zone in the corner to give a region of high strain rate above the die entry. The remaining material appears to be 'static' but is moving—albeit slowly—as a result of the high limiting viscosity predicted by (39) at yield. This illustrates one of the shortcomings of the regularisation approach.
- It is noticeable that there is little influence of β at high Bn.

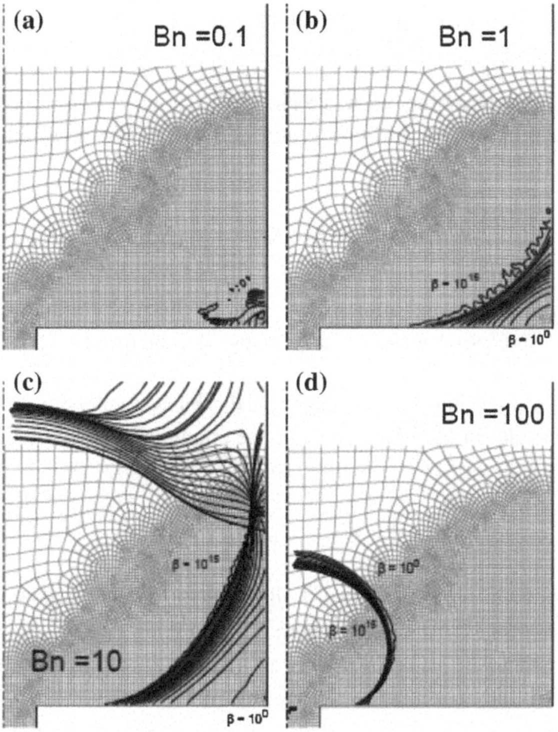

Fig. 15 Effect of Bingham number and wall slip coefficient β on location of deformation boundaries in extrusion of a Bingham fluid through a square-ended cylindrically axisymmetric die. Grey pattern shows local mesh used in the FEM calculations. Reproduced from Bryan et al. (2015)

These results illustrate the challenge in modelling forming processes. Pressure and flow rate data sets can be fitted by several combinations of model parameters: internal data (local velocity, static zone shape or local pressure measurements) are required to verify and/or compare simulations.

3 YSFs, Heat Transfer and Structure Development

We can consider two applications of heat transfer involving YSFs as

(a) Heat transfer **through** a YSF. This can be divided into cases where the properties of the fluid are not affected significantly, and ones where they are changed.
(b) Heat transfer **to** a YSF, where changes in thermal energy content modify the structure of the fluid, its flow behaviour, and thus the flow itself. The Bingham number then varies *locally*. Examples include casting of metals or thermosetting

suspensions; flow of waxy crude oil along a subsea pipeline; and slow surface flows of magma.

Convective heat transfer with YSFs. Up to now we have been looking for solutions to the momentum and continuity equations. We now have to include a statement of conservation of thermal energy, such as

$$\rho \frac{Dh}{Dt} = -\nabla \cdot (-\lambda \nabla T) + \tau : \nabla \mathbf{u} + Q \tag{40}$$

Here h is the enthalpy per unit mass, λ is the thermal conductivity and Q is a source term: the first term on the RHS describes conduction and the second viscous dissipation.

In practice, Eq. (40) is then simplified on the basis of physics of the case being studied. For example, in the absence of phase change, the specific enthalpy can be expressed in terms of a reference temperature, T_0, such that

$$h = h_0 + C_p (T - T_0) \tag{41}$$

where C_p is the specific heat capacity. For flow at steady state, with negligible viscous dissipation and source terms (40) becomes

$$\nabla \cdot (-\lambda \nabla T) = -\rho C_p \mathbf{u} \cdot \nabla T \tag{42}$$

For uniform λ, this becomes

$$\nabla^2 T = \frac{\rho C_p}{\lambda} \mathbf{u} \cdot \nabla T \tag{43}$$

Dimensionless form

Introducing a characteristic velocity V, length scale L, and apparent viscosity η_a, allows us to write (43) in dimensionless form, where the<> brackets indicate a scaled quantity;

$$\begin{aligned} \langle \nabla^2 T \rangle &= \frac{\eta_a C_p}{\lambda} \frac{\rho V L}{\eta_a} \langle \mathbf{u} \rangle \cdot \langle \nabla T \rangle \\ &= Pr\, Re \langle \mathbf{u} \rangle \cdot \langle \nabla T \rangle \end{aligned} \tag{44}$$

Here *Pr* is the Prandtl number and *Re* is the Reynolds number. The thermal energy equation is often written in terms of *Pe,* the Peclét number for heat transfer, as

$$\begin{aligned} \langle \mathbf{u} \rangle \cdot \langle \nabla T \rangle &= \frac{1}{Pr \times Re} \langle \nabla^2 T \rangle \\ &= \frac{1}{Pe} \langle \nabla T \rangle \end{aligned} \tag{45}$$

For a Bingham fluid, the Bingham viscosity term is often used for η_a in the Prandtl and Reynolds numbers. Equation (45) is the equation to be solved for heat transfer within YSFs. Coupling between (45) and the momentum equation can arise at several levels:

(a, i) Uniform fluid properties

If the parameters in the rheological model are independent of temperature—the simplest case—then (42) (or (45)) can be solved once the velocity field has been calculated.

Many fluid properties are sensitive to temperature so the accuracy will be limited. However, the assumption of uniform fluid properties can be used to generate an initial estimate of the velocity and temperature fields for subsequent iteration. It also provides a method for checking the calculations.

The heat transfer calculation is still not straightforward: we expect there to be non-deforming regions in a YSF flow and this introduces a bifurcation in the heat transfer calculations:

- In flowing regions, one solves (42), as in any other forced convection or free convection problem (note that we haven't said where the motive force comes from yet!).
- In the unyielded region(s), for steady state, (42) collapses to Laplace's equation ($\nabla^2 T = 0$), i.e. steady state conduction.
- Locating the yield surface again becomes very important.

If we use regularisation approaches, there will be a small but finite velocity within the unyielded region. Equation (44) allows us to gauge the deviation of the solution of the temperature field in the unyielded region from the Laplace solution.

For fluids with small Prandtl number, e.g. metals, with a high thermal conductivity, the RHS and thus the deviation, will be small.

For fluids with high η_a, and therefore small *Re*, the RHS will again be small. For cases with high *Pe*, the contribution from 'creep' in the unyielded region can be significant, particularly if the material is at the heat transfer surface.

(a, ii) Temperature dependent properties

Whereas thermal conductivity and specific heat are weak functions of temperature, viscous effects are usually strongly sensitive to temperature. Depending on the nature of the YSF, the yield strength may also be temperature sensitive. The momentum and enthalpy equations are then coupled via the apparent viscosity and have to be solved simultaneously.

Forced convective heat transfer through a YSF: example. In heat transfer calculations, we often wish to express the heat flux across an interface, q, as the product of a heat transfer coefficient, U, and a temperature difference driving force:

$$q = U\,(T_s - T_{bulk}) \tag{46}$$

Here, T_s is the temperature at the surface and T_{bulk} is either the temperature of the surroundings (in an unconfined flow) or a flow averaged value (the 'mixing cup' temperature in duct flows).

Chemical engineers will then express this as a dimensionless group, the Nusselt number, defined

$$Nu = \frac{LU}{\lambda} \tag{47}$$

Note that workers differ in the choice of length scale for L, particularly for cylinders and spheres where radius and diameter are both used.

Two forms of Nusselt number can be generated: a local value, and one averaged over the surface of interest, i.e.

$$Nu_{overall} = \frac{1}{A}\int_A Nu \cdot dA \tag{48}$$

Solving (45) requires specification of the initial condition (often an inlet temperature) and the boundary conditions, the most common being

Set temperature (Dirichlet condition): calculation of the temperature field allows q to be evaluated and U identified from (46).
Specified heat flux (Neumann condition): T_s is calculated as part of the temperature field etc.

Nirmalkar et al. (2013a) considered convective heat transfer from a hot sphere moving at velocity V relative to a pool of a Bingham fluid. Figure 16 shows the geometry and key dimensions. They have also presented the results for a Herschel-Bulkley fluid (Nirmalkar et al. 2013b). They considered the case of uniform fluid properties and presented their results in terms of modified forms of the Reynolds, Prandtl and Bingham numbers in which the apparent viscosity used in the definition was that evaluated at the characteristic shear rate, given by V_o/L. For a Bingham fluid, the groups are

$$Re_B^* = \frac{Re_B}{1 + Bn} \tag{49}$$

$$Pr^* = \frac{\mathrm{Pr}}{1 + Bn} \tag{50}$$

$$Bn^* = \frac{Bn}{1 + Bn} \tag{51}$$

They used Papanastasiou regularisation to describe the YSF and obtained solutions using the COMSOL Multiphysics finite element software. Viscous dissipation was ignored and there was no slip at the surface of the sphere.

Let us think about what we expect to see in the solutions:

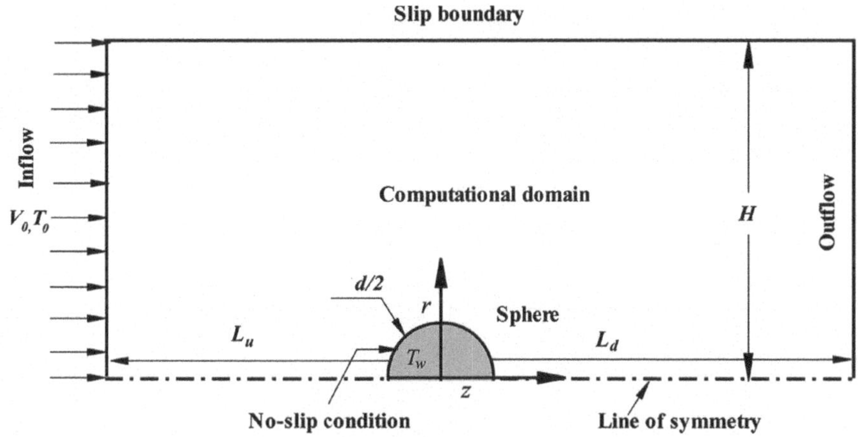

Fig. 16 Schematic of geometry and computational domain for heat transfer from a sphere to a pool of YSF. Reprinted with permission from Nirmalkar et al. (2013a) © 2013 American Chemical Society

(i) The presence of a yield stress means that we expect to see little deformation distant from the sphere: we expect to see a somewhat circular 'ring of shearing' around the sphere.
(ii) Heat transfer by advection should only arise in the shearing band: beyond this, heat transfer will be dominated by conduction.
(iii) The larger the Bingham number, the narrower the shearing band will be. The shear rate at the sphere surface will increase with *Bn*, increasing advection and thus the rate of heat transfer, the heat transfer coefficient, and the Nusselt number. This will be accompanied by increased wall friction and thus a higher drag coefficient, C_D, defined as

$$C_D = \frac{F_{drag}/\frac{\pi}{4}d^2}{\frac{1}{2}\rho V_0^2} \tag{52}$$

The samples of streamlines evaluated in creeping and faster flows in Fig. 17 show that the yield stress suppresses the wake observed at $Re = 10$ and 100. The drag coefficient results, presented in Fig. 18, show that the modified Reynolds number incorporates the influence of *Bn*.

Figure 19 shows that the size of the shearing zone decreases modestly with *Bn*.

The isotherm contours in Fig. 20 show several noteworthy features:

- The near sphericity at $Re^* = 1$, $Pr^* = 1$;
- The spread downstream with increasing Re^* and Pr^* owing to advection of the solid body that is the unyielded material distant from the sphere (the heat source);
- The narrowing of the contours with increasing *Bn*;
- Cases with equal *Bn* and *Pe* give similar patterns.

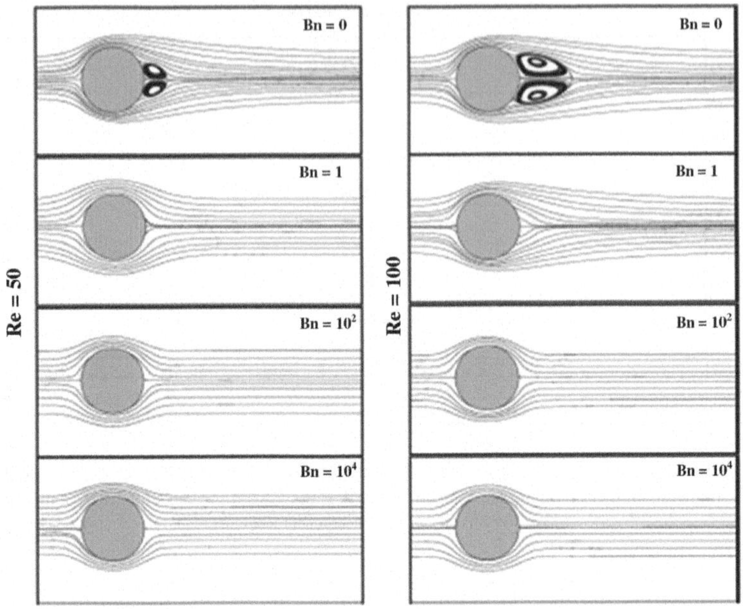

Fig. 17 Streamline contours (flow from left). Reprinted with permission from Nirmalkar et al. (2013a) © 2013 American Chemical Society

The suppression of the wake in the yield stress is manifested in the local heat transfer results in Fig. 21. Increasing *Bn* leads to a general, modest, increase in local *Nu* and a noticeably wider region of high heat transfer rate at the equator.

The overall Nusselt number is expressed in the form of the Chilton-Colburn j-factor in Fig. 22.

$$j_H = \frac{Nu_{overall}}{Re^* Pr^{*1/3}} \tag{52}$$

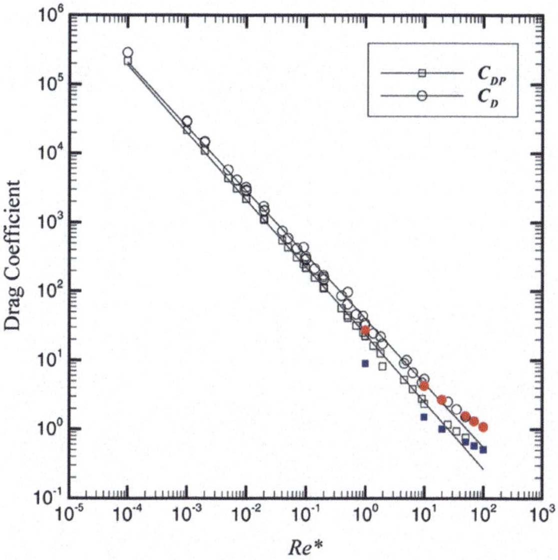

Fig. 18 Effect of modified Reynolds number, Eq. (49), on drag coefficient. C_{DP} is the contribution to C_D from normal forces. Solid symbols indicate results for a Newtonian fluid. Reprinted with permission from Nirmalkar et al. (2013a) © 2013 American Chemical Society

The use of the modified Reynolds and Prandtl numbers collapses the results for different *Bn* values to a single curve.

Natural convection

In natural convection the motion of the fluid arises from the presence of a heated or cooled surface. There must be a variation in fluid density and this is often modelled using the Boussinesq approximation, $\Delta\rho = \rho_0\beta_B(T - T_0)$, where β_B is the coefficient of thermal expansion.

An example of the above modelling approach, with other properties remaining constant, is given by Sairamu et al. (2013).

Heat transfer to a YSF: **evolution of structure**. As noted above, heat transfer creates a temperature distribution: many rheological properties are temperature dependent and this introduces coupling of the momentum and thermal energy equations.

In several cases, the rheological properties are determined by physical features which are determined by the amount of heat transferred.

Consider a highly filled suspension of particles in a wax continuous phases. The shear viscosity of such a material is often described by the Krieger-Dougherty equation

$$\eta_a = \eta_{a,cont}\left(1 - \phi/\phi_{\max}\right)^{-k_D\phi_{\max}} \tag{54}$$

Here $\eta_{a,cont}$ is the apparent shear viscosity of the continuous phase at the overall shear rate and temperature under consideration, and $\phi_{\max}$ is the maximum volume

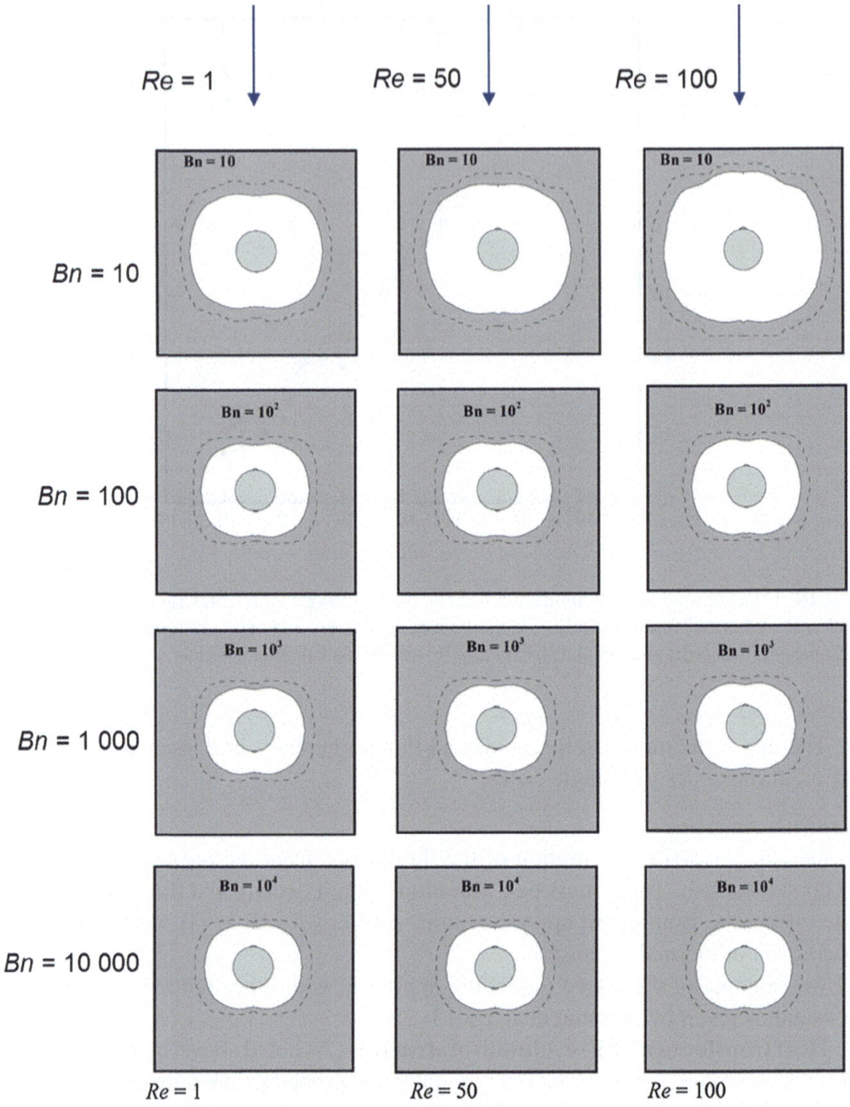

Fig. 19 Influence of *Bn* on the shape of yielded zone boundary for a Papanastasiou growth rate parameter, *m*, of 10^5. The dashed line shows the results obtained from a biviscous model with $\mu_y/\mu_B = 10^5$. The flow enters from the top. In their formulation of the Papanastasiou expression the shear rate is scaled by the characteristic shear rate, V_0/d. Reprinted with permission from Nirmalkar et al. (2013a) © 2013 American Chemical Society

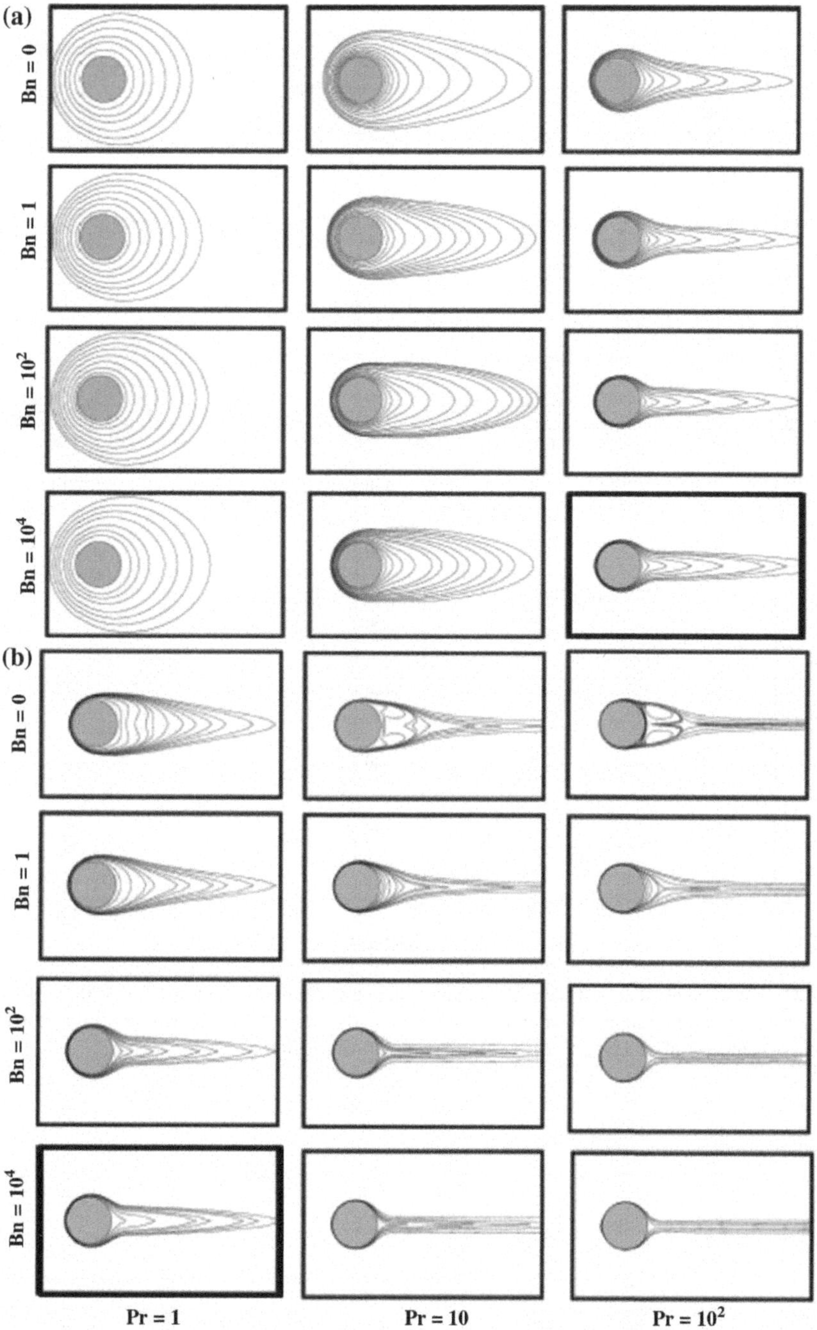

Fig. 20 Representative isotherm contours for flow, from right to left, with $Re^* = 1$ and **b** $Re^* = 100$. Bold boxes identify cases with equal *Bn* and *Pe*. Reprinted with permission from Nirmalkar et al. (2013a) © 2013 American Chemical Society

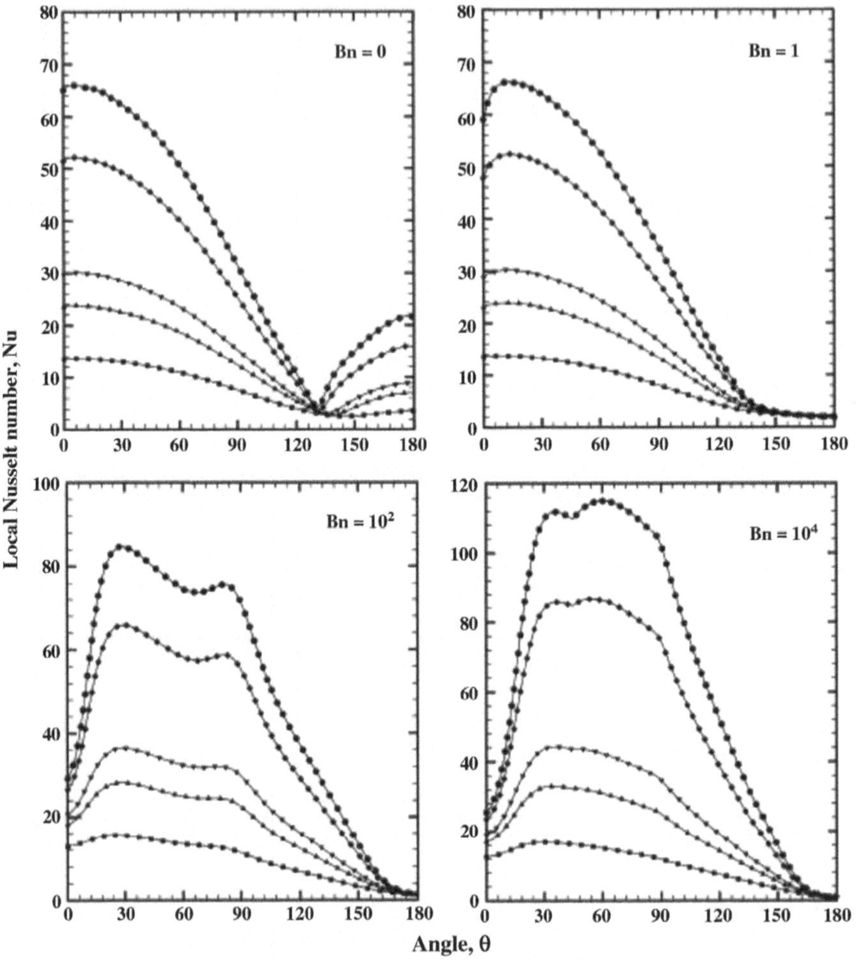

Fig. 21 Variation in local heat transfer coefficient over the sphere surface for $Re^* = 100$. $\theta = 0$ corresponds to the leading stagnation point. Note that the y-axis scales differ. Reprinted with permission from Nirmalkar et al. (2013a) © 2013 American Chemical Society

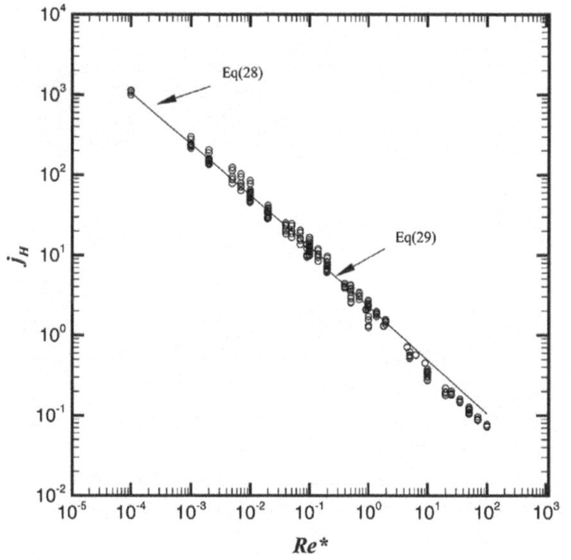

Fig. 22 Relationship between the overall heat transfer coefficient, expressed as j_{H}, and the modified Reynolds number. Equation [28] is $j_H = 2.3/\mathrm{Re}^{*2/3}$, $10^{-4} < Re^* < 0.199$; Equation [29] is $j_H = 2.27/\mathrm{Re}^*$, $0.199 < Re^* < 100$. Reprinted with permission from Nirmalkar et al. (2013a) © 2013 American Chemical Society

fraction for the solids. As ϕ approaches $\phi_{\max}$ the apparent viscosity increases and yield stress behaviour will increase. In the melt, ϕ is determined by the volume of particles but as the melt is cooled, higher melting point components of the wax will crystallise out, increasing ϕ and promoting yield strength behaviour. Eventually most of the wax will have crystallised and the material will have solidified.

The temperature range over which the continuous phase solidifies is called the *mushy zone* in applications such as casting of alloys and flow of magmas.
Consider two variants:
Casting of pure material: there is a single, well defined solidification temperature and latent heat is consumed at a solidification front, giving rise to a sharp delineation between soft (molten or semi-solid material) and a solid (with large yield stress). Latent heat contributions are often significantly larger than sensible heat, so the problem becomes one of heat transfer to the solidification front. The rate of transfer of latent heat determines the location of the solid-liquid interface. This is a moving boundary problem and is an example of the *Stefan problem*.

Sketch

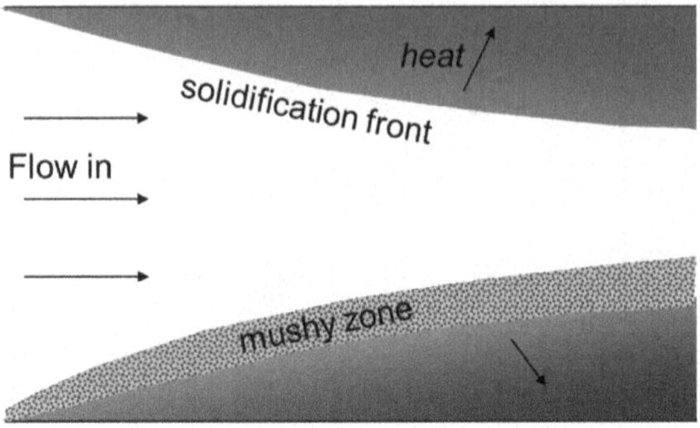

Casting an alloy or solidification of a mixture: the presence of more than one component means that there is a range of solidification temperatures, and thermal energy passes through the crystallising zone. The problem becomes one of tracking the local extent of solidification.

More on mixtures
The degree of solidification of a mixture is not expected be linear in temperature. The melting point depends on the mole (or mol) fraction of solidifying species, and this changes as higher melting point components crystallise and leave the solution.

Different examples:

(a) Wax formation in a subsea oil pipeline: the solution consists of crystallising species (the waxy components) as well as non-crystallising ones (the other oil components). Solidification of the wax species forms a particulate *gel*, with oil held within the framework of crystals.
(b) Casting of waxes/alloys: all the components solidify. The higher melting point components will solidify first, forming crystals of their respective species (or mixed phase, depending on the chemistry). On further cooling the continuous phase will reach the eutectic point at which it all solidifies, giving a microstructure somewhat like cement (aggregates in a cement matrix).
(c) Chocolate: this is a multiphase system and the composition depends on type and country of origin. The dispersed phase is a mixture of sugar, milk solids and cocoa solids (or a combination of these in crumb chocolate) and the continuous phase is a blend of different fats (cocoa butter). The latter melt or solidify during the cooling and tempering stages of production.

Modelling considerations
It is convenient to work in terms of a microstructural parameter X, which is the fraction of solidified material, where $0 \leq X \leq 1$.

The rheology of the YSF is intimately linked to X. Heat transfer calculations will determine X: the challenge is to determine the form of the YSF constitutive relationship incorporating X.

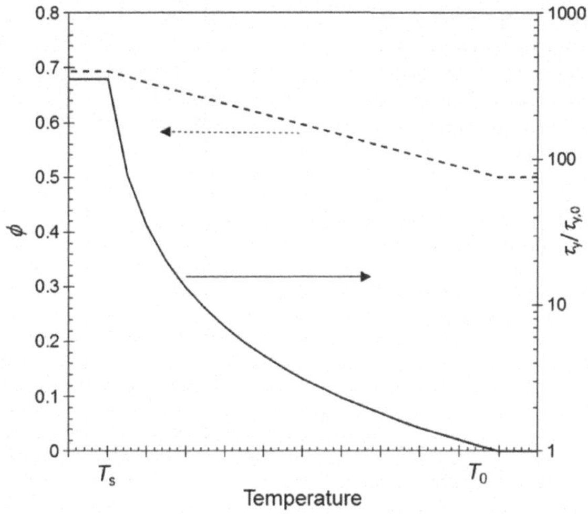

Fig. 23 Illustration of the effect of solidification of the continuous phase in a dense suspension from $\phi_p = 0.50$ to 0.99 ϕ_{max}, where $\phi_{max} = 0.70$, over the temperature range T_s (solidified) to T_0 (melt). Note that real solidification profiles are unlikely to be linear

As an example, consider a dense suspension where the effect of solidification on the yield stress follows a Krieger-Dougherty relationship

$$\tau_y = \tau_{y,0}\left(\frac{1-\phi/\phi_{max}}{1-\phi_p/\phi_{max}}\right)^{-2.5\phi_{max}} \tag{55}$$

Let the initial volume fraction of (non-solidifying) particles be ϕ_p. Consider the case where solidification of the continuous phase causes ϕ to increase linearly with temperature over the range T_0–T_s, reaching 0.99 ϕ_{max}:

$$\phi = \phi_p + (0.99\phi_{max} - \phi_p)\left(\frac{T-T_0}{T_s-T_0}\right) \quad T_s \le T \le T_0 \tag{56}$$

Figure 23 shows the resultant effect on yield stress. It should be noted that making rheological measurements in the mushy region is not straightforward (see Barker and Wilson 2008).

Wall slip

Solidification is also accompanied by more significant slip effects. Figure 24a shows the effect of temperature (and partial solidification) on the wall shear stress associated with wall slip of the 50 vol.% ZnO/wax suspension whose melt behaviour is shown in Fig. 24b. These data were obtained by pushing a plug of material along a duct at controlled speed and hydrostatic pressure.

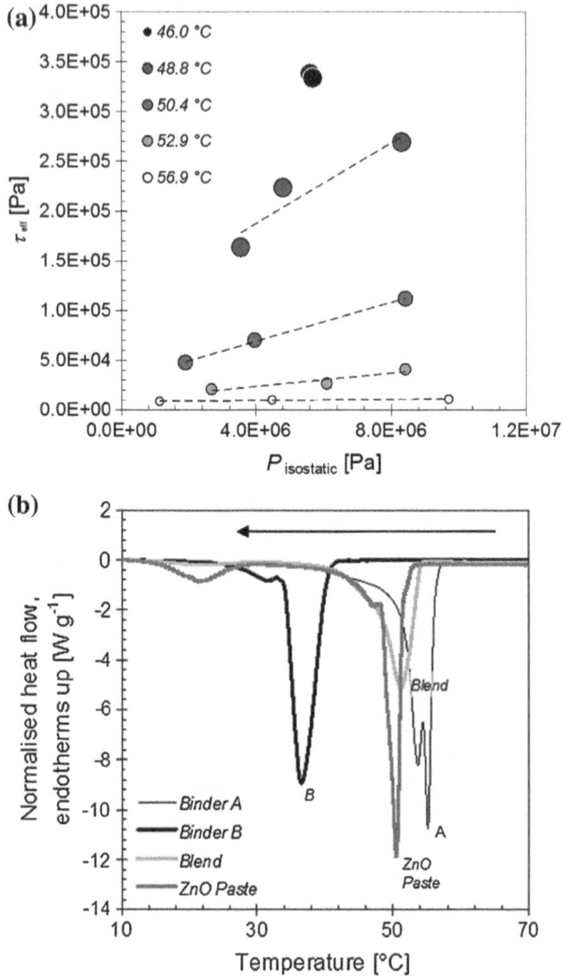

Fig. 24 **a** Effect of temperature and isostatic pressure on wall shear stress for a 50 vol.% ZnO-wax suspension with melting range 54–40 °C as determined by DSC (**b**). Reproduced from Barker and Wilson (2008)

Points to note include

(i) Decreasing the temperature increases the solids volume fraction and the wall shear stress associated with slip.
(ii) In the solidification region, the wall shear stress shows a linear dependence on isostatic pressure, indicating a contribution from Coulombic friction. This contribution becomes more significant at higher solids volume fraction.
(iii) Wall slip in these systems is complex, as viscous dissipation at the wall is likely to cause some local melting of the continuous phase. Heat transfer through the wall also needs to be considered.

Calculation methods[3]

Having established that the extent of solidification is important, we note that in some cases heat transfer is the controlling factor but in others crystallization kinetics and diffusion effects can arise.

Stefan originally considered the solidification of sea ice, but the problem of locating the solidification front has been found to arise in other systems. In most cases reported to date the mobile phase is treated as a Newtonian fluid and the solidified material a simple solid.

A good review of the mathematical techniques used for phase-change modelling was given by Hu and Argyropoulos (1996). Techniques generally fall into one of two classifications: multi-domain or continuum formulations.

Multi-domain methods (also termed variable grid methods)

The solidus and liquid domains are modelled separately using

$$\nabla \cdot (-\lambda \nabla T) = -\rho C_p \mathbf{u} \cdot \nabla T \tag{57}$$

The two domains are coupled at the solidification interface via the Stefan condition (an energy balance), essentially

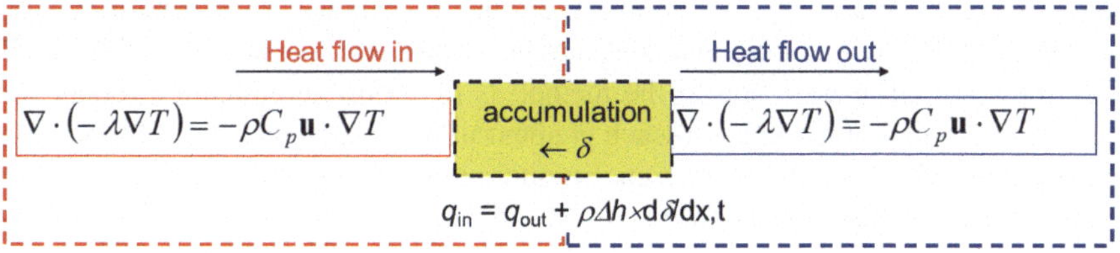

heat flux across boundary = rate of growth of solid region + conduction through solid.

The moving boundary is tracked explicitly in such a way that it falls on a grid point at each evaluated time step. The grid may either be interface fitting, in which case the time step is adjusted so the boundary coincides with a grid point; or dynamic, where the grid is modified to coincide with the boundary for a given time step.

Variable grid techniques are best suited to sharp interface problems such as arise with the phase-change of pure metals (see Jana et al. 2007), as the smearing of the phase-change process over a temperature interval which is encountered in alloys and organic mixtures is more problematic, since the formulation inherently needs to locate the precise position of the solid/liquid boundary. In many ways this replicates the challenge of locating the yielding zone in YSF flows.

Continuum formulations (weak methods)

There are many similarities with YSF flows and regularisation methods here. A single set of equations are applied over the entire computational domain. The position of the phase-change boundary is not tracked explicitly: this information is extracted a

[3] Much of this section is based on the PhD dissertation of Dean Barker (Univ. Cambridge 2008)

posteriori by recovering the temperature field and hence locating the position of the solidification region.

The latent heat is incorporated into the mathematical formulation to release or absorb energy (for solidification or melting events, respectively) over the same temperature interval. Methods to do this include.

Apparent heat capacity method
The latent heat is included as an extra component to the sensible heat capacity, C_p, in (42) over the phase-change interval. This results in an augmented heat capacity over the mushy transition at each computational node within the phase change interval. One drawback of this technique is that if a node passes through the phase transition temperature range in a single time step, the latent heat contribution to the total heat capacity is missed, leading to an error in the total heat balance.

Effective heat capacity method
The 'skipping problem' in solving (42) is countered by assuming a temperature profile between adjacent nodes, and an effective C_p is computed based on an integration of the entire element.

Heat integration method
A heat balance calculation is performed to track the evolution of latent heat during phase-change. For example, for solidifying a pure substance, a temperature below the solidification temperature indicates that the control volume has entered the melt state transition. The next step in the method is to set up an enthalpy account for each element, then to add heat to each element until it reassumes the solidification temperature, whilst removing the same amount of heat from the enthalpy account to capture the latent heat contribution.

Source-based methods
Equation (45)—i.e. (40) without the viscous dissipation term in this case—is used, including the source term Q via which heat can either be added or removed from the system. A kinetic scheme is required to limit the latent heat contribution to the temperature range of interest. For example, level set methods can be used to describe the extent of solidification, thereby linking the change in rheology explicitly to the latent heat contribution.

$$\rho \frac{Dh}{Dt} = -\nabla \cdot (-\lambda \nabla T) + Q \tag{58}$$

The enthalpy method is similar, with the enthalpy h framed to include the latent heat as a function of temperature rather than including the source term.

Industrial application
The Time Capsule® is a New Zealand pharmaceutical product for herbivores (sheep, horses, cattle) to combat facial eczema. It is manufactured by continuous casting of a 50 vol.% ZnO/wax suspension. The paste passes along a cylinder with water-cooled walls which promote the solidification of the wax matrix. The challenge is to solidify enough material so that it retains its shape once it leaves the cooler section.

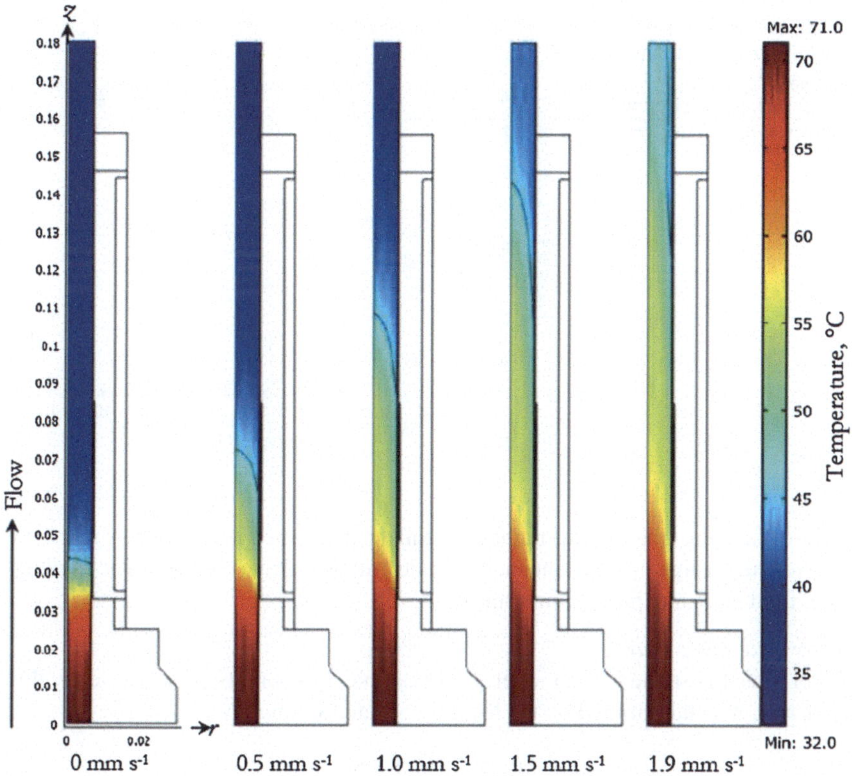

Fig. 25 Temperature distribution in the continuous casting of a 50 vol.% ZnO/wax paste for the inlet temperature 71 °C case. Values of mean velocity reported. Black line denotes the solidification front. Reproduced with permission from Barker (2008)

Figure 25 shows the effect of flow rate on the temperature distribution. The rheology of the material changes from that of a dense suspension (where it is modelled as a temperature dependent Carreau fluid) to a YSF and thence a Coulombic solid (see Figs. 3 and 24b). This is an example of a combined Graetz and Stefan problem.

More on microstructural models. The use of a microstructural parameter to describe the evolution of properties of a complex fluid is well-established in rheology. The Cross model for a Generalised Newtonian fluid describes the shear viscosity in terms of the number of linkages, N, between units in the fluid. The number of linkages is determined by competing steps of shear-induced breakage, and continuous reformation. Cross proposed a kinetic scheme of the form

$$\frac{dN}{dt} = k_r\left(N_0 - N\right) - k_d N\dot{\gamma}^m \tag{59}$$

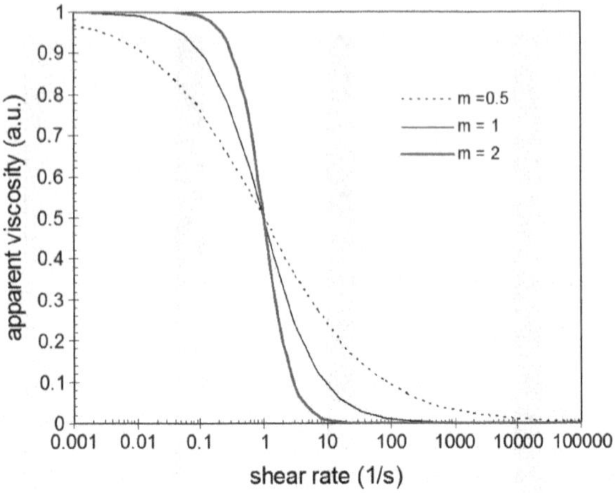

Fig. 26 Plot of the Cross model with $K_{dr} = 1$ (units will vary with m), $\eta_o = 1$ and $\eta_\infty = 0$, for a selection of values of m

Here N_0 is the (maximum) number of linkages, attained at rest (when $\dot{\gamma} = 0$); k_r is the restructuring rate constant and k_d the destructuring rate constant. The rheology of the fluid is then expressed in terms of N.

Steady state behaviour

We continue to use the Cross model as an example. For conditions when steady flow is reached, N is determined by the local shear rate. Setting $dN/dt = 0$ gives

$$\frac{n}{n_0} = \frac{1}{1 + k_d \dot{\gamma}^m / k_r} \tag{60}$$

Now for the important bit—linking N to rheology: the Cross model is a Generalised Newtonian fluid so N determines the apparent viscosity.

At high shear rates, as $\dot{\gamma} \to \infty$, $N \to 0$ and the apparent viscosity of the material will approach that of the base (with no linkages), η_∞. Similarly, at very low shear rates, $N \to n_0$, and the viscosity approaches its maximum value, η_0. The increase in viscosity in the micro-structured fluid is directly proportional to the number of interactions:

$$\frac{\eta(\dot{\gamma}) - \eta_\infty}{\eta_0(\dot{\gamma} = 0) - \eta_\infty} = \frac{N}{N_0} \tag{59}$$

Combining (59) and (60) gives

$$\eta = \eta_\infty + \frac{\eta_0 - \eta_\infty}{1 + K_{dr}\dot{\gamma}^m} \tag{61}$$

where $K_{dr} = k_d/k_r$. Figure 26 shows the form of (61). The Cross model predicts shear-thinning behaviour.

It is hopefully obvious that the same approach can be applied to YSFs, where an expression analogous to (62) can be used to describe the effect of structure parameter N, X, etc. on τ_y and the viscous parameters in the constitutive equation.

Unsteady state behaviour
A second feature of (59) is that it is time-dependent: it can be used to describe thixotropy by evaluating $N(t)$ with respect to $N_{-\infty}$ (the starting state), from

$$N = N_{-\infty} + \int_{-\infty}^{t} k_r \left(N_0 - N'\right) - k_d N' \dot{\gamma} \left(t'\right)^m dt' \tag{62}$$

In simulations of unsteady flow simulations of YSFs, such as start-up or interrupted flows, the parameter equivalent to N has to be evaluated as a function of the shear history of the element, in a similar manner to the treatment of viscoelastic materials.

YSF Microstructure Models. Several workers have developed expressions for the time dependent rheology of YSFs, wherein the choice of variable controlling the microstructural parameter (be it strain, strain rate or shear stress) depends on the physics. Examples of those available include

(a) The general elasto-YSF model of De Souza Mendes and Thompson (2013), with microstructural parameter Λ (they use λ, which is used in these lectures for thermal conductivity):

$$\frac{d\Lambda}{dt} = \frac{1}{t_{eq}} \left[\left(\frac{1}{\Lambda} - \frac{1}{\Lambda_0} \right)^a - f(\tau)\Lambda^b \right] \tag{64}$$

(b) Putz and Burghalea (2009) model for the observed solid-fluid transition in gels, including Carbopol.

$$\frac{d\Phi}{dt} = R_d(\Phi, t, \Gamma) + R_r(\Phi, t, \Gamma) + \delta_N \tag{65}$$

in which Φ is the fraction of fluid phase; Γ is the stress-based forcing parameter, R_d and R_r are the destruction and recombination functions, respectively, and δ_N is the contribution from thermal noise.

(c) De Kee et al. (1983) model for gel breakup, applied by Zhao et al. (2012) to waxy crude oils:

$$\frac{d\Lambda}{dt} = -a\dot{\gamma}^b (\Lambda - \Lambda_e)^n \tag{66}$$

where Γ_e is the equilibrium value eof the structure parameter in th slurry flow state.

(d) PTFE extrusion: the 'mozzarella model'. This was developed by Hatzikiriakos and co-workers (see Patil et al. 2006) for the extrusion forming of polytetrafluroethylene pastes. PTFE cannot be processed in the melt so it is extruded as a highly-filled particulate pastes, with an organic liquid binder. During extension PTFE particles contact under compressive stress and can fuse, such that fibrils of PFTE are formed between particles which move apart. The material exhibits elasto-visco-plastic rheology and fibrillation causes strain hardening of the material. They presented a constitutive model with shear-thinning and shear-thickening terms, the relative contribution of each to the stress being determined by a structural parameter ξ:

$$\tau = (1 - \xi)\,\eta_1\dot{\gamma} + \xi\eta_2\dot{\gamma} \tag{67}$$

ξ represents the fraction of the domains of the paste that are fibrillated: its evolution follows

$$u \cdot \nabla\xi = \alpha\,|\dot{\gamma}|\,\sqrt{\psi} - \beta\,|\dot{\gamma}|\,\xi \tag{68}$$

where α and β are constants and ψ is a parameter quantifying the extent of strain and rotation in the flow.

4 Phase Motion in Multiphase YSFs

The previous lecture considered phenomena which affect the microstructure of a YSF and thus its rheology, including temperature- and flow-driven phenomena. Here we consider flow-induced meso- and macro-structural changes which can occur with multiphase materials.

Liquid phase migration (LPM). Highly-filled suspensions are examples of multiphase materials which often exhibit YSF behaviour. As the volume fraction of solids approaches the maximum packing fraction they exhibit plasticity and there is much overlap with the behaviour of soils. They can experience meso- or even macro-scale rearrangement as a result of the stresses imposed on the different phases during flow. The liquid/binder phase can move relative to the solids/dispersed phase, causing a change in the local composition of the material and thus its rheology. In soil mechanics this is termed drainage: in paste extrusion is called liquid phase migration or dewatering.

LPM is observed in forming operations for many different materials (see Martin et al. 2004), including catalyst pastes (Bardsley and Bridgwater 2012), pharmaceutical doughs (Rough et al. 2002) and vertebrae repair cements (O'Neill et al. 2016).

When a stress is imposed on the material, part of it is imposed on the solids and part on the liquid phase. Terzaghi's principle of soil mechanics states that the imposed stress can be expressed as

$$\sigma_{total} = \sigma_{solids} + P_L \tag{67}$$

In soil mechanics the solids stress, σ_{solids}, is that imposed on the solids network and contains all deviatoric components, while the pore pressure P_L is a hydrostatic stress and is that imposed on the liquid. The pore pressure can support a significant fraction of the stress, particularly if the liquid cannot drain away. It can, however, drain away and in the limiting case of rapid drainage the total stress is soon imposed fully on the solids phase, causing it to fail if the yield criterion is met. Figure 27 shows where LPM can arise in extrusion.

Filtration. In the barrel, the billet of material is exposed to the full driving stress, the extrusion pressure P. The resultant pore pressure will promote liquid filtration though the solids network. This is usually modelled using Darcy's law,

$$\nabla P_L = K_{DL}\mu u^* \tag{70}$$

where u^* is the superficial velocity of liquid, K_{DL} is the permeability and μ is the (Newtonian) liquid phase viscosity. There are several expressions for the permeability of a particulate network: the Carman-Kozeny result for flow in packed beds is:

$$\nabla P_L = k_t \frac{S^2}{\varepsilon^3}\mu u^* \tag{71}$$

Here k_t is the tortuosity, S is the wetted surface area per unit volume, and ε is the local voids volume fraction. This form of Darcy's law is used as LPM involves changes in ε and thus non-uniform permeability. For an assembly of uniform spheres of diameter d, S can be calculated using

$$S = 6\frac{(1-\varepsilon)}{d} \tag{72}$$

Applying continuity to the liquid in an element of paste gives

$$\nabla \cdot u^* + \frac{d\varepsilon}{dt} = 0 \tag{73}$$

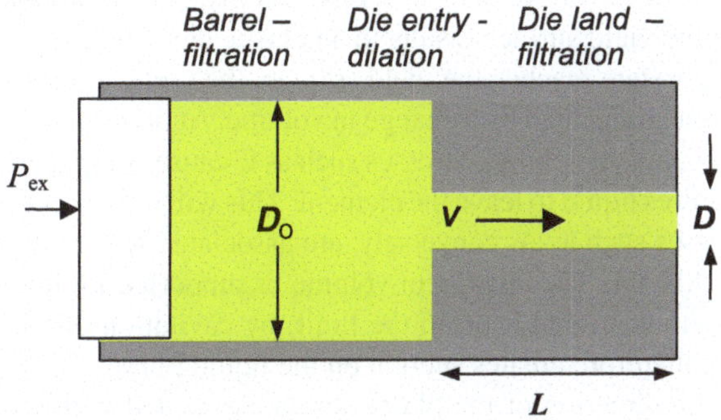

Fig. 27 Schematic of ram extrusion showing regions where LPM can occur

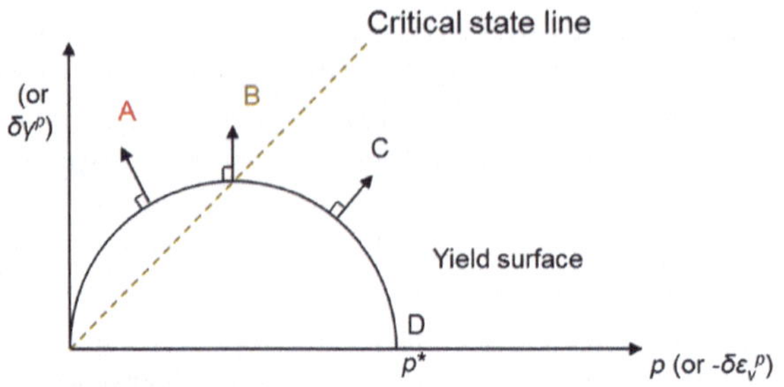

Fig. 28 Two-dimensional representation of the modified Cam-Clay yield surface in deviatoric (Mises) stress—solids effective stress space. γ^{p} and $\varepsilon_{\mathrm{v}}^{\mathrm{p}}$ are increments in plastic strain and volumetric plastic strain, respectively. Reproduced from Patel et al. (2007)

If the pore pressure is known, Eq. (73) can be integrated to give the evolution of liquid content of a packet of material over time.

A similar analysis can be applied to the material in the die land. The pore pressure at the exit will be close to zero (there may be a small component induced by surface tension) but the pore pressure at the inlet needs to be calculated. The key question is how to estimate the pore pressure, i.e. separate the components in (69).

Dilation. The concept of dilation arises in soil mechanics. In soils, which can be considered to be suspensions well into the jammed state, the particulate phase is modelled as a porous phase with voids ratio e ($= \varepsilon^{-1} - 1$) and can either deform elastically (for stress loadings up to some yield value) or plastically when the yield value is reached. The yield surface is three-dimensional, and various models for yield behaviour exist, including the Cam-Clay and Drucker-Praeger formulations. Figure 28 illustrates the possible behaviours in the Cam-Clay model.

The solids stress is expressed as the sum of the deviatoric (or von Mises) stress and the effective stress, which is the arithmetic average of the three principal stresses. These provide the axes on Fig. 28. In the Cam-Clay model, the yield surface is circular and the diameter, p^*, is the set by the maximum effective stress that the material has been subjected to with a deviatoric stress of zero. This is associated with a particular value of ε or e. The material 'remembers' this quantity. Any combination of stresses which lie below the yield surface results in an elastic deformation.

When the stress state reaches the yield surface, the yield surface will move, and this will often be accompanied by a change in volume. At point B, on the critical state line, there is no change in volume. Stresses such as C cause compaction, reducing the voidage, and requires liquid to leave the element. This will only occur if the liquid can drain away. Stresses such as A, conversely, are associated with increases in volume and are termed dilatant. The change in volume is supported by liquid entering the element from regions around it or, in the limit, by cavitation of the liquid already present. Dilation therefore creates suction on the liquid phase.

In all cases, the direction of the plastic strain associated with moving the yield locus are set by a flow rule. The directions shown in Fig. 28 are those for an associated

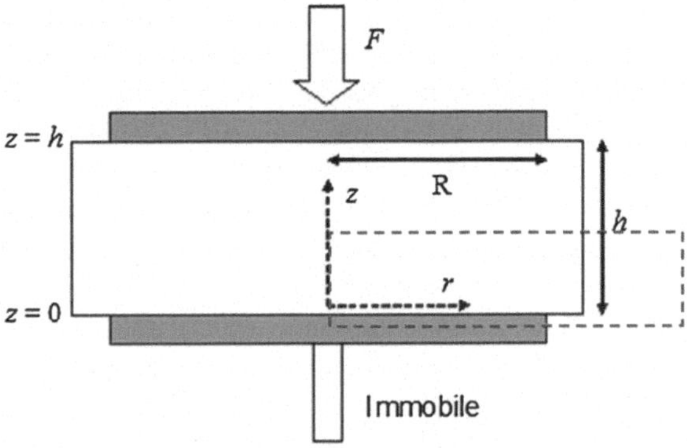

Fig. 29 Schematic of the squeeze flow test. Dashed quadrant indicates the control volume in the Nikkoo and Gadala-Maria (2014) model

flow rule. At D, the direction has no plastic component (the element does not change shape, just volume). Two points to take from Fig. 28 are

(i) *Filtration* corresponds to point D: there is little shearing imposed on the material in the barrel and little further shear imposed in the die land once the flow is developed, so the von Mises stress is zero and any volumetric strain will be compactive.
(ii) In the *die entry*, the material is subject to extension and thus deviatoric stresses. These can result in yield in the dilatant region (i.e. at A) so will give rise to suction in the liquid phase: the shearing of the solids as they are extruded draws liquid in from slower moving regions, such as elements upstream or the static zones formed in square entry dies. This results in progressively stronger static regions.

Incorporating the solids framework behaviour into flow of a yield stress fluid requires combining the above relationships with the equations of motion for an extruding solid. The solids behaviour is expressed in terms of strain rather than strain rate, so favours a Lagrangian approach. Patel et al. (2007) reported results for extrusion of a modified Cam Clay using the ABAQUS platform for their calculations (see later).

Fluid-based models. Several workers have reported models describing LPM in the squeeze flow of yield stress fluids (e.g. Sherwood 2002; Roussel and Lanos 2004a, b) and there are several experimental studies of LPM in the squeeze flow of concentrated suspensions.

Squeeze flow generates a cylindrically axisymmetric flow pattern. The geometry in Fig. 29 is that of a hamburger: two circular plattens of radius R are moved towards each other, forcing the disc of material sandwiched between them to flow outwards. In practice one platen is moved, subject to either a constant force, F, or constant speed ($-\mathrm{d}h/\mathrm{d}t$). In squeeze flow the disc initially fills the gap between the plattens: the case where the disc radius is initially smaller than that of the plattens is known as *upsetting*.

The model reported by Nikkoo and Gadala-Maria (2014) is presented here. It accompanied an experimental study of squeeze flow of glass ballotini in viscous silicone oils, where they used a novel technique to record the normal stress distribution on the plattens (Nikkoo et al. 2013, 2014). Their experimental results showed that large normal stresses were generated in regions where the particles reached the maximum packing fraction and *jamming* occurred: significant stresses were then transmitted through the particle network. The pore pressure problem (so to speak) is circumvented by assuming that the solids stress is zero until the particles reach the jammed state. This assumption is widely used in the fluids modelling approach.

The solid and liquid phases are not assumed to be in mechanical equilibrium, so that their local superficial velocities, u^*_i, can differ. There is no net liquid motion in the z direction (the plattens are impermeable) so (70) can be written in terms of the relative velocity u^*_r as

$$\frac{\partial P}{\partial r} = K_{DL}\mu\left(u_L^* - u_s^*\right) = K_{DL}\mu u_r^* \tag{74}$$

where P is the local pressure in the suspension. The equations of motion are written in 2-D, with volume averaged velocities u_i:

$$\rho\left(\frac{\partial u_z}{\partial t} + u_r\frac{\partial u_z}{\partial r} + u_z\frac{\partial u_z}{\partial r}\right) = -\frac{\partial P}{\partial z} + \frac{1}{r}\frac{\partial\left(r\sigma_{rz}\right)}{\partial r} + \frac{\partial \sigma_{zz}}{\partial r} \tag{75}$$

$$\rho\left(\frac{\partial u_r}{\partial t} + u_r\frac{\partial u_r}{\partial r} + u_z\frac{\partial u_r}{\partial r}\right) = -\frac{\partial P}{\partial r} + \frac{1}{r}\frac{\partial\left(r\sigma_{rr}\right)}{\partial r} + \frac{\partial \sigma_{zz}}{\partial r} - \frac{\sigma_{\theta\theta}}{r} \tag{76}$$

and the density of the suspension is given by

$$\rho = \rho_s\phi + \rho_L(1-\phi) \tag{77}$$

Conservation of volume for solids is written as

$$\begin{aligned}\frac{\partial\phi}{\partial t} + u\cdot\nabla\phi &= -\nabla\cdot N_s\\ &= -\nabla\cdot\left[u_r^*\phi\left(1-m_s\right)\right]\end{aligned} \tag{78}$$

where m_s is the local solids mass fraction (evaluated from ϕ and the ratio of densities).

The rheology of the suspension was modelled as a viscous fluid using the Krieger-Dougherty expression with ϕ_{max} =0.68 for ϕ <0.50, and as a YSF for ϕ >0.50, via

$$\eta_a = \mu_L\left(1-\frac{\phi}{\phi_{\max}}\right)^{-2.5\phi_{\max}} + \frac{\tau_0}{\dot\gamma}\left(1-\exp\left[-1000\dot\gamma\right]\right) \tag{79}$$

They also included wall slip, which was evident at very high solids volume fraction, using a slip factor S:

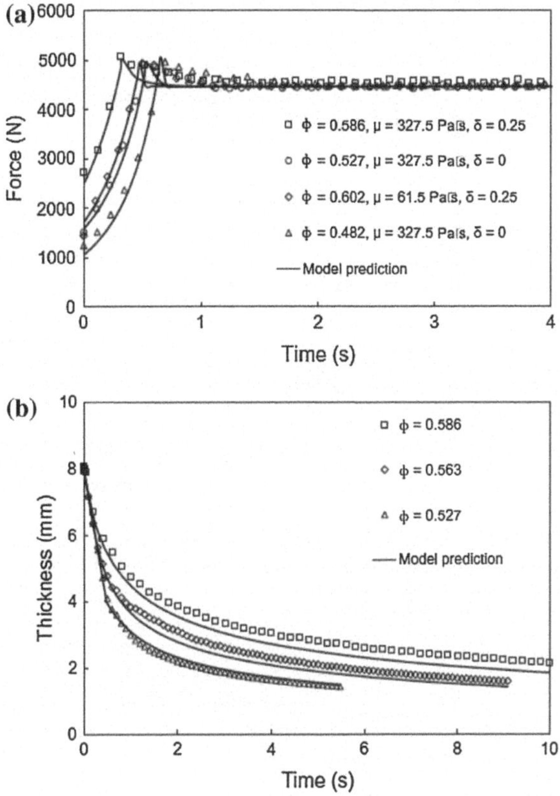

Fig. 30 Evolution of **a** force and **b** gap height during squeeze flow of dense suspensions with different initial solids volume fraction and liquid viscosity. The top platen was moved at constant velocity until a squeezing force of 4450 N was reached, and the platen position was subsequently adjusted to maintain F at this value. The term δ in the legend is the slip factor, S. Reproduced from Nikkoo and Gadala-Maria (2014)

$$S = \frac{h}{r}\frac{v_s}{-dh/dt} \tag{80}$$

Equation (80) arises from a simple geometrical argument which gives $S = 0.5$ for perfect slip and $S = 0$ for no-slip. They determined S for different volume fractions by fitting the model to subsets of data (but it only applied to the highest values of ϕ). The model is solved for the dashed quadrant marked on Fig. 29 with boundary conditions of axisymmetry, impermeable walls, identical values of the slip velocity on the bottom platen, and atmospheric pressure at the exit.

The force acting on the platens is given by

$$F = \int_0^R 2\pi r \sigma_{zz} dr \tag{81}$$

Figures 30, 31 and 32 show some of the model predictions alongside experimental data. The agreement is good and indicates that LPM can be modelled up to the jamming transition.

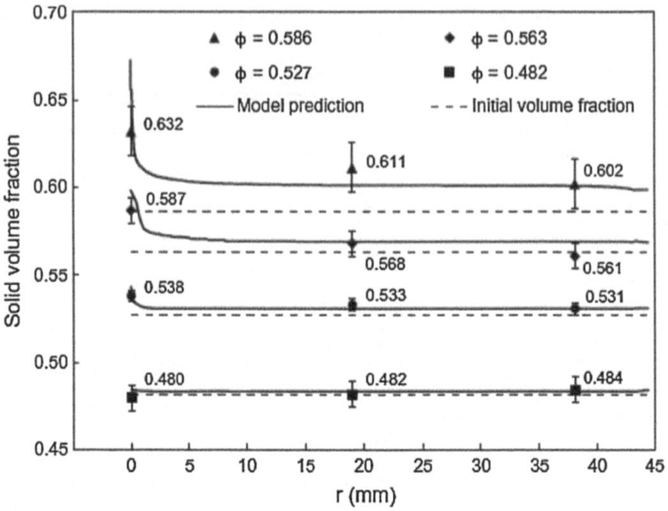

Fig. 31 Comparison of measured and predicted solids distributions for suspensions with different initial composition. Experimental conditions in Fig. 30. Reproduced from Nikkoo and Gadala-Maria (2014)

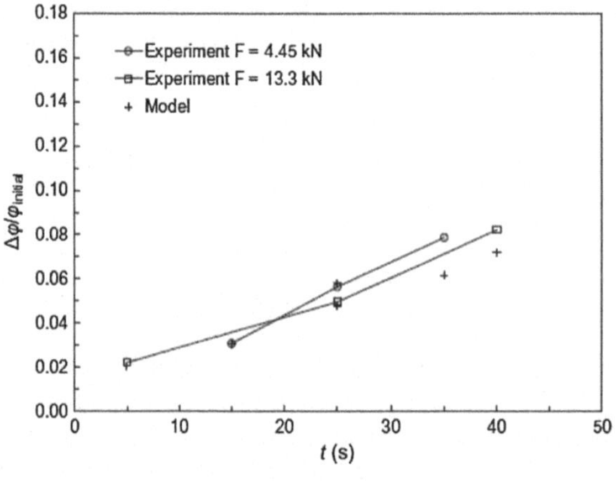

Fig. 32 Fractional change in ϕ in the central region for a suspension with initial $\phi = 0.586$ for squeeze times and squeeze forces indicated in the legend. Reproduced from Nikkoo and Gadala-Maria (2014)

Soil mechanics-based models. Nikkoo and Gadala-Maria's model describe LPM causing a progressively more viscous YSF. It extended up to the jamming transition, in which state a significant fraction of the stress will be borne by the solids network. At this point the material is acting as a soft solid, and there is lively conjecture

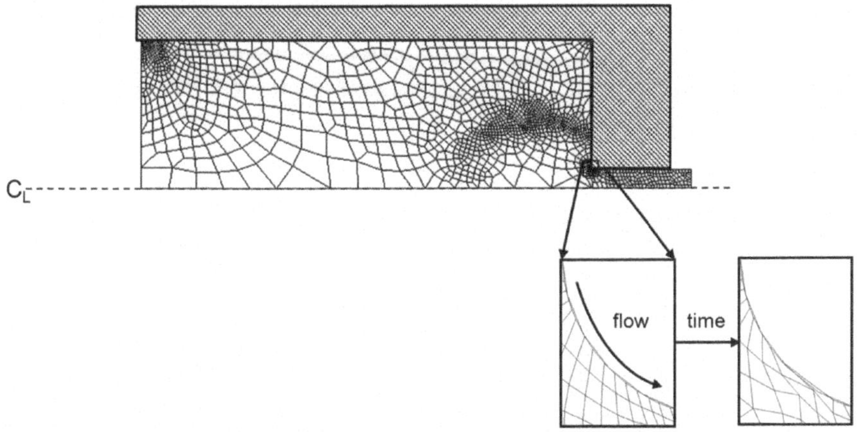

Fig. 33 Mesh generated for ram extrusion of a highly-filled suspension through a square-ended die with smooth walls (flow is from left to right). Reproduced from Patel et al. (2007)

as to whether fluid mechanics approaches (typically related to strain rate) or solid mechanics ones (typically related to strain) should be used.

We saw in the discussion of Fig. 28 that extension of a solids network can give rise to dilation and suction. Extension-driven suction cannot be modelled by the fluid mechanics approaches above as the stress in the fluid is assumed to be taken up by the liquid phase. Approaches similar to those used in soil mechanics are then necessary. The pore pressure and solids stress have to be modelled independently, coupled by conservation of volume (and thus filtration).

Numerical solutions are necessary, in which the strain field (for solids behaviour) must be evaluated in tandem with the pore pressure (for liquid). In a Lagrangian framework this requires extensive remeshing as elements distort (see Fig. 33). Patel et al. (2007) reported results obtained for ram extrusion through smooth-walled square-entry and conical-entry dies. Patel (2008) also obtained results for extrusion for rough walls and squeeze flow configurations (papers in preparation). The mesh does not match the region of high plastic strain in Fig. 10 (flow of a perfectly plastic material through a similar geometry): the concentration of mesh elements around the die entry are required to evaluation dilation effects.

Figure 34 shows the evolution of total (extrusion) pressure on the ram, P_e, during extrusion as well as the contributions from the solid and liquid phases. The latter dominates, as the liquid cannot drain away quickly enough to impose large stresses on the solids. This result also indicates that the stresses measured in the system will be nearly isotropic as the stress is chiefly borne by the liquid.

The initially high solids stress in Fig. 34 is related to the initial development of high solids content zones in the corners which are evident in Fig. 35: dilation at the die entry gives rise to inhomogeneity in the composition of the material.

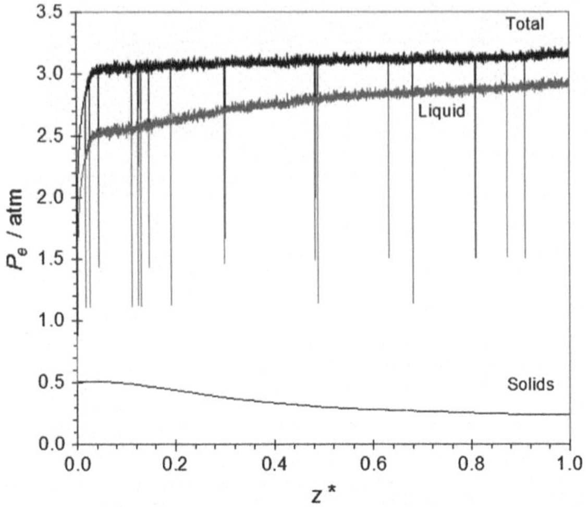

Fig. 34 Contributions to the stress during ram extrusion of a highly-filled suspension through the geometry in Fig. 37. Spikes are numerical artefacts. Reproduced from Patel et al. (2007)

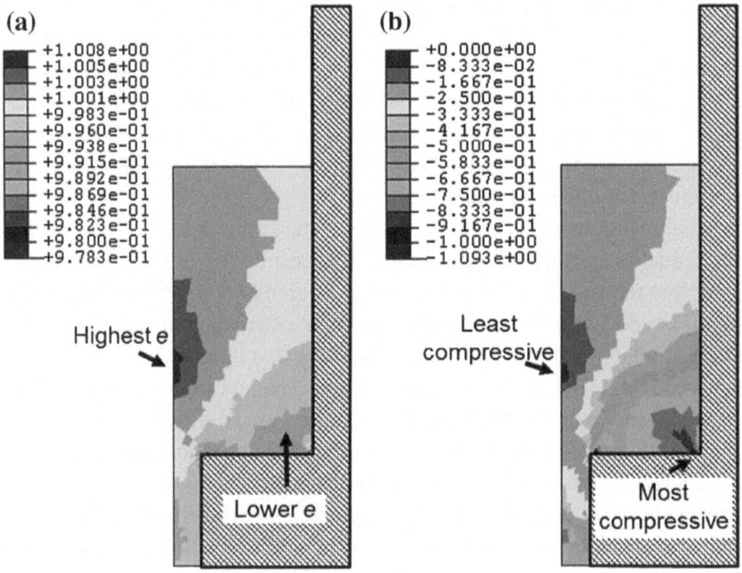

Fig. 35 Distribution of **a** voids ratio ($e_{initial} = 1$), and **b** axial effective stress (units of atm, positive when tensile) at the end of the ram extrusion simulation in Fig. 34. Reproduced from Patel et al. (2007)

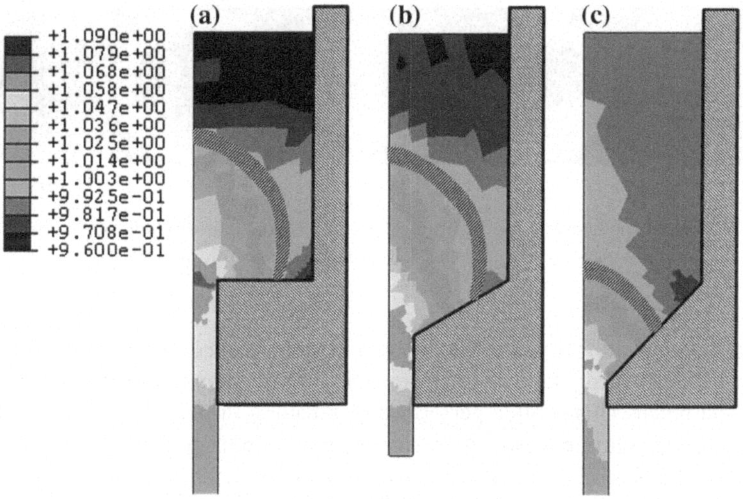

Fig. 36 Distribution of voids ratio for die entry angles of **a** 90°, **b** 60° and **c** 45°, for a ratio of die land length/barrel radius, $L/R_b = 1$, $V/R_b = 0.002\ \mathrm{s}^{-1}$ and no wall friction. The initial voids ratio, $e_{initial}, = 1$: hatched pattern indicates regions with e close to $e_{initial}$. Reproduced from Patel et al. (2007)

Figure 36 shows that the distribution of liquid is influenced by the die geometry. In this case, using a conical die reduces the maximum strain experienced by the material in the corners (which are smooth so can still deform) and thus the dilation there.

The distribution of liquid is also determined by the timescale, in that no LPM is predicted at higher flow rates because the liquid phase will not have time to redistribute. The initial force required to drive the ram is likely to be larger but it will remain constant over time as the composition of the material does not change. When LPM occurs, the solid volume fraction in the barrel increases over time (see Fig. 4), increasing the yield strength of the material and thus the extrusion force (see Rough et al. 2002; O'Neill et al. 2016).

This is expected from considerations of the Deborah number, as the process timescale is set by the geometry (size of the deformation zone) and the flow rate, whereas the LPM timescale is dominated by the permeability of the solids and the extent of dilation. Martin et al. (2004) showed that a simple scaling result from soil mechanics could describe the onset of LPM in several pastes.

5 Cleaning—The Dirty Side of YSFs

Many products in the food, pharmaceutical and fast moving consumer good sectors are yield stress fluids: the list includes gels, toothpaste, ointments and chocolate.

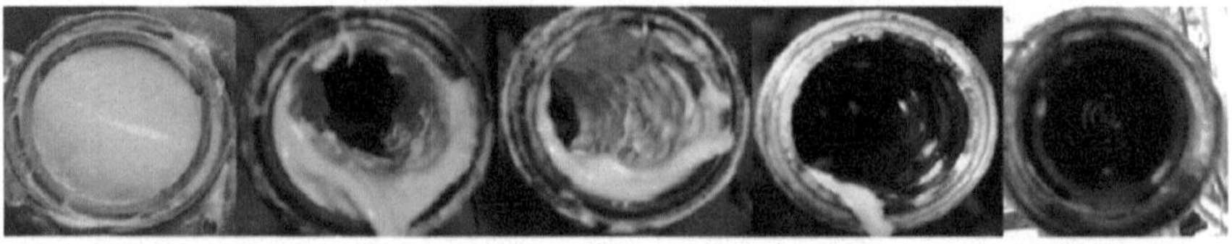

Fig. 37 Extent of toothpaste removal from a 25.4 mm i.d. pipe at different times during water purging at 50 °C and mean velocity 1.1 ms^{-1} Reproduced from Cole et al. (2010)

They are often manufactured in batches and in multi-product plant. Production lines have to be cleaned regularly in order to avoid cross-contamination by one material in the next batch. This is often performed by cleaning-in-place (CIP) operations, in which cleaning solution is passed through the lines or units to (i) purge material, and (ii) remove any soiling layers generated on process surfaces. The aim of this section is to introduce cleaning operations as YSF flow problems (or research opportunities).

Cleaning-in-place (CIP) mechanisms. The three components involved are

(i) The soil
(ii) The substrate (the surface of the equipment)
(iii) The solution

The cleaning solution is usually an aqueous solution. This gives rise to several possible cleaning mechanisms. If the YSF is water soluble, dissolution and diffusion can play different roles

(i) *Total dissolution*—the majority of the components of the soil dissolve and the soil layer is removed in a steady fashion. This is essentially an interfacial mass transfer problem, where the rate is enhanced by motion of the soil layer.
(ii) *Partial dissolution* (erosion)—some components of the YSF dissolve, or cleaning agent penetrates into the layer (e.g. as a result of swelling promoted by pH changes). Both of these result in progressive weakening of the cohesive strength of the soil and increased propensity for removal by hydraulic action in addition to dissolution.
(iii) *Adhesive removal* (peeling)—ingress of cleaning agent results in weakening of the soil-substrate bond. The soil detaches from the substrate under hydraulic action.

If the YSF is immiscible with water, the cleaning mechanism is driven by hydraulic action and results in a coupled flow problem.

Example 1: Purging. Palabiyik et al. (2014) reported an experimental study of the removal of toothpaste from initially filled pipework by water flushing. Shear rheometry of the undiluted toothpaste indicated Herschel-Bulkley behaviour with a yield stress of 92 Pa at 30 °C. The series of photographs in Fig. 37 illustrates the three stages of removal which are evident in the removal profiles in Fig. 38.

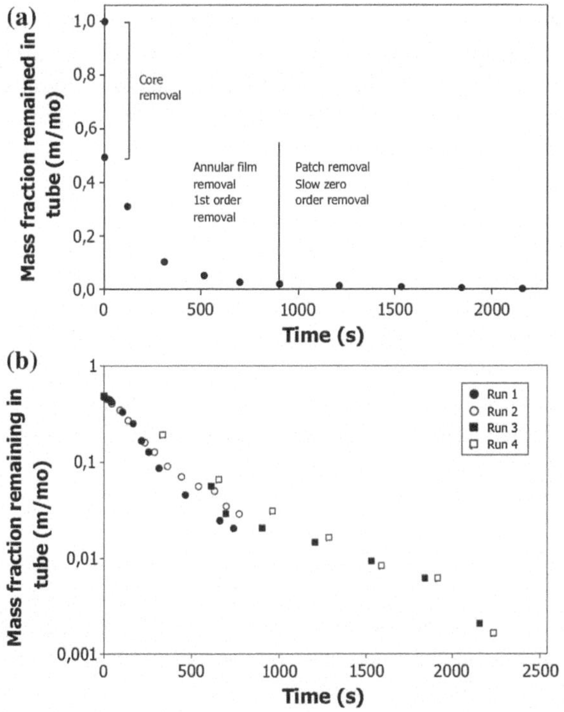

Fig. 38 Progress of removal in water purging of toothpaste from a 47.7 mm i.d. pipe at **a** 50 °C, mean velocity 0.55 ms^{-1}; and **b** 50 °C and 0.55 ms^{-1} (note different ordinate axis scale). Reproduced from Palabiyik et al. (2014)

(i) *Core removal stage*

The water breaks through the YSF by pushing out the core. In the experiments the mass fraction of material removed was found to lie consistently around 0.5, as shown in Fig. 38a. Cox (1962) reported values of 0.4–0.6 for the purging of viscous Newtonian liquids from long tubes. This is a rapid event—the characteristic time of the flow, given by tube length/mean velocity, is 2 s—and gives the initial condition in Fig. 38b.

Dimakopoulos and Tsamopoulos (2003) modelled the transient displacement of a Bingham fluid from a pipe by air and similar approaches, modified to include the inertia of the flushing liquid could be applied here to estimate the fraction of material purged in the flushing stage.

(ii) *Annular film removal*

Subsequent cleaning occurs by the thinning of the annular film of the YSF. The mass of material decays exponentially (see Fig. 38b) until about 900 s. The photographs in Fig. 37 at $t = 1$ s and 15 s show that the interface is wavy. Alba and Frigaard (2016) have recently presented an analysis of the removal of YSFs from pipes which includes the wavy regime. Applying their results to this example is confounded by the fact that the shear stress exerted by the purging liquid flow is noticeably less than the measured yield stress of the toothpaste: for a mean velocity of 0.55 ms^{-1}, the Reynolds number in a 25.4 mm i.d. pipe is around 25,000 and the wall shear stress for a smooth tube is approximately 1 Pa. Surface roughness will enhance this but the factor of 90 difference is unlikely to arise from the wall waviness evident in the picture.

The long timescale of this regime, being several hundred times the characteristic timescale in (*i*), also indicates that the controlling mechanism may not be coupled flow. Inspection of the patent literature reveals that many commercial toothpastes employ water soluble binders such as glycerol. Dissolution will increase the liquid content of the material and reduce its yield strength: its rheology is therefore time dependent, determined by the ingress of solution into the paste. This stage would appear to be a coupled diffusion/hydraulic problem, and the timescales are therefore considerably longer.

(iii) *Patch removal*

In due course the wavy interface penetrates through to the substrate and the mechanism changes from removal of a continuous film to erosion of residual islands of soil, termed 'patches'. The rate of removal is no longer first order in remaining mass fraction and features a longer timescale. The last part of the soil takes a long time to remove if the cleaning conditions are kept constant.

These observations are consistent with the coupled diffusion/hydraulic mechanism in (ii). The initial purging stage leaves behind a layer of uneven thickness: if this is eroded away at a uniform rate per unit area, dry patches will form when the solution breaks through to the surface. The area available for erosion will be smaller and the overall rate will then decrease. A relationship between the mass of residue remaining and the area available for erosion is needed in order to generate a quantitative model.

Example 2: Cleaning soiled walls by impinging jets. Jets of liquid created by nozzles and spray balls are widely used to clean the internal surfaces of, and fittings within, vessels and storage tanks. Figure 39a shows a coherent jet of water impinging perpendicularly on a vertical wall which is coated with petroleum jelly, which can be considered to be a YSF with a yield stress of approximately 12 Pa. The flow of water away from the point of impingement creates a circular cleaned region, of radius a (Fig. 39b), which increases with time. O is the point of impingement. The model for $a(t)$ by Wilson et al. (2014), adapted for a YSF by Glover et al. (2016), is presented here.

The rate of peeling is assumed to be proportional to the flow of momentum per unit width, M, viz.

$$\frac{da}{dt} = k'M \tag{82}$$

Fig. 39 Cleaning of a soil layer from a vertical wall by an impinging water jet. **a** photograph of removal of a 200 μm thick petroleum jelly layer by a 2 mm diameter water jet: **b** model

where k' is a cleaning rate constant. Calculating M requires a model for the hydrodynamics of the thin film of water flowing radially outwards. Wilson et al. employed an approximate model for the water flow, where it was assumed to behave as a Nusselt film until the point where surface tension creates a feature resembling a hydraulic jump (which in their experiments was always beyond a). This gave

$$M = \frac{3\dot{m}}{5\pi}\frac{V}{r} \tag{83}$$

where $\dot{m}$ is the mass flow rate in the jet and V is the mean velocity in the film at radius r.

The velocity profile in the film at radius r is given by

$$\frac{1}{V} - \frac{1}{V_o} = \frac{10\pi^2\rho\mu}{3\dot{m}^2}\left[r^3 - r_o^3\right] \tag{84}$$

where V_0 and r_0 are the velocity and radius of the jet, respectively.

Assuming that $1/V_0$ is small, and that $a \gg r_0$, allows M at the cleaning front to be calculated. Equation (82) becomes

$$\begin{aligned}\frac{da}{dt} &= \frac{3k'\dot{m}^3}{\pi c}\frac{1}{5a^4}\\ &= K_c^5\frac{1}{5a^4}\end{aligned} \tag{85}$$

Here c is a constant determined by liquid properties ($c = 10\pi^2\rho\mu/3$), and K_c is a flow rate dependent cleaning rate constant. Integrating (85) from the point where a circular cleaning front is first observed, with size a_0, at time t_0 gives

Fig. 40 Glover et al. (2016) model for cleaning of a YSF soil layer of thickness δ

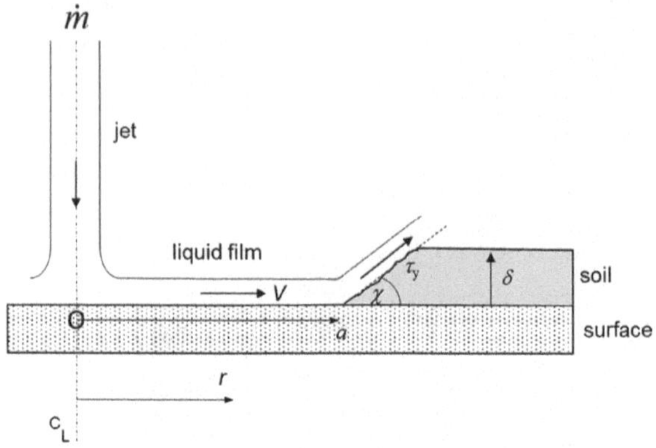

$$a^5 - a_o^5 = K_c^5 (t - t_o)$$
$$= K_c^5 \Delta t \tag{86}$$
$$\text{or} \quad a \approx K_c \Delta t^{1/5} \tag{87}$$

Equation (87) was developed to describe removal of simple soiling layers. It gives a good description of the initial stage of cleaning of petroleum jelly layers (see Fig. 41) but asymptotic behaviour is observed at longer times. The asymptotic behaviour was attributed to the presence of the yield stress.

Glover et al. adapted the model of Wilson et al. by assuming that the cleaning front took the form of a ramp with height δ inclined to the substrate at angle χ (see Fig. 40). A momentum balance is used to determine the effect of the ramp. Beyond a the liquid flows upwards so that a fraction of its momentum flux is no longer horizontal. The change in horizontal momentum, i.e. the difference between M and $M\cos\chi$, provides the driving force for cleaning in Eq. (82).

Asymptotic limit
When the cleaning front reaches the asymptotic limit, at $r = a_{\max}$, the net momentum flux is equal to the force required to overcome the shear yield stress of the layer and induce motion.

The area of the yielding region for a complete circle of radius $a_{\max}$, is approximately $2\pi a_{\max}\delta/\sin\chi$: the length of the ramp is assumed to be small compared to $a_{\max}$. A force balance in the horizontal direction at $a_{\max}$ gives

$$\frac{6}{5}\left(\dot{m}V - \dot{m}V\cos\chi\right) = \tau_y\left(\frac{2\pi a_{\max}\delta}{\sin\chi}\right)\cos\chi \tag{88}$$

The coefficient 6/5 arises from treating the flow as a Nusselt film. Substituting for V from (84), with $1/V_o$ and r_o again both small and $r = a_{\max}$, yields

$$a_{\max} = \left(\frac{3\dot{m}^3}{5\pi c}\frac{1}{\tau_y\delta}\left[\tan\chi - \sin\chi\right]\right)^{1/4} \tag{89}$$

The rate of cleaning is now given by

$$\frac{da}{dt} = k'(M - M_Y) \quad M > M_Y \tag{90}$$

$$\frac{da}{dt} = 0 \quad M \leq M_Y \tag{91}$$

where M_Y is the momentum flux required to cause yield. For a flat shear plane on the soil, with area per unit width δ cosecχ and inclined an angle χ, the momentum balance gives

$$M_Y - M_Y \cos\chi = \tau_y (\delta \operatorname{cosec}\chi) \cos\chi \tag{92}$$

which gives

$$M_Y = \frac{\tau_y \delta}{(\tan\chi - \sin\chi)} \tag{93}$$

Some further rearrangement, substituting for $\tau_y\delta$ from (89) gives

$$M_Y = 3\dot{m}^3 / 5\pi c a_{\max}^4 \tag{94}$$

Integrating Eq. (90) then yields

$$t - t_o = \frac{5}{K^5} \int_{a_o}^{a} \frac{a^4}{1 - (a/a_{\max})^4} da \tag{95}$$

For the case where $a = 0$ when $t = 0$, the result is

$$t = \frac{5}{4}\left(\frac{a_{\max}}{K}\right)^5 \left\{ \ln\left(\frac{1 + a/a_{\max}}{1 - a/a_{\max}}\right) - 4\left(\frac{a}{a_{\max}}\right) + 2\tan^{-1}\left(\frac{a}{a_{\max}}\right) \right\} \tag{96}$$

Figure 41 shows that Eq. (96) gives a reasonable description of the experimental data. The values of χ extracted from the data ranged from 10–25°, which could be compared with detailed measurements of cleaning front shape.

Cleaning is often a complex process. Simple models such as those outlined above, based on momentum balances, can often provide valuable insight into the process and could provide sufficient resolution that more detailed modelling is not required.

6 Closing Remarks

This chapter has presented a number of practical applications where yield stress fluids are employed or where yield stress behaviour arises. The existence of a yield stress creates significant differences in flow behaviour from other types of fluid which can

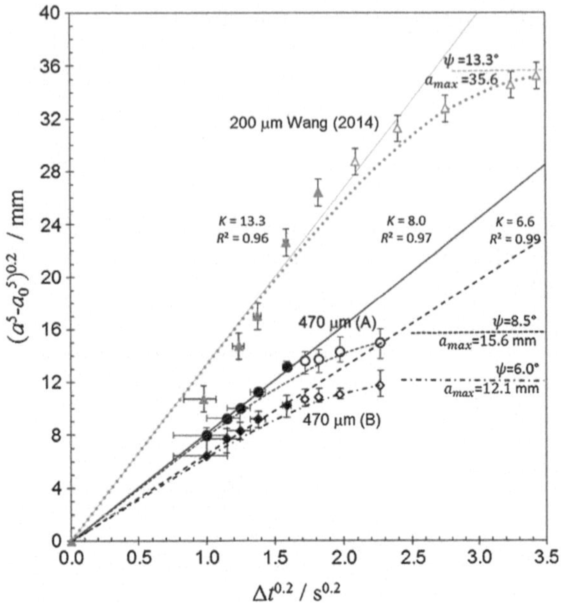

Fig. 41 Growth of cleared region in petroleum jelly layers for 2 mm nozzle at a water flow rate of 35 gs^{-1} at 20 °C. Symbols: circles, diamonds, Perspex substrate, $\delta = 470 \pm 50$ μm; triangles, glass substrate, $\delta = 200 \pm 30$ μm, reported by Wang (2014). Solid loci show fit of initial data (solid symbols) to (87); dashed loci show fit for all data in a series to (96). Reproduced from Glover et al. (2016)

either be exploited, or need to be understood, in order to design or operate the system to best effect.

With any real material, there is likely to be some uncertainty arising from the validity of a yield stress fluid model to describe its behaviour. Since yield stress behaviour is associated with solid-like behaviour, one needs to consider whether the purpose is best suited to mathematical descriptions based on fluid mechanics (and strain rate) or solid (strain) mechanics approaches.

Yield stress behaviour arises from microstructural phenomena, and at some material-dependent timescale or stress level the assumption of yield stress behaviour is likely to be inaccurate. It is important to establish those timescales, strain and/or stress levels and the nature of the deviation which accompanies the breakdown of the assumption.

Similarly, the microstructural features will differ between materials, and will be sensitive both to the composition and how the material is prepared (together often referred to as its formulation). The ability of the model to describe features of a flow will depend on the reliability of the input parameters, and there is likely to be some uncertainty in predictions arising from this 'natural' variation in the material. This variation applies both to the bulk material and its interaction with boundaries.

In short, yield stress models can give useful results, and reasonable quantitative predictions. Be very happy if the results fit the observations well.

References

Aeschlimann, J. M., & Beckett, S. T. (2000). International inter-laboratory trials to determine the factors affecting the measurement of chocolate viscosity. *Journal of Texture Studies, 31*(5), 541–576.

Alba, K., & Frigaard, I. A. (2016). Dynamics of the removal of viscoplastic fluids from inclined pipes. *Journal of Non-Newtonian Fluid Mechanics, 229,* 43–58.

Amarasinghe, A. D. U. S., & Wilson, D. I. (1998). Interpretation of paste extrusion data. *Chemical Engineering Research and Design, 76*(1), 3–8.

Assaad, J. J., Harb, J., & Maalouf, Y. (2016). Effect of vane configuration on yield stress measurements of cement pastes. *Journal of Non-Newtonian Fluid Mechanics, 230,* 31–42.

Balmforth, N. J., Frigaard, I. A., & Ovarlez, G. (2014). Yielding to stress: recent developments in viscoplastic fluid mechanics. *Annual Review of Fluid Mechanics, 46,* 121–146.

Banfill, P.F.G. (2006) Rheology of fresh cement and concrete, *Rheology Reviews*, 61–130.

Bardsley, M. A., & Bridgwater, J. (2012). Evaluation of liquid phase migration in pastes and gels. *Industrial and Engineering Chemistry Research, 51,* 1774–1781.

Barker, D. A., & Wilson, D. I. (2008). Rheology of a thermoplastic paste: from the melt to the mushy state. *Chemical Engineering Science, 63,* 1438–1448.

Barker, D.A. (2008) Thermal processing of highly filled suspensions, PhD Dissertation, University of Cambridge.

Basterfield, R. A., Adams, M. J., & Lawrence, C. J. (2005). On the interpretation of orifice extrusion data for viscoplastic materials. *Chemical Engineering Science, 60,* 2599–2607.

Benbow, J.J. and Bridgwater, J. (1993) *Paste flow and extrusion*, publ. OUP.

Bryan, M. P., Rough, S. L., & Wilson, D. I. (2015). Investigation of static zone formation and wall slip through sequential ram extrusion of contrasting micro-crystalline cellulose-based pastes. *Journal of Non-Newtonian Fluid Mechanics, 220,* 57–68.

Cole, P. A., Asteriadou, K., Robbins, P. T., Owen, E. G., Montague, G. A., & Fryer, P. J. (2010). Comparison of cleaning of toothpaste from surfaces and pilot scale pipework. *Food Bioproducts Processing, 88,* 392–400.

Cox, B. G. (1962). On driving a viscous fluid out of a tube. *Journal of Fluid Mechanics, 14,* 81–96.

Curran, S. J., Hayes, R. E., Afacan, A., Williams, M. C., & Tanguy, P. A. (2002). Properties of carbopol solutions as models for yield-stress fluids. *Journal of Food Science, 67,* 176–180.

Kee, D., Code, R. K., & Turcotte, G. (1983). Flow properties of time-dependent foodstuffs. *Journal of Rheology, 27,* 581–604.

Dimakopoulos, Y., & Tsamopoulos, J. (2003). Transient displacement of a viscoplastic material by air in straight and suddenly constricted tubes. *Journal of Non-Newtonian Fluid Mechanics, 112,* 43–75.

De Souza Mendes, P. R., & Thompson, R. L. (2013). A unified approach to model elasto-viscoplastic yield-stress materials and apparent yield-stress fluids. *Rheologica Acta, 52,* 673–694.

Elhweg, B., Burns, I. W., Chew, Y. M. J., Martin, P. J., Russell, A. B., & Wilson, D. I. (2009). Viscous dissipation and apparent wall slip in capillary rheometry of ice cream. *Food Bioproducts Processing, 87,* 266–272.

Ferstl, H., Barbist, R., Rough, S. L., & Wilson, D. I. (2012). Influence of visco-elastic binder properties on ram extrusion of a hardmetal paste. *Journal of Materials Science, 47*(19), 6835–6848.

Glover, H. W., Brass, T., Bhagat, R. K., Davidson, J. F., Pratt, L., & Wilson, D. I. (2016). Cleaning of complex soil layers on vertical walls by fixed and moving impinging liquid jets. *Journal of Food Engineering, 178,* 95–109.

Gray, J. M. N. T., & Edwards, A. N. (2014). A depth-average $\mu(I)$ rheology for shallow granular free-surface flows. *Journal of Fluid Mechanics, 755,* 503–544.

Horrobin, D. J., & Nedderman, R. M. (1998). Die entry pressure drops in paste extrusion. *Chemical Engineering Science, 53*(18), 3215–3225.

Horrobin, D. J. (1999) Theoretical aspects of paste extrusion, PhD Dissertation, University of Cambridge.

Hu, H., & Argyropoulos, S. A. (1996). Mathematical modelling of solidification and melting: A review. *Modelling and Simulation Materials Science and Engineering, 4,* 371–396.

Jana, S., Ray, S., & Durst, F. (2007). A numerical method to compute solidification and melting processes. *Applied Mathematical Modelling, 31,* 93–119.

Jastrzebski, Z. D. (1967). Entrance effects and wall effects in an extrusion rheometer during the flow of concentrated suspensions. *Industrial and Engineering Chemistry Fundamentals, 6*(3), 445–454.

Jop, P., Forterre, Y., & Pouliquen, O. (2006). A constitutive relation for dense granular flows. *Nature, 44,* 727–730.

Martin, P. J., Wilson, D. I., & Bonnet, P. (2004). Rheological study of a talc-based paste for extrusion-granulation. *Journal of the European Ceramic Society, 24*(10–11), 3155–3168.

Martin, P. J., & Wilson, D. I. (2005). A critical assessment of the Jastrzebski interface condition for the capillary flow of pastes, foams and polymers. *Chemical Engineering Science, 60,* 493–502.

Mitsoulis, E. (2007). Flows of viscoplastic materials: Models and computations. *Rheology Reviews, 64,* 137–178.

Mitsoulis, E., & Hatzikiriakso, S. G. (2009). Steady flow simulations of compressible PTFE paste extrusion under severe wall slip. *Journal of Non-Newtonian Fluid Mechanics, 157,* 26–33.

Nikkoo, M., & Gadala-Maria, F. (2014). Modelling radial filtration in squeeze flow of highly concentrated suspensions. *Rheologica Acta, 53,* 607–619.

Nikkoo, M., Khodabandehlou, K., Brozovsky, L., & Gadala-Maria, F. (2013) Normal stress distribution in highly concentrated suspensions undergoing squeeze flow, *Rheologica Acta*, 52, 155–163.

Nikkoo, M., Hofman, A., & Gadala-Maria, F. (2014). Correlation between radial filtration and normal stress distribution in highly concentrated suspensions undergoing constant-force squeeze flow. *Rheologica Acta, 53,* 303–314.

Nirmalkar, N., Chhabra, R. P., & Poole, R. J. (2013a). Numerical predictions of momentum and heat transfer characteristics from a heated sphere in yield stress fluids. *Industrial and Engineering Chemistry Research, 52,* 6848–6861.

Nirmalkar, N., Chhabra, R. P., & Poole, R. J. (2013b). Effect of shear-thinning behavior on heat transfer from a heated sphere in yield-stress fluids. *Industrial and Engineering Chemistry Research, 52,* 13490–13504.

O'Neill, R., McCarthy, H. O., Cunningham, E., Montufar, E., Ginebra, M-P., Wilson, D.I., Lennon, A., & Dunne, N. (2016). Extent and mechanism of phase separation during the surgical injection (extrusion) of calcium phosphate pastes, *JoM: Materials in Medicine*, 27, 29.

Ortega-Avila, J. F., Perez-Gonzalez, J., Marin-Santibanez, B. M., Rodriguez-Gonzalez, F., Aktas, S., Malik, M., et al. (2016). Axial annular flow of a viscoplastic microgel with wall slip. *Journal of Rheology, 60,* 503–515.

Palabiyik, I., Olunloyo, B., Fryer, P. J., & Robbins, P. T. (2014). Flow regimes in the emptying of pipes filled with a Herschel-Bulkley fluid. *Chemical Engineering Research and Design, 92,* 2201–2212.

Patil, P. D., Feng, J. J., & Hatzikiriakos, S. G. (2006). Constitutive modeling and flow simulations of polytetrafluoroethylene (PTFE) paste extrusion. *Journal of Non-Newtonian Fluid Mechanics, 134,* 44–53.

Patel, M. J., Blackburn, S., & Wilson, D. I. (2007). Modelling of paste flows subject to liquid phase maldistribution. *International Journal for Numerical Methods in Engineering, 72*(10), 1157–1180.

Patel, M.J. (2008). Theoretical aspects of paste formulation for extrusion, PhD Dissertation, Univeristy of Cambridge.

Pérez-González, J., López-Durán, J. L., Marín-Santibáñez, B. M., & Rodríguez-González, F. (2012). Rheo-PIV of a yield-stress fluid in a capillary with slip at the wall. *Rheologica Acta, 51,* 937–946.

Powell, J., & Blackburn, S. (2010). Co-extrusion of multi-layered ceramic micro-tubes for use as solid oxide fuel cells. *Journal of the European Ceramic Society, 30,* 2859–2870.

Putz, A. M. V., & Burghalea, T. I. (2009). The solid-fluid transition in a yield stress shear thinning physical gel. *Rheologica Acta, 48,* 673–689.

Rahman, L., Rowe, P., Cheyne, A., & Wilson, D. I. (2001). Ram extrusion of potato starch dough through multi-holed dies. *Food and Bioproducts Processing, 80*(1), 12–19.

Rough, S. L., Wilson, D. I., & Bridgwater, J. (2002). A model describing liquid phase migration within an extruding microcrystalline cellulose paste. *Chemical Engineering Research and Design, 80,* 701–714.

Roussel, N., & Lanos, C. (2004a). Particle fluid separation of dense suspensions: experimental measurement on squeezed clay paste. *Journal of Rheology, 51,* 493–515.

Roussel, N., & Lanos, C. (2004b). Particle fluid separation in shear flow of dense suspensions: Experimental measurements on squeezed clay pastes. *Applied Rheology, 14,* 256–265.

Sairamu, M., Nirmalkar, N., & Chhabra, R. P. (2013). Natural convection from a circular cylinder in confined Bingham plastic fluids. *International Journal of Heat and Mass Transfer, 60,* 567–581.

Sherwood, J. D. (2002). Liquid-solid relative motion during squeeze flow of pastes. *Journal of Non-Newtonian Fluid Mechanics, 104,* 1–32.

Wang, T. (2014). PhD Dissertation, University of Cambridge.

Wilmot, R. B., Barbist, R., Ferstl, H., Gruber, C., Branstetter, W., Patel, M. J. & Wilson, D. I. (2009). Characterisation and design of wax-based hard metal extrusion processes, Plansee Seminar 2009, Vol. 9, Paper HM15, Plansee Group, Reutte, Austria.

Wilson, D. I., & Rough, S. L. (2012). Paste engineering: Multi-phase materials and multi-phase flows. *Canadian Journal of Chemical Engineering, 90*(2), 277–289.

Wilson, D. I., Atkinson, P., Köhler, H., Mauermann, M., Stoye, H., Suddaby, K., et al. (2014). Cleaning of soft-solid soil layers on vertical and horizontal surfaces by coherent impinging liquid jets. *Chemical Engineering Science, 109,* 183–196.

Yoshimura, A., & Prud'homme, R. K. (1988). Wall slip corrections for Couette and parallel disk viscometers. *Journal of Rheology, 32*(1), 53–67.

Zhang, M., Mascia, S., Rough, S. L., Ward, R., Seiler, C., & Wilson, D. I. (2013). A novel lab-scale screen extruder for studying extrusion-spheronisation. *International Journal of Pharmaceutics, 455,* 285–297.

Zhao, U., Kumar, L., Paso, K., Ali, H., Safieva, J., & Sjöblom, J. (2012). Gelation and breakage behaviour of model wax-oil systems: rheological properties and model development. *Industrial and Engineering Chemistry Research, 51,* 8123–8133.

The manufacturer's authorised representative in the EU is Springer Nature Customer Service Centre GmbH, Europaplatz 3, 69115 Heidelberg, Germany. If you have any concerns regarding our products, please contact ProductSafety@springernature.com

Printed and bound by CPI Group (UK) Ltd, Croydon, CR0 4YY
05/10/2026
02228139-0005